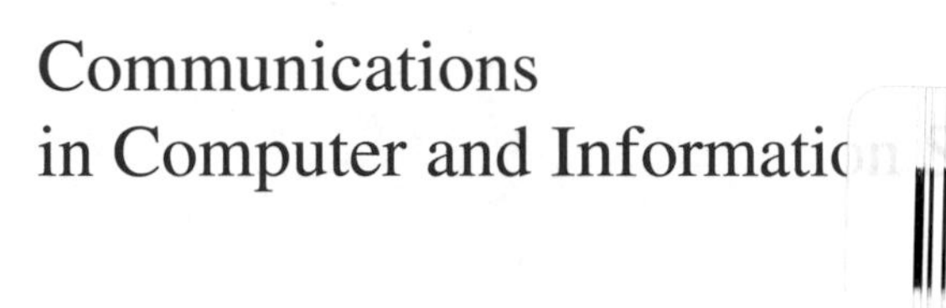
Communications
in Computer and Informatic

T0092459

Kang Li Xin Li Shiwei Ma
George W. Irwin (Eds.)

Life System Modeling and Intelligent Computing

International Conference on Life System Modeling
and Simulation, LSMS 2010
and International Conference on Intelligent Computing
for Sustainable Energy and Environment, ICSEE 2010
Wuxi, China, September 17-20, 2010
Proceedings, Part II

Volume Editors

Kang Li
Queen's University Belfast
Belfast, UK
E-mail: k.li@qub.ac.uk

Xin Li
Shanghai University, China
E-mail: su_xinli@yahoo.com.cn

Shiwei Ma
Shanghai University, China
E-mail: swma@mail.shu.edu.cn

George W. Irwin
Queen's University Belfast
Belfast, UK
E-mail: g.irwin@ee.qub.ac.uk

Library of Congress Control Number: 2010934237

CR Subject Classification (1998): J.3, H.2.8, F.1, I.6.1, I.6, I.2.9

ISSN 1865-0929
ISBN-10 3-642-15858-7 Springer Berlin Heidelberg New York
ISBN-13 978-3-642-15858-2 Springer Berlin Heidelberg New York

springer.com

Printed in Germany

Typesetting: Camera-ready by author, data conversion by Scientific Publishing Services, Chennai, India
Printed on acid-free paper 06/3180 5 4 3 2 1 0

Preface

The 2010 International Conference on Life System Modeling and Simulation (LSMS 2010) and the 2010 International Conference on Intelligent Computing for Sustainable Energy and Environment (ICSEE 2010) were formed to bring together researchers and practitioners in the fields of life system modeling/simulation and intelligent computing applied to worldwide sustainable energy and environmental applications.

A life system is a broad concept, covering both micro and macro components ranging from cells, tissues and organs across to organisms and ecological niches. To comprehend and predict the complex behavior of even a simple life system can be extremely difficult using conventional approaches. To meet this challenge, a variety of new theories and methodologies have emerged in recent years on life system modeling and simulation. Along with improved understanding of the behavior of biological systems, novel intelligent computing paradigms and techniques have emerged to handle complicated real-world problems and applications. In particular, intelligent computing approaches have been valuable in the design and development of systems and facilities for achieving sustainable energy and a sustainable environment, the two most challenging issues currently facing humanity. The two LSMS 2010 and ICSEE 2010 conferences served as an important platform for synergizing these two research streams.

The LSMS 2010 and ICSEE 2010 conferences, held in Wuxi, China, during September 17–20, 2010, built upon the success of two previous LSMS conferences held in Shanghai in 2004 and 2007 and were based on the Research Councils UK (RCUK)-funded Sustainable Energy and Built Environment Science Bridge project. The conferences were jointly organized by Shanghai University, Queen's University Belfast, Jiangnan University and the System Modeling and Simulation Technical Committee of CASS, together with the Embedded Instrument and System Technical Committee of China Instrument and Control Society. The conference program covered keynote addresses, special sessions, themed workshops and poster presentations, in addition to a series of social functions to enable networking and foster future research collaboration.

LSMS 2010 and ICSEE 2010 received over 880 paper submissions from 22 countries. These papers went through a rigorous peer-review procedure, including both pre-review and formal refereeing. Based on the review reports, the Program Committee finally selected 260 papers for presentation at the conference, from amongst which 66 were subsequently selected and recommended for publication by Springer in two volumes of *Communications in Computer and Information Science* (CCIS). This particular volume of *Communications in Computer and Information Science* (CCIS) includes 34 papers covering 5 relevant topics.

The organizers of LSMS 2010 and ICSEE 2010 would like to acknowledge the enormous contributions from the following: the Advisory and Steering Committees for their guidance and advice, the Program Committee and the numerous referees worldwide for their significant efforts in both reviewing and soliciting the papers, and the Publication Committee for their editorial work. We would also like to thank

Alfred Hofmann, of Springer, for his continual support and guidance to ensure the high-quality publication of the conference proceedings. Particular thanks are of course due to all the authors, as without their excellent submissions and presentations, the two conferences would not have occurred.

Finally, we would like to express our gratitude to the following organizations: Chinese Association for System Simulation (CASS), IEEE SMCS Systems Biology Technical Committee, National Natural Science Foundation of China, Research Councils UK, IEEE CC Ireland chapter, IEEE SMC Ireland chapter, Shanghai Association for System Simulation, Shanghai Instrument and Control Society and Shanghai Association of Automation.

The support of the Intelligent Systems and Control research cluster at Queen's University Belfast, Tsinghua University, Peking University, Zhejiang University, Shanghai Jiaotong University, Fudan University, Delft University of Technology, University of Electronic Science Technology of China, Donghua University is also acknowledged.

July 2010

Bohu Li
Mitsuo Umezu
George W. Irwin
Minrui Fei
Kang Li
Luonan Chen
Xin Li
Shiwei Ma

LSMS-ICSEE 2010 Organization

Advisory Committee

Kazuyuki Aihara, Japan
Shun-ichi Amari, Japan
Erwei Bai, USA
Zongji Chen, China
Peter Fleming, UK
Sam Shuzhi Ge, Singapore
Guo-sen He, China
Huosheng Hu,UK
Tong Heng Lee, Singapore
Frank L. Lewis, USA
Stephen K.L. Lo, UK
Okyay Kaynak, Turkey
Marios M. Polycarpou, Cyprus
Zhaohan Sheng, China
Peter Wieringa, The Netherlands
Olaf Wolkenhauer, Germany
Cheng Wu, China
Yugeng Xi, China
Minlian Zhang, China
Guoping Zhao, China

Steering Committee

Sheng Chen, UK
Kwang-Hyun Cho, Korea
Xiaoguang Gao, China
Tom Heskes, The Netherlands
Shaoyuan Li, China
Liang Liang, China
Zengrong Liu, China
Sean McLoone, Ireland
Robert Harrison, UK
MuDer Jeng, Taiwan, China
Xiaoyi Jiang, Germany
Da Ruan, Belgium
Kay Chen Tan, Singapore
Kok Kiong Tan, Singapore
Stephen Thompson, UK
Haifeng Wang, UK
Tianyuan Xiao, China
Jianxin Xu, Singapore
Guangzhou Zhao, China
Donghua Zhou, China
Quanmin Zhu, UK

Honorary Chairs

Bohu Li, China
Mitsuo Umezu, Japan

General Chairs

George W. Irwin, UK
Minrui Fei, China

International Program Committee

IPC Chairs

Kang Li, UK
Luonan Chen, Japan

IPC Regional Chairs

Haibo He, USA
Amir Hussain, UK
Guangbin Huang, Singapore
Wen Yu, Mexico
John Morrow, UK
Qiguo Rong, China
Shiji Song, China
Taicheng Yang, UK
Jun Zhang, USA
Xingsheng Gu, China
Yongsheng Ding, China
Zhijian Song, China
Ming Chen, China
Feng Ding, China
Weidong Chen, China

IPC Members

Maysam F. Abbod, UK
Peter Andras, UK
Costin Badica, Romania
Vitoantonio Bevilacqua, Italy
Uday K. Chakraborty, USA
Tianlu Chen, China
Yuehui Chen, China
Xinglin Chen, China
Weidong Cheng, China
Minsen Chiu, Singapore
Michal Choras, Poland
Tommy Chow, Hong Kong, China
Kevin Curran, UK
Mingcong Deng, Japan
Frank Emmert-Streib, UK
Jianbo Fan, China
Haiping Fang, China
Jiali Feng, China
Wai-Keung Fung, Canada
Houlei Gao, China
Huijun Gao, China
Xiao-Zhi Gao, Finland
Lingzhong Guo, UK
Xudong Guo, China
Aili Han, China
Minghu Ha, China
Haibo He, USA
Pheng-Ann Heng, China
Laurent Heutte, France
Fan Hong, Singapore
Xia Hong, UK
Wei-Chiang Hong, China
Yuexian Hou, China
Jiankun Hu, Australia
Xiangpei Hu, China
Guangbin Huang, Singapore
Peter Hung, Ireland
Amir Hussain, UK
MuDer Jeng, Taiwan, China
Xiaoyi Jiang, Germany
Pingping Jiang, China
Yasuki Kansha, Japan
Tetsuya J. Kobayashi, Japan
Aim`e Lay-Ekuakillel, Italy
Gang Li, UK
Xiaoou Li, Mexico
Xuelong Li, UK
Yingjie Li, China
Paolo Lino, Italy
Tim Littler, UK
Hongbo Liu, China
Hua Liu, China
Wanquan Liu, Australia
Zhi Liu, China
Sean McLoone, Ireland
Marion McAfee, UK
Fenglou Mao, USA
Kezhi Mao, Singapore
Guido Maione, Italy
John Morrow, UK
Wasif Naeem, UK
Mark Price, UK
Donglian Qi, China
Feng Qiao, China
Alexander Rotshtein, Ukraine
Chenxi Shao, China
Jiafu Tang, China
David Wang, Singapore
Haiying Wang, UK
Hongwei Wang, China
Hui Wang, UK
Kundong Wang, China
Ruisheng Wang, USA
Shujuan Wang, China
Wenxing Wang, China
Yong Wang, Japan
Zhuping Wang, China
Zhengxin Weng, China
Lisheng Wei, China
Ting Wu, China
WeiQi Yan, UK
Rongguo Yan, China
Lianzhi Yu, China
Wen Yu, Mexico
Zhang Yuwen, USA
Hong Yue, UK
Peng Zan, China
Guofu Zhai, China
An Zhang, China
Degan Zhang, China
Qing Zhao, Canada
Lindu Zhao, China
Huiru Zheng, UK
Liangpei Zhang, China
Qingchang Zhong, UK
Huiyu Zhou, UK
Shangming Zhou, UK

Secretary-General

Xin Sun, China
Ping Zhang, China
Huizhong Yang, China

Publication Chairs

Xin Li, China
Wasif Naeem, UK

Special Session Chairs

Xia Hong, UK
Li Jia, China

Organizing Committee

OC Chairs

Shiwei Ma, China
Yunjie Wu, China
Fei Liu, China

OC Members

Min Zheng, China
Yijuan Di, China
Qun Niu, UK
Banghua Yang, China
Weihua Deng, China
Xianxia Zhang, China
Yang Song, China
Tim Littler, UK

Reviewers

Renbo Xia, Vittorio Cristini, Aim'e Lay-Ekuakille, AlRashidi M.R., Aolei Yang, B. Yang, Bailing Zhang, Bao Nguyen, Ben Niu, Branko Samarzija, C. Elliott, Chamil Abeykoon, Changjun Xie, Chaohui Wang, Chuisheng Zeng, Chunhe Song, Da Lu, Dan Lv, Daniel Lai, David Greiner, David Wang, Deng Li, Dengyun Chen, Devedzic Goran, Dong Chen, Dongqing Feng, Du K.-L., Erno Lindfors, Fan Hong, Fang Peng, Fenglou Mao, Frank Emmert-Streib, Fuqiang Lu, Gang Li, Gopalacharyulu Peddinti, Gopura R. C., Guidi Yang, Guidong Liu, Haibo He, Haiping Fang, Hesheng Wang, Hideyuki Koshigoe, Hongbo Liu, Hongbo Ren, Hongde Liu, Hongtao Wang, Hongwei Wang, Hongxin Cao, Hua Han, Huan Shen, Hueder Paulo de Oliveira, Hui Wang, Huiyu Zhou, H.Y. Wang, Issarachai Ngamroo, Jason Kennedy, Jiafu Tang, Jianghua Zheng, Jianhon Dou, Jianwu Dang, Jichun Liu, Jie Xing, Jike Ge, Jing Deng, Jingchuan Wang, Jingtao Lei, Jiuying Deng, Jizhong Liu, Jones K.O., Jun Cao, Junfeng Chen, K. Revett, Kaliviotis Efstathios, C.H. Ko, Kundong Wang, Lei Kang,

Leilei Zhang, Liang Chen, Lianzhi Yu, Lijie Zhao, Lin Gao, Lisheng Wei, Liu Liu, Lizhong Xu, Louguang Liu, Lun Cheng, Marion McAfee, Martin Fredriksson, Meng Jun, Mingcong Deng, Mingzhi Huang, Minsen Chiu, Mohammad Tahir, Mousumi Basu, Mutao Huang, Nian Liu, O. Ciftcioglu, Omidvar Hedayat, Peng Li, Peng Zan, Peng Zhu, Pengfei Liu, Qi Bu, Qiguo Rong, Qingzheng Xu, Qun Niu, R. Chau, R. Kala, Ramazan Coban, Rongguo Yan, Ruisheng Wang, Ruixi Yuan, Ruiyou Zhang, Ruochen Liu, Shaohui Yang, Shian Zhao, Shihu Shu, Yang Song, Tianlu Chen, Ting Wu, Tong Liang, V. Zanotto, Vincent Lee, Wang Suyu, Wanquan Liu, Wasif Naeem, Wei Gu, Wei Jiao, Wei Xu, Wei Zhou, Wei-Chiang Hong, Weidong Chen, WeiQi Yan, Wenjian Luo, Wenjuan Yang, Wenlu Yang, X.H. Zeng, Xia Ling, Xiangpei Hu, Xiao-Lei Xia, Xiaoyang Tong, Xiao-Zhi Gao, Xin Miao, Xingsheng Gu, Xisong Chen, Xudong Guo, Xueqin Liu, Yanfei Zhong, Yang Sun, Yasuki Kansha, Yi Yuan, Yin Tang, Yiping Dai, Yi-Wei Chen, Yongzhong Li, Yudong Zhang, Yuhong Wang, Yuni Jia, Zaitang Huang, Zhang Li, Zhenmin Liu, Zhi Liu, Zhigang Liu, Zhiqiang Ge, Zhongkai Li, Zilong Zhao, Ziwu Ren.

Table of Contents – Part II

The First Section: Advanced Evolutionary Computing Theory and Algorithms

The Second Section: Advanced Neural Network Theory and Algorithms

The Third Section: Innovative Education in Systems Modeling and Simulation

The Fourth Section: Intelligent Methods in Developing Vehicles, Engines and Equipments

The Fifth Section: Fuzzy, Neural, and Fuzzy-Neuro Hybrids

Table of Contents – Part I

The First Section: Intelligent Modeling, Monitoring, and Control of Complex Nonlinear Systems

The Second Section: Modeling and Simulation of Societies and Collective Behaviour

The Third Section: Advanced Theory and Methodology in Fuzzy Systems and Soft Computing

The Fourth Section: Biomedical Signal Processing, Imaging, and Visualization

The Fifth Section: Computational Intelligence in Utilization of Clean and Renewable Energy Resources

The Sixth Section: Innovative Education for Sustainable Energy and Environment

The Seventh Section: Intelligent Methods in Power and Energy Infrastructure Development

Co-Evolutionary Cultural Based Particle Swarm Optimization Algorithm

Yang Sun, Lingbo Zhang, and Xingsheng Gu

Research Institute of Automation, East China University of Science and Technology, 200237, Shanghai, China
xsgu@ecust.edu.cn

Abstract. Particle swarm optimization (PSO), cultural algorithm (CA) and co-evolutionary algorithm (CEA) are all research hotspots in the field of intelligent computing. In order to apply their advantages, a hybrid algorithm CECBPSO is proposed in this paper. In the hybridization, PSO is introduced into the framework of CA, and then a co-evolutionary mechanism between two cultural based PSO algorithms is established. In this way, useful experiences can be exchanged among the populations, and randomly reinitialized particles are introduced into the algorithm. Both of them can help the algorithm improving the efficiency and escape the local optima when the particles get premature. The performance is evaluated on five test functions. Simulation results show that the hybridizing of the three algorithms greatly improves the performance.

Keywords: Global optimization, particle swarm optimization, cultural algorithm, co-evolutionary algorithm

1 Introduction

Intelligent heuristic algorithms are more effective to solve the complicated, nonlinear, discrete optimization problems compared with the traditional mathematic methods. Particle swarm optimization (PSO)[1,2] is a kind of heuristic evolutionary algorithm based on swarm intelligence which was proposed by Kennedy and Eberhart in 1995. It was inspired by the social behavior of birds foraging. Because of its simplicity and effectivity, PSO has attracted more and more attention, and has been applied in many areas[3-5]. However, basic PSO still can be improved[6-7]. Cultural algorithm (CA) was firstly purposed by Reynolds[8] in 1994. CA simulates the social and cultural changes: biological evolution in micro level and cultural evolution in macro level. The dual inheritance enhances the efficiency of CA. Co-evolutionary algorithm (CEA) [9-10] is a kind of new evolution algorithm on the base of co-evolutionary theory.

In this paper, PSO is introduced into the framework of CA, and a new algorithm called CBPSO is put forward. Then a hybridization of PSO, CA and CEA called CECBPSO is proposed by establishing a co-evolutionary mechanism between two CBPSO algorithms in order to fully use the advantages of the basic algorithms.

This paper is organized as follows: Section 2 briefly describes the concepts of PSO, CA and CEA. The proposed algorithms CBPSO and CECBPSO are detailedly

K. Li et al. (Eds.): LSMS/ICSEE 2010, Part II, CCIS 98, pp. 1–7, 2010.

presented in Section 3. A series of experiments on testing the performances are done in Section 4. Finally, Section 5 comes to a conclusion based on the above results.

2 Brief Description of PSO, CA and CEA

Particle Swarm Optimization (PSO). In the basic PSO proposed by Kennedy and Eberhart, each solution is considered as an individual in a group of "birds" (i.e. particles). Initially, particles are assigned a group of random velocities and positions. To achieve the optimum, particles update their velocities and positions as follows:

$$v_{id}^{k+1} = wv_{id}^{k} + c_1 r_1 (p_{id}^{k} - x_{id}^{k}) + c_2 r_2 (p_{gd}^{k} - x_{id}^{k}) \quad (1)$$

$$x_{id}^{k+1} = x_{id}^{k} + v_{id}^{k+1} \quad (2)$$

where v_{id}^{k} and x_{id}^{k} are the velocity and position of the *i*th particle, w is inertia weight, p_{id}^{k} represent *pbest* of *i*th particle, p_{gd}^{k} denotes *gbest*, c_1 and c_2 are acceleration factors, r_1 and r_2 are random numbers distributed in (0,1).

Cultural Algorithm (CA). The individuals of cultural algorithm are divided into two parts: population space and belief space. In the process of evolution, the two spaces first evolve respectively, then they communicate with each other through specific protocols: accept operation and affect operation. In this way, knowledge of the two spaces interacts with each other and supports each other.

Co-Evolutionary Algorithm (CEA). In the co-evolutionary model, the population is divided into several sub-groups. All the sub-groups evolve respectively. At the same time, they also evolve together based of knowledge sharing and information transfer. This is the main idea of the co-evolutionary theory. But, currently, there hasn't been a unified framework of co-evolutionary algorithm. Researchers in various fields set up their own models and algorithms according to their own ideas. The mechanism of coordination in this paper will be shown later in Section 3.2.

3 Co-Evolutionary Cultural Based Particle Swarm Optimization

In order to fully use the advantages of PSO, CA and CEA, a hybridization of them is considered to enhance the performances of algorithms, called CECBPSO (**C**o-**E**volutionary **C**ultural **B**ased **P**article **S**warm **O**ptimization). The hybridization is established in two steps:

Step 1: Introduce PSO into the framework of CA, build a new method called CBPSO (**C**ultural **B**ased **P**article **S**warm **O**ptimization).

Step 2: Establish co-evolutionary mechanism between two or more CBPSO algorithms, get CECBPSO.

3.1 Cultural Based Particle Swarm Optimization

There are two spaces in the framework of CA. Here we use PSO in both of the two spaces. Through the dual inheritance and interaction between the two spaces, CBPSO has better global search capability. CBPSO algorithm is constructed as follows:

1) Design of the population space and the belief space
Assume that there are N particles searching in a D-dimensional space. Divide them into two parts: N_1 particles searching in the population space (PS); and N_2 particles searching in the belief space (BS). The ratio between the number of the particles in PS and BS is represented as BR, generally set at 0.4. The algorithm derives *gbest* of all the N particles, called *ggbest*, according to Eq. (5).

$$N = N_1 + N_2 \tag{3}$$

$$BR = N_2 / N_1 \tag{4}$$

$$ggbest = \text{best}(gbest_P, gbest_B) \tag{5}$$

2) Accept operation
In each generation, the algorithm performs accept operation according to the accept operation probability, called *Acp*. If $rand \leqslant Acp$, the best particle in PS will replace the worst particle in BS.

3) Affect operation
In each generation, the CBPSO also performs affect operation according to the affect operation probability, called *Afp*. The number of the particles joining the affect operation is called *Afn*. If $rand \leqslant Afp$, the top *Afn* particles in BS will replace the bottom *Afn* particles in PS.

Following is the brief procedure of CBPSO:

Step 1: Initiation. In generation k: =1, establish two spaces. Initialize all the particles and evaluate the fitness values. Calculate *ggbest* according to Eq. (5).

Step 2: Termination. Stop the algorithm if the optimal solution is satisfied, otherwise go to Step 3.

Step 3: Updating. Set k: =k+1, update each particle in PS and BS as basic PSO according to Eq. (1) and (2) respectively, and evaluate the fitness.

Step 4: Accept operation.

Step 5: Affect operation.

Step 6: Statistics. Calculate *ggbest* according to Eq. (5). Return to Step 2.

3.2 Co-Evolutionary Cultural Based Particle Swarm Optimization

In this step, a co-evolutionary mechanism between two CBPSO algorithms is established to take full advantage of CEA and CBPSO. As shown in Fig. 1, in CECBPSO, there are two CBPSO populations and a shared global belief space. In each generation, the two populations execute their own PSO operations and cultural operations respectively. After these operations, the shared global belief space begins to play a role. The particles of the two sub-belief spaces are collected into the shared global belief space. Excellent particles will be reserved. The bad ones will be abandoned, and replaced by reinitialized particles. Then the shared global belief space will implement affect operation to population space 1 and 2. The two sub-population spaces also exchange their experiences in each generation. Through the coordination mechanism, the algorithm has a higher probability of jumping out of local optima and the whole swarm can move to the global optimal solution more quickly.

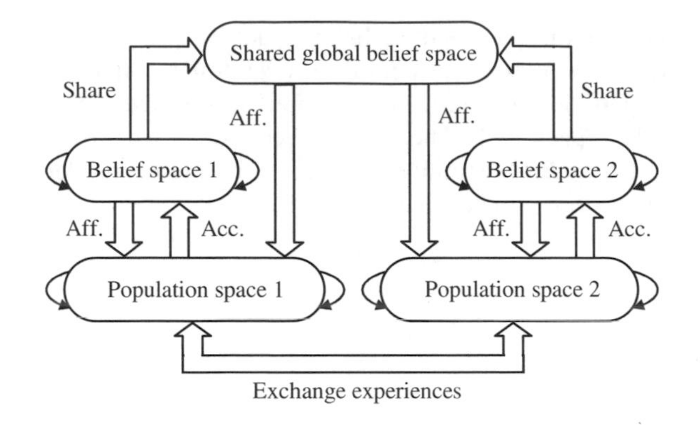

Fig. 1. CECBPSO algorithm model

CECBPSO algorithm is constructed as follows:

1) Design of the two CBPSO populations
Assume that there are $2N$ particles searching in a D-dimensional space. Divide them into four parts: N_{11} particles searching in the population space 1 (P1); N_{12} particles searching in the belief space 1 (B1); N_{21} particles searching in the population space 2 (P2); N_{22} particles searching in the belief space 2 (B2). Define BR_1, BR_2 as Eq. (7) and (8). The best particle of all the $2N$ particles is also called *ggbest*.

$$N = N_{11} + N_{12} = N_{21} + N_{22} \tag{6}$$

$$BR_1 = N_{12} / N_{11} \tag{7}$$

$$BR_2 = N_{22} / N_{21} \tag{8}$$

$$ggbest = \text{best}(gbest_{P1}, gbest_{B1}, gbest_{P2}, gbest_{B2}) \tag{9}$$

2) Design of the shared global belief space
The shared global belief space (SGBS) is composed by N_0 particles.

$$N_0 = N_{12} + N_{22} \tag{10}$$

3) Design of the update of SGBS
In each generation, the algorithm updates the SGBS after the two CBPSO populations' own operations. In each update of SGBS, there are two steps: firstly, all the particles in B1 and B2 compose the particles in SGBS; secondly, reserve the excellent particles and reinitialize the bad ones. The threshold value of reserving and abandoning is called TV, defined as Eq. (11). For each particle in SGBS, if $fitness(x_i) \leqslant TV$, the particle will be reserved, or it will be reinitialized.

$$TV = \sum_{i=1}^{N_0} fitness(x_i^k) / N_0 \tag{11}$$

4) Affect operation from SGBS to P1 and P2
In each generation, the algorithm performs affect operation from SGBS to P1 and P2 according to the affect operation from SGBS probability, called *Afp_SGBS*. There are

other two parameters in this operation, called Afn_1_SGBS and Afn_2_SGBS. The operation is implemented as: If $rand \leqslant Afp\_SGBS$, the Afn_1_SGBS number reinitialized particles in SGBS will replace the bottom Afn_1_SGBS particles in P1 and the bottom Afn_1_SGBS particles in P2; the Afn_2_SGBS number excellent particles in SGBS will replace the other bottom Afn_2_SGBS particles in P1 and the other bottom Afn_2_SGBS particles in P2.

5) Exchange experiences between P1 and P2
In each generation, after the affect operation from SGBS to population spaces, the operation of exchanging experiences between the two population spaces will be implemented. In this operation, two parameters are defined: the probability of exchanging experiences, called *EEp*, and the number of particles joining this operation, called *EEn*. The operation is implemented as: If $rand \leqslant EEp$, the *EEn* excellent particles in P1 will replace the bottom *EEn* particles in P2, and also, the *EEn* excellent particles in P2 will replace the bottom *EEn* particles in P1.

Following is the brief procedure of CECBPSO:

Step 1: Initiation. In generation k: =1, establish four spaces. Initialize all the particles and evaluate the fitness of each particle. Establish the shared global belief space (SGBS). Calculate the *ggbest* of the spaces according to Eq. (9).

Step 2: Termination. Stop the algorithm if the optimal solution is satisfied, otherwise go to Step 3.

Step 3: Updating. Set k: =k+1, update the velocity and position of each particle in P1, B1, P2 and B2 as basic PSO, and evaluate the fitness values.

Step 4: Accept operation from P1 to B1 and affect operation from B1 to P1.

Step 5: Accept operation from P2 to B2 and affect operation from B2 to P2.

Step 6: Update SGBS.

Step 7: Affect operation from SGBS to P1 and P2.

Step 8: Exchange experiences between P1 and P2.

Step 9: Statistics. Calculate the *ggbest* according to Eq. (9). Return to Step 2.

4 Simulations

Five well-known test functions are used in the performance test: Sphere, Rosenbrock, Rastrigin, Schaffer and Ackley. The formulas and parameters can be seen in [7].

To examine the performance of CBPSO and CECBPSO, the two algorithms are compared with other three algorithms: basic PSO, CPSO[6] and PSOPDE[7].These methods have the same total particles number of 60, and maximum generation Max_{gen}=1000. Each one of them is calculated for 60 times. For the basic PSO and each improved PSO, the inertia weight w linearly declines from w_{max}=0.95 to w_{min}=0.4, and the acceleration factors c_1 and c_2 are compromisingly set to c_1=c_2=1.5. Other parameters in CPSO and PSOPDE are set according to [6] and [7]. In CBPSO, parameters are set as: N_1=43, N_2=17, BR=0.39, Acp=0.1, Afp=0.4, and Afn=1. In CECBPSO, N_{11}=N_{21}=21, N_{12}=N_{22}=9, BR_1=BR_2=0.43, Acp_1=Acp_2=0.1, Afp_1=Afp_2=0.4, Afn_1=Afn_2=1, Afp_SGBS=0.8, Afn_1_SGBS=Afn_2_SGBS=5, EEp=0.8, EEn=5.

Table 1. Results of five algorithms on all test functions

Fun.	Item	Basic PSO	CPSO	PSOPDE	CBPSO	CECBPSO
f_1	Best	5.3956×10^{-8}	3.8194×10^{-13}	2.8707×10^{-22}	1.2628×10^{-29}	9.1779×10^{-48}
	Worst	5.3000×10^{-3}	4.9924×10^{-6}	3.4741×10^{-12}	5.7217×10^{-21}	3.0716×10^{-41}
	Mean	2.1961×10^{-4}	1.0507×10^{-7}	6.4784×10^{-14}	1.5760×10^{-22}	1.2571×10^{-42}
	Std	7.3759×10^{-4}	$6.4823\times10^{--7}$	4.4863×10^{-13}	8.0771×10^{-22}	4.1763×10^{-42}
	Rate	1.0000	1.0000	1.0000	1.0000	1.0000
	Itera	293.4000	605.9500	462.3167	454.4167	188.1667
f_2	Best	0.9927	5.8822	16.8873	0.1214	0.0034
	Worst	129.3654	169.2358	183.3045	116.1462	21.6303
	Mean	54.7950	47.9790	66.2669	30.6171	13.2810
	Std	32.7101	34.2021	39.4233	25.4905	3.9466
	Rate	0.9667	0.9500	0.8667	0.9833	1.0000
	Itera	499.9483	472.4386	264.8654	464.9492	110.9333
f_3	Best	22.8840	17.9093	30.8563	26.8639	9.9496
	Worst	94.5209	68.6521	74.6360	80.5914	44.7731
	Mean	53.0810	34.9230	47.0675	47.6087	21.4082
	Std	15.7868	9.5428	11.8403	12.1095	6.6794
	Rate	1.0000	1.0000	1.0000	1.0000	1.0000
	Itera	318.6000	478.4167	70.1500	285.8667	128.5333
f_4	Best	0	0	0	0	0
	Worst	0.0097	0.0097	0	0	0
	Mean	0.0015	0.0008	0	0	0
	Std	0.0035	0.0027	0	0	0
	Rate	0.8500	0.9167	1	1	1
	Itera	113.5490	166.0909	338.2833	167.7333	34.3333
f_5	Best	2.9850×10^{-7}	1.6251×10^{-8}	1.9450×10^{-7}	1.5072×10^{-12}	7.9936×10^{-15}
	Worst	2.2223	2.0119	2.1201	1.8997	2.9310×10^{-14}
	Mean	0.5587	0.4869	0.2189	0.2683	1.4921×10^{-14}
	Std	0.7032	0.6991	0.5119	0.5604	5.1265×10^{-15}
	Rate	0.1667	0.5333	0.7500	0.8000	1
	Itera	762.3000	920.5625	953.5333	664.4167	426.4333

Table 1 shows the performance of the five algorithms. Best, Worst, Mean, Std, Rate and Itera represent the best minimum, the worst minimum and the mean minimum in the 60 trials, the standard deviation, the success rate and the average iterations of having reached the goal.

Seen form Table 1, all the five algorithms can get the sole optimum in the test of Sphere function. Only CECBPSO can find the greatest fitness in 1000 generations. In the test of Rosenbrock function, CECBPSO has the success rate of 100%, and has the highest precision and efficiency. For the Rastrigin function, all the five algorithms have the 100% success rate. Basic PSO finds the worst solution, and CPSO has the lowest optimal efficiency. PSOPDE and CBPSO find almost the same fitness value. Comparatively, CECBPSO finds the best solution. For Schaffer function, PSOPDE, CBPSO and CECBPSO find the global minimum in all the 60 trials. Comparatively, CECBPSO has the greatest performance because of its efficiency. The results of basic PSO and CPSO are not satisfactory. Like Rastrigin and Schaffer function, Ackley function is also a multimodal function which has a lot of local minima. Obviously,

CECBPSO has an excellent performance in each aspect. The other four algorithms find close fitness value. Basic PSO has the lowest success rate.

5 Conclusion

Considering the features of basic PSO, CA and CEA, a hybrid algorithm CECBPSO is proposed in this paper. Based on the hybridization, multi-populations are established. The mechanism of sharing knowledge and experiences can help the algorithm improving the efficiency. The randomly reinitialized particles make the algorithm easier to jump out of local minima. The performance of CECBPSO is evaluated on five test functions. The results show that CECBPSO has much higher accuracy and efficiency than the other four algorithms, and the hybridizing of the three algorithms greatly improves the performance.

Acknowledgments. We are very grateful to the editors and anonymous reviewers for their valuable comments and suggestions to help improve our paper. This work is supported by National High Technology Research and Development Program of China (863 Program) (No. 2009AA04Z141), Shanghai Commission of Science and Technology (Grant no. 08JC1408200), Shanghai Leading Academic Discipline Project (Grant no. B504).

References

1. Eberhart, R.C., Kennedy, J.: A new optimizer using particle swarm theory. In: Proceedings of the sixth international symposium on micro machine and human science, pp. 39–43. IEEE Press, Nagoya (1995)
2. Kennedy, J., Eberbart, R.C.: Particle swarm optimization. In: Proceedings of the IEEE international conference on neural networks, vol. IV, pp. 1942–1948. IEEE Press, Piscataway (1995)
3. Shi, Y., Eberhart, R.C.: Fuzzy adaptive particle swarm optimization. In: Proceedings of the 2001 Congress on Evolutionary Computation, pp. 101–106. IEEE Press, Seoul (2001)
4. Hu, X.H., Eberhart, R.C.: Multi-objective optimization using dynamic neighborhood particle swarm optimization. In: Proceedings of the 2002 Congress on Evolutionary Computation, pp. 1677–1681. IEEE, Honolulu (2002)
5. Juang, C.F.: A hybrid of genetic algorithm and particle swarm optimization for recurrent network design. J. IEEE Trans. Syst. Man and Cybernetics: Part B-Cybernetics 34, 997–1006 (2004)
6. Liu, B., Wang, L., Jin, Y.-H., et al.: Improved Particle Swarm Optimization Combined with Chaos. J. Chaos, Solitons and Fractal 25, 1261–1271 (2005)
7. Xu, W., Gu, X.: A hybrid particle swam optimization approach with prior crossover differential evolution. In: Proceedings of the 1st ACM/SIGEVO Summit on Genetic and Evolutionary Computation, Shanghai, pp. 671–677 (2009)
8. Reynolds, R.G.: An Introduction to Cultural Algorithms. In: Proceedings of the Third Annual Conference on Evolutionary Programming, pp. 131–139. World Scientific River Edge, New Jersey (1994)
9. Potter, M.A.: The design and analysis of a computational model of cooperative coevolution. Ph. D. dissertation, George Mason University (1997)
10. Potter, M.A., De Jong, K.A.: A cooperative coevolutionary approach to function optimization. In: Proceedings of the Third Conference on Parallel Problem Solving From Nature, New York, pp. 249–257 (1994)

Non-cooperative Game Model Based Bandwidth Scheduling and the Optimization of Quantum-Inspired Weight Adaptive PSO in a Networked Learning Control System

Lijun Xu[1,*], Minrui Fei[1,**], and T.C. Yang[2]

[1] Shanghai Key Laboratory of Power Station Automation Technology,
Shanghai University, Shanghai 200072, China
[2] University of Sussex, UK
ruby_mickey@sina.com, mrfei@staff.shu.edu.cn,
taiyang@sussex.ac.uk

Abstract. In this paper, under a framework of Networked two-layer Learning Control Systems (NLCSs), optimal network scheduling is studied. Multi networked feedback control loops called subsystems in a NLCS share common communication media and therefore there is a competition for available bandwidth and data rate. A non-cooperative game(NG) model is first formulated for the problem studied. The existence and uniqueness of Nash Equilibrium point is proved. Subsequently, the utility function of subsystems is designed, taking account of both transmission data rate and control sampling period according to the feature of scheduling pattern and network control. Following this, a quantum-inspired weight adaptive particle swarm optimization algorithm is developed to obtain an optimal solution. Simulation results presented in the paper have demonstrated the effectiveness of the proposed theoretical approach and the algorithm developed.

Keywords: NCS, Scheduling, Non-cooperative Game, PSO, Data rate.

1 Introduction

While majority of researches are carried out on single-layer networked control system (NCS) [1], little has been done to investigate nonlinear with more complex NCS architecture for nonlinear plants. We have already proposed two-layer networked learning control system (NLCS[2]) architecture suitable for complex plants.

* This work is supported by National Natural Science Foundation of China under Grant 60834002 and 60774059, the Excellent Discipline Head Plan Project of Shanghai under Grant 08XD14018 and Mechatronics Engineering Innovation Group project from Shanghai Education Commission.

** Corresponding author.

K. Li et al. (Eds.): LSMS/ICSEE 2010, Part II, CCIS 98, pp. 8–15, 2010.

In real-time NLCS, problems of limited network resources are catastrophic. The quality for performances of network service (QoS) and control system (QoC) rely on not only the design of system architecture and control algorithm, but also scheduling of network information to reduce information transmission collision and implement resource allocation[3] of network nodes. Therefore, it is necessary to introduce new methods and theories to scheduling and optimization of NLCS. Game Theory and Intelligent Optimization methods provide efficient mathematical foundations.

Lots of successful applications of non-cooperative Game(NG)[4] have been reported in many complex optimization and decision problems[5]. NG emphasizes individual rational in which the equilibrium solutions of competition satisfying individual rational are defined by Nash equilibrium (NE)[6]. By designing a rational competitive mechanism according to scheduling in NLCS, a satisfactory solution concurrently guaranteeing individual and collective requirements can be induced.

Particle swarm optimization (PSO)[7] is a class of stochastic optimization technique inspired by the behavior of bird flocks. To improve performance of original PSO algorithm and avoid trapping to local excellent situations, we construct the new quantum solutions expression and adaptive weight for multi-objective optimization particle swarm to seek the optimal decision of network resources allocation.

The rest of the paper is organized as follows: Section 2 presents a NLCS and NG based scheduling optimization scheme. Section 3 introduces available resources changes into solving the equilibrium point and utility function. Section 4 presents Quantum-Inspired Weight Adaptive PSO to solve the problem efficiently. Section 5 provides the simulation results. Finally, a brief conclusion is given in Section 6.

2 Non-cooperative Game Model for Scheduling in NLCS

NLCS contains many subsystems which may complete a series of different tasks. Under the limitations of communications and network resources, subsystems will compete for consuming network resources according to their own needs. Consequently, the mathematical description of two-layer NLCS is demonstrated: The number of subsystems is n and the objective set is $L = \{l_i | 1 \le i \le n\}$.

Because of the diversity of data formats in system, different methods and coding techniques are used to achieve higher coding efficiency. The encoded and compressed data are transmitted in a shared network. To ensure a stable transmission data rate, a rate control algorithm dynamically adjust parameters of encoders and varying sampling strategies are applied to afford the target data rates. Optimal data rate set in NLCS is $D^{\max} = \{\delta_i^{\max} | 1 \le i \le n\}$. The minimum of data rate is $\delta_i^{\min}$ for subsystem to ensure the most basic work. Available data rate is $\delta_i \in [\delta_i^{\min}, \delta_i^{\max}]$, and the available data rate set of all subsystems in NLCS is $D = \{\delta_i | 1 \le i \le n\}$. The characteristic of subsystem is described as a probability distribution and probability density function, obtaining the mean value of occupied bandwidth for a subsystem.

$$f_i(\delta_i)=\begin{cases}0 & ,\delta_i<\delta_i^{\min}\\ (\frac{\Delta}{\Delta_i})^c, & \delta_i^{\min}\le\delta_i\le\delta_i^{\max}\\ 1 & ,\delta_i>\delta_i^{\max}\end{cases},\quad g_i(\delta_i)=\begin{cases}\frac{c(\Delta)^{c-1}}{(\Delta_i)^c}, & \delta_i^{\min}\le\delta_i\le\delta_i^{\max}\\ 0 & ,\delta_i<\delta_i^{\min}\, or\, \delta_i>\delta_i^{\max}\end{cases}$$

$$Ef_i(\delta_i)=\int_{-\infty}^{+\infty}\delta_i g_i(\delta_i)\mathrm{d}\delta_i=\int_{\delta_i^{\min}}^{\delta_i^{\max}}\delta_i c(\Delta)^{c-1}/(\Delta_i)^c\,\mathrm{d}\delta_i=\frac{c\delta_i^{\max}+\delta_i^{\min}}{c+1} \tag{1}$$

All the subsystems of the system have formed a NG. The NG model is a triple, namely $G(L,S,u_i)$, where S is the set of strategies including transmission data rate, allotted bandwidth and sampling period for subsystems, that is, $S=\{s_i|s_i=(b_i,\delta_i,t_i),i\in[1,L]\}$. $b_i\in[\delta_i^{\min},U_d]$ is the pre-allocated data rate to the subsystem l_i and t_i is the sampling period. Maximum bandwidth of the network is U_d and $\delta_i^{\max}\le U_d$. u_i is the utility function of l_i, and utility function u_i is the mapping of S to the set of real numbers $u: S\to R$. If and only if $u_i(s_i^*)>u_i(s_i)$, characteristic of s_i^* is better than s_i for subsystem l_i.

3 Nash Equilibrium in Scheduling Strategy of NLCS

Nash Equilibrium (NE) is an important analysis means of NG theory[8]. NE describes that all the participants face a situation of not changing his own strategy either when other people will not change, that is, his strategy is optimal at this moment.

NE of NG for scheduling in NLCS is defined: n control subsystems participated in NG. In case that allocated transmission data rate and sampling period strategy profile $\{s_1^*,s_2^*,\cdots,s_n^*\}$ satisfy that s_i^* of each subsystem l_i is the optimal one or at least not inferior to $\{s_1^*,\cdots,s_{i-1}^*,s_{i+1}^*,\cdots,s_n^*\}$ of the other $n-1$ subsystems.

Above-mentioned utility function u_i describes the Nash equilibrium as:

$$\forall s_i\in S\quad u_i\left(s_1^*,\cdots,s_{i-1}^*,s_i^*,s_{i+1}^*,\cdots,s_n^*\right)\ge u_i\left(s_1^*,\cdots,s_{i-1}^*,s_i,s_{i+1}^*,\cdots,s_n^*\right)$$

$$\text{Denoted of}\qquad u_i\left(s_i^*,s_{-i}\right)\ge u_i\left(s_i,s_{-i}\right) \tag{2}$$

Theorem: The Nash equilibrium exists and is unique if every player's best response function $s_i^*=r(s_{-i}), i\in[1,L]$ such that:

(1) $r(s_{-i})$ is a differentiable function to each strategy $s_j, j \neq i, j \in [1, L]$;

(2) $\sum_{j=1,\ j\neq i}^{n} \left|\partial r_i(s_{-i})/\partial s_j\right| \leq \lambda < 1$.

According to the above theorem, we structured the utility function $u_i\left(\delta_i, b_i, t_i\right) =$ (3)

$$\begin{cases} \left[\|e_i\|^2 - (\delta_i - Ef_i)^2 - \gamma_i(\alpha_i + \beta_i t_i)\right]\left(\dfrac{b_i}{Ef_i}e\right)^{\mu} / e^{\mu\frac{b_i}{Ef_i}} \|e_i\|^3, & 0 \leq b_i \leq Ef_i \\ \left[1 - \left(\dfrac{b_i - Ef_i}{Ef_i - \delta_i^{\max}}\right)^{\nu}\right] \dfrac{\left[\|e_i\|^2 - (\delta_i - Ef_i)^2 - \gamma_i(\alpha_i + \beta_i t_i)\right]}{\|e_i\|^3}, & Ef_i \leq b_i \leq \delta_i^{\max} \end{cases}$$

Rights Law compound data rate δ_i and sample period t_i, where γ_i is weight coefficient, which balance two of them. $\|e_i\|$ is maximum mathematical expectation deviation between optional data rate and data rate distribution of each subsystem, that is $\|e_i\| = \max\left(\left|\delta_i^{\min} - Ef_i\right|, \left|\delta_i^{\max} - Ef_i\right|\right)$. And μ, ν are the empirical constants may be set for specific subsystems to adjust rate of change of utility functions.

4 Quantum-Inspired Weight Adaptive PSO for NG Scheduling

In the analysis of network load distribution, the key is to find a optimal allocation scheme for subsystems. To improve performance of original PSO[9] and avoid trapping to local excellent situations, we construct a new quantum solutions expression and adaptive weight for MOO particle swarm to seek the optimal decision.

From the perspective of quantum mechanics, a new PSO algorithm model based on the DELTA potential is proposed, whose particles have quantum behavior. The new PSO algorithm is simple and easy to achieve and have advantages of less adjustable parameters, with good stability and convergence.

The basic unit of information in quantum computation [10] is qubit. A qubit is a two-level quantum system and it can be represented by a unit vector of a two dimensional Hilbert space $(\alpha, \beta \notin \mathbb{C})$. Inspired by the concept of quantum computing, PSO is designed with a novel Q-bit representation, a Q-gate as a variation operator, and an observation process. The following rotation gate is used in PSO:

$$\begin{bmatrix} \alpha_i^* \\ \beta_i^* \end{bmatrix} = U(\Delta\theta_i) \begin{bmatrix} \alpha_i \\ \beta_i \end{bmatrix} = \begin{bmatrix} \cos(\Delta\theta_i) & -\sin(\Delta\theta_i) \\ \sin(\Delta\theta_i) & \cos(\Delta\theta_i) \end{bmatrix} \begin{bmatrix} \alpha_i \\ \beta_i \end{bmatrix} \tag{4}$$

$$U(\Delta\theta)=\begin{bmatrix}\cos(\Delta\theta) & -\sin(\Delta\theta)\\ \sin(\Delta\theta) & \cos(\Delta\theta)\end{bmatrix} \tag{5}$$

$$\theta_i = s(\alpha_i,\beta_i)\times\Delta\theta_i \tag{6}$$

The basic idea would take each individual as a particle which has no size and no quality in a N-dimensional space. In N-dimensional search space, $X^i=\left(x_{i,1},x_{i,2},\cdots x_{i,N}\right)$ and $V^i=\left(v_{i,1},v_{i,2},\cdots v_{i,N}\right)$ are respectively the position and speed of particle i. Each iteration, particles update by tracking two optimal solutions. One is individual optimal solution *pbest* $P^i=\left(p_{i,1},p_{i,2},\cdots p_{i,N}\right)$; the other is the global optimal solution *gbest* P_g. All particles in population will be seen as smart group. We find partial quantum optimal angle and the overall quantum optimal angle in each iteration process. According to quantum revolving door dynamic adjust quantum angle. Its specific operations: the speed, location, the best individual and overall position respective is $v_{i,j}$, $\theta_{i,j}$, $\theta_{i,j}^{pbest}$ and $\theta_{i,j}^{gbest}$. Speed and location of the iterative formula are as follows:

$$\begin{gathered} v_{i,j}(t+1)=wv_{i,j}(t)+c_1r_1\left(\theta_{i,j}^{pbest}-\theta_{i,j}(t)\right)+c_2r_2\left(\theta_{i,j}^{gbest}-\theta_{i,j}(t)\right)\\ \theta_{i,j}(t+1)=\theta_{i,j}(t)+v_{i,j}(t+1)\end{gathered} \tag{7}$$

Where c_1,c_2 are positive learning coefficients. r_1,r_2 are random numbers in $[0,1]$. To balance global search capability and capacity of local improvements of Quantum-inspired PSO algorithm, nonlinear dynamic inertia weight w is presented, whose value automatically changes with particle objective function:

$$w=\begin{cases} w_{\max} & ,f>f^{avg}\\ w_{\min}-(w_{\max}-w_{\min})(f-f^{\min})/(f^{avg}-f^{\min}), & f\le f^{avg}\end{cases} \tag{8}$$

Where $w_{\max}$ and $w_{\min}$ are extremums of w. $f=\sum u_i$ is the current value of particle objective function. f^{avg} is the average value and $f^{\min}$ is the minimum one.

5 Simulation Analysis

We present results of a simulation to operate and evaluate NG model of scheduling strategy and show validity and solution of Quantum-Inspired Weight Adaptive PSO of NLCS in a particular network environment. We run program on Intel(R) Core(TM) 2 Quad CPU and 4G DDR3 RAM, compiled environment of Matlab 7.0.

Table 1. Range of the output data rates of subsystems for two distributions

Distribution of Output Data Rate	Discrete Normal Distribution		Stochastic Distribution	
Range of Data Rate	$\delta_i^{\min}$	$\delta_i^{\max}$	$\delta_i^{\min}$	$\delta_i^{\max}$
Subsystems l_i $i=1,2,\cdots,10$	42.49	94.95	28	44
	64.83	105.04	68	92
	76.43	114.32	96	134
	87.91	122.52	18	26
	99.28	129.48	101	119
	98.12	131.32	118	134
	89.55	124.75	33	42
	79.35	116.89	88	118
	67.79	107.92	21	32
	54.84	98.04	305	401

In the test environment, we set the number of subsystems $n = 11$, where the 11th subsystem is used as the disturbance node, and required range of output data rate distribution of them as $\left(\delta_1^{\min},\delta_1^{\max}\right)\rightarrow\left(\delta_{10}^{\min},\delta_{10}^{\max}\right)$ in Table 1 .We prepared two different distributions of data rates which are Discrete Normal Distribution(DND) and Stochastic Distribution(SD) aiming to the simulations are of universal significance. The empirical constants are $c=4,\mu=6,\nu=4$. The pre-allocation of network resources to the system is $U_d(1-\omega)=1000\text{Kbps}$.

When the overall fitness value stabilized, we get the results by earlier described optimization calculation. $\{b_1,b_2,\cdots b_{10}\}=\{73.72, 89.04, 99.37, 110.21, 119.38, 119.72, 112.15, 103.12, 92.85, 80.44\}$ of DND and {41, 82, 129, 23, 112, 125, 38, 107, 28,315} of SD, that are the Nash equilibriums separately. Figure 1 and 2 are the proportion of statistical output data rates lower limit, the reserved data rate and output data limit, showing the ratio between reserved data rates and limits within the requirements of the two distributions of data rates. It is obvious that game model based optimize the network resource allocation in NLCS is fair which can be found from the results, and able to find a reasonable solution under the constraints states. The overall performance of the system remained stable by effectively restricting large flow subsystem of network resource utilization.

Figure 3 is the ratio diagram of scheduling adjusted sampling period of subsystem and original sampling period t_i^* / t_i, showing that the lower ratio, the Optimizing is better and the optimization of the sampling period after the adjustment within the requirements of the two data rates distributions. When DND data rates is used, the larger probability distribution data rate is of better optimization effect, indicating that this optimization method is adequate for requirements of most data rate. With SD, the ratio mean is 77% with the same effect of discrete normal distribution. Thus, game theory based resource scheduling has good effects on sampling period optimization.

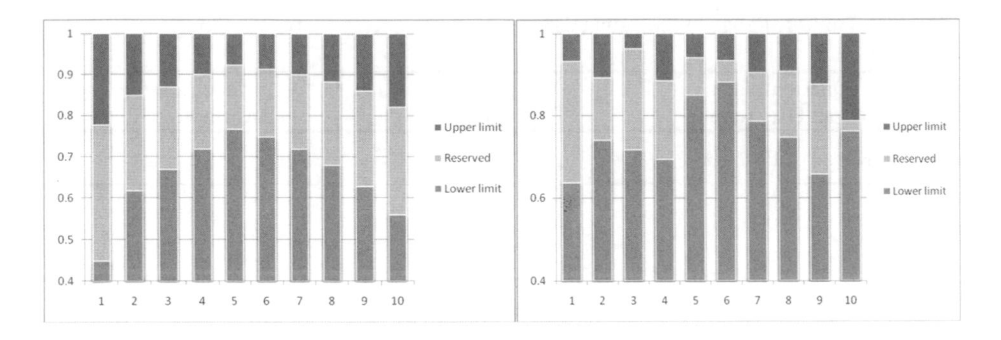

Fig. 1 and 2. Data rate percentage plot of DND and SD

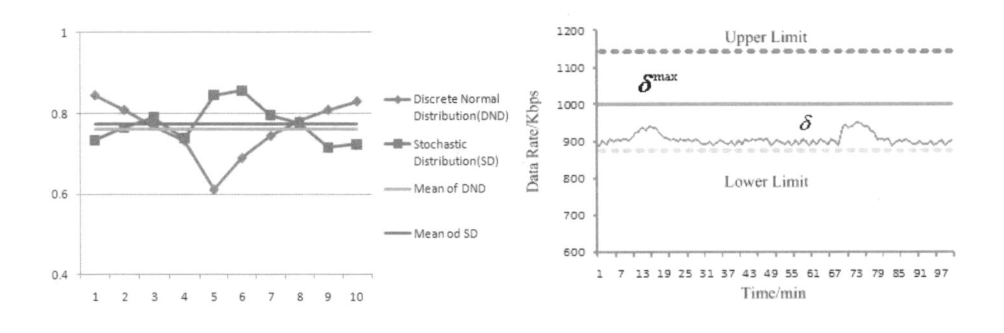

Fig. 3. t_i^* / t_i proportion

Fig. 4. Total data rate variation simulation in NLCS

In order to obtain the better observation of the actual effect, we established a pilot network. Each subsystem has its end to end QoS control and rate control. The total desired bandwidth is 1000Kbps. We obtained the network resource allocation scheme by calculating, and then set the QoS policy on the server.

The subsystems with random distribution data rate was tested on the anti-interference. We plus period of 30s and amplitude of 10K of the noise in [10,20] and [70, 80] minutes. Interference node is on the role, who constantly send specified sized UDP packets to be bandwidth-intensive. We simulate the following three conditions: 1) Control network topology changes, a sudden increase in the number of nodes; 2) Increased demand for network resources of nodes; 3) The emergence of external interference. Experimental simulation runs 100mins; data acquisition sampling period is 1min. Statistics server records amount of data transmission of each subsystem, accessing the control network total transfer data rate of change is shown in graph 4.

Through the experimental simulation, we found that game model based optimize resource scheduling strategy can effectively meet the needs of limited resources control network. Under the condition of periodic and large random disturbances, system can still work stably between minimum bandwidth and the designed bandwidth to guarantee the fairness of using limited bandwidth of various subsystems. Scheduling results are much better than traditional First In First Out(FIFO), Priority queue scheduling(PQ), and weighted fair queuing scheduling(WFQ) etc..

6 Conclusion

Resource allocation problem in NLCS is a multi-constrained nonlinear optimization problem. We established a non-cooperation game model and the distribution of available resources changed into solving the equilibrium point. Game model based optimize resource scheduling strategy of quantum-inspired weight adaptive PSO algorithm can effectively meet the needs of limited resources control network. But the performances of the algorithm to some extent rely on empirical constants, needing to manually be set. Consequently, our next steps are proposing more efficient algorithm and introducing some uncertainties in non-cooperation game model intending to solve more complex problems as multi-layer hybrid scheduling in NLCS.

References

1. Yang, T.C.: Networked control system: a brief survey. J. IEE Proc. Control Theory Appl. 153(4), 403–412 (2006)
2. Du, D.J., Fei, M.R., Li, K.: A two-layer networked learning control system using actor-critic neural. J. Applied Maths and Computation 205(1), 26–36 (2008)
3. Salles, R.M.: Lexicographic maximin optimisation for fair bandwidth allocation in computer networks. J. European Journal of Operational Research 185, 778–794 (2008)
4. Nash, J.F.: NGs. J. Annals of Mathematics 54(2), 286–295 (1951)
5. Ganesh, A.: Congestion Pricing and Noncooperative Games in Communication Networks. J. Operations Research 55(3), 430–438 (2007)
6. Nash, J.F.: Equilibrium Points in n-Person Games. J. Proceedings of the National Academy of Sciences of the United States of America 36(1), 48–49 (1950)
7. Goh, C.K., Tan, K.C., Liu, D.S., Chiam, S.C.: A competitive and cooperative co-evolutionary approach to multi-objective particle swarm optimization algorithm design. J. European Journal of Operational Research 202, 42–54 (2010)
8. Jiang, Y.B.: Non-cooperative network scheduling game model of streaming media based surveillance system and its optimization based on genetic algorithm. J. Chinese Journal of Scientific Instrument 28(10), 1800–1805 (2007)
9. Chao-Tang: A discrete particle swarm optimization for lot-treaming flowshop scheduling problem. J. European Journal of Operational Research 191, 360–373 (2008)
10. Malossini, A., Blanzieri, E., Calarco, T.: Quantum Genetic Optimization. J. IEEE Trans. on evolutionary computation 12(2), 231–241 (2008)

Modified Bacterial Foraging Optimizer for Liquidity Risk Portfolio Optimization

Ben Niu[1,2], Han Xiao[2], Lijing Tan[3], Li Li[2], and Junjun Rao[2]

[1] Hefei Intelligent Computing Lab, Hefei Institute of Intelligent Machines, Chinese Academy of Science, Hefei 230031, China
[2] College of Management, Shenzhen University, Shenzhen 518060, China
[3] Measurement Specialties Inc., Shenzhen 518107, China
drniuben@gmail.com

Abstract. Recently, bacterial foraging optimizer (BFO) is gaining popularity in the community of researchers because of its efficiency in solving some real-world optimization problems. But very little research work has been undertaken to deal with portfolio optimization problem using BFO approach. This article comes up with a novel approach by involving a linear variation of chemotaxis step in the basic BFO for finding the optimal portfolios. Our proposed approach is evaluated on application on an improved portfolio optimization model considering both the market and liquidity risk. The experimental results demonstrate the positive effects of the strategy.

Keywords: Liquidity risk, portfolio optimization, bacterial foraging.

1 Introduction

In recent years, for solve the complex non-linear questions, swarm intelligent optimization algorithms as a new kind of computational intelligence have been proposed. But most of swarm intelligent optimization algorithms are inspired by the behavior of animals with higher complexity. Genetic algorithms (GA) [1-2] simulate the crossover and mutation of the natural selection and genetic replication. Particle swarm optimization (PSO) [3] was gleaned ideas from swarm behavior of bird flocking or fish schooling. However the states of the above mentioned animals are more complex and their behaviors are difficult to describe qualitatively. As prokaryote, bacteria behave in a simple pattern which can be easily described. Inspired by the foraging behavior of *Escherichia coli* (E.coli) in human intestines, Passino proposed an optimization algorithm known as bacterial foraging optimization (BFO) recently [4].

But, as a new algorithm, BFO still has many disadvantages. For instance, when chemotaxis step length (C) was too large bacteria failed to locate the global optimum by swimming without stop. While C was too small, it took a long time for the swarm to find the global optimum. But in the original bacterial foraging algorithm the

K. Li et al. (Eds.): LSMS/ICSEE 2010, Part II, CCIS 98, pp. 16–22, 2010.

chemotaxis step length C is a constant. In this paper, we proposed an improved BFO model with linear decreasing chemotaxis step (BFO-LDC). The purpose of this paper is to apply the proposed BFO-LDC technique to the portfolio optimization problem.

Modern portfolio analysis started from pioneering research work of Markowitz [5] who proposed the original mean–variance model. Sharpe [6] developed the mean-variance theory, formed the capital asset pricing model (CAPM) that showed an asset's expected return is proportional to the risk that is not dispersed. JP Morgan Company developed a new method namely value-at-risk to measure risk approach, VAR is a widely used risk measure of the risk of loss on a specific portfolio of financial assets. Bangia [7] improved a new portfolio optimization model considering both the liquidity risk and the market risk by VAR. Alexander and Baptista [8] established a new model to measure portfolio risk, using VAR replaced the variance. Consigi [9] and Berkowitz [10] studied the mean-VAR model. In the paper, we improve a new model using VAR for measuring both market and liquidity risk. The proposed BFO-LDC algorithm is devised to obtain the solutions of the improved portfolio optimization model. The results of the proposed method are also compared with genetic algorithm, particle swarm optimization and basic bacterial foraging optimization.

2 The Improved Model—Liquidity Risk Portfolio Optimization

Liquidity is another important feature of the securities beside the price volatility. It is generally believed a price balance ability when a lot of trading in securities. Liquidity risk can be divided into two categories according to their source: exogenous liquidity risk and endogenous liquidity risk.

1) Exogenous liquidity risk is the result of market characteristics. It is common to all market players and unaffected by the actions of any one participant. It is determined by the market. Small and stable bid-ask spreads, stable and high levels of quote depth, which make exogenous liquidity good.
2) Endogenous liquidity risk is specific to one's position in the market, varies across market participants, and the exposure of any one participant is affected by her actions, the larger the size, the greater the endogenous liquidity risk.

As the BDSS model, it focused on methods for quantifying exogenous rather than endogenous liquidity risk, it used the index to measure the exogenous risk as form:

$$ECL = \frac{1}{2} p_t (\bar{s} + \alpha\bar{\delta}) \tag{1}$$

$\bar{s}$ is the average relative spread ($\bar{s} = \frac{1}{k}\sum_{i=1}^{k} \frac{(ask_i - bid_i)}{(ask_i + bid_i)/2}$).

$\bar{\delta}$ is the volatility of relative spread.

α is the scaling factor such that we achieve roughly a 99% probability coverage.

P_t is today's mid-price for the asset or instrument.

But, Amihund [6] said: the intraday data is not easy to get, and the data is not important. Beside Chinese securities market is different from the foreign, it is auction system, and so we make these indexes to measure the liquidity risk: max daily price, min price, close price and the turnover rate.

$$V = \frac{p_h - p_l}{p_c} \frac{1}{to} \tag{2}$$

p_h, p_l, p_c is the max, min and close daily price, to is the relation turnover rate.

The V index doesn't contain the trading volume and size. So the index can not measure the endogenous risk. We use the daily volume to weight the max, min and close price. And then the new index can contain the two liquidity risks.

$$P_i = \sum_{n-1}^{3} p_i vol(n) / [vol(1) + vol(2) + vol(3)] \quad i = h, l, c \tag{3}$$

$$ECL = \frac{1}{2} P_t (W\bar{A}V + \alpha' \delta') \tag{4}$$

$vol(n)$ is the daily volume $n = 1, 2, 3$.

$W\bar{A}V$ is the average relative price , is defined as $WAV = \frac{P_h - P_l}{P_c} \frac{1}{to}$. As the mean-VAR model, the market price risk is $\Phi^{-1}(c)\delta_x - R_x$ (δ_x is the standard deviation, R_x is the expected yield). In the proposed new model, λ is the risk-averse factor, which distribute in [0, 1]. Smaller λ represents the investor could bear larger risk. We assume that the c is 99%, so the $\Phi^{-1}(c) = 2.33$.

$$\begin{aligned} \min F(x) &= \lambda[\Phi^{-1}(c)\delta_x - R_x + 1/2(W\bar{A}V + \alpha' \delta')] - (1-\lambda)\sum_{i=1}^{n} x_i r_i \\ &= \lambda[2.33\sqrt{\sum_{i=1}^{n}\sum_{j=1}^{n} x_i x_j \sigma_{ij}} - \sum_{i=1}^{n} x_i r_i + 1/2(W\bar{A}V + 2.33\sum_{i-1}^{n}\sum_{j=1}^{n} x_i x_j \sigma'_{ij})] \\ &\quad - (1-\lambda)\sum_{i=1}^{n} x_i r_i \end{aligned} \tag{5}$$

$$\text{s.t} \begin{cases} \sum_{i=1}^{n} x_i = 1 \\ x_i \geq 0 \end{cases}$$

3 The Proposed Bacterial Foraging Optimization

The Bacterial foraging optimization mimics the foraging behavior of Escherichia coli bacteria that live in human intestine. The foraging strategy of Escherichia coli bacteria present in human intestine can be explained by four processes, namely chemotaxis swarming, reproduction and elimination dispersal. Any initial population started by random, and then through the four processes, the method can get the optimal solution. Due to the limited page space, the original BFO can refer to literature [4]. In the original bacterial foraging algorithm the chemotaxis step length C is a constant. By this way, it is hard to keep a right balance between global search and local search ability and thus influence accuracy and speed of the search.

We proposed a simple scheme to modulate the chemotaxis step size with a view to improving its convergence behavior without imposing additional requirements in terms of numbers of evaluations. The size of the step length adjusted in the reproduction and elimination-process, which ensures the bacteria moving global optimum quickly at the beginning, and the global optimum accurately in the end.

In this proposed method we used a linearly varying chemotaxis step length over iterations, in which the chemotaxis step length starts with a high value $C_{\max}$ and linearly decreases to $C_{\min}$ at the maximal number of iterations. The mathematical representations of the BFO method are given as shown in

$$C_j = C_{\min} + \frac{iter_{\max} - iter}{iter_{\max}}(C_{\max} - C_{\min}) \quad (6)$$

where $iter_{\max}$ the maximal number of iterations is, $iter$ is the current number of iterations, j is the jth chemotaxis step. with $C_{\max} = C_{\min}$, the system becomes a special case of fixed chemotaxis step length, as the original proposed BFO algorithm. From hereafter, this BFO algorithm will be referred to as bacterial foraging optimizer with linear decreasing chemotaxis step (BFO-LDC).

4 Illustrative Examples

We choose four assets as the sample, and they are from different industry, different place, where two of them from Shanghai A share are PuFa Bank (600000) and JiangXi Copper Industry (600362), ShangHai Automotive Industry (600124) and China Petrochemical Corporation (600028). The basic data about the assets were from January 1st in 2009 to December 30th in 2009. The relation number is as follows:

$\lambda = (0.15, 0.85)$ $\quad r = (0.003496, 0.006156, 0.007198, 0.002933)$

$\sigma = [0.000821, 0.00061, 0.000352, 0.000398; 0.00061, 0.001828, 0.000597, 0.000513; 0.000352, 0.000597, 0.001174, 0.000282; 0.000398, 0.000512, 0.000282, 0.000582]$

$\sigma' = [0.000126, 0.000125, 0.000088, -0.0000142; 0.0000125, 0.0000149, 0.000056, 0.000042; 0.000088, 0.0000556, 0.001203, 0.0007658; -0.0000142, 0.000042, 0.0000765, 0.00713]$

$\overline{WAV} = (0.0355, 0.00956, 0.0831, 0.10047)$

For comparison purposes, four swarm based algorithms were used, PSO, GA, BFO and BFO-LDC. The parameters used in PSO, GA, BFO and BFO-LDC are listed in next paragraph. A total for 15 runs for the PSO, GA and BFO experimental setting are performed.

In PSO, the inertia weight is 0.9 to 0.4, c_1 and c_2 are 2, the iterations is 800. In the GA, the number of population is 800, mutation children random and Gaussian is 20 and elitism children are 2. In the BFO and BFO-LDC, the data of bacteria in the population is 50, and chemotactic steps are 80, the number of elimination-dispersal events (Ned) is 2, the number of reproduction steps (Nre) is 5, the c_{max} and c_{min} are 0.2 and 0.01. Because the values of all the data are small, all the results are multiplied by 100 in the following section.

In order to apply the above mentioned algorithms to the improved portfolio optimization model, We use the real-number encoding method to construct their indexes, where each variable represents the holdings of each asset. The position of the bacteria (particles in PSO or individuals in GA) presents the proportion of each assest. The objective function is the fitness function designed in Eq (5).

4.1 Experimental Results

Numerical results with different λ obtained by the standard PSO, the GA and the BFO are showed in the Table 1 and the Table 2. The final portfolio selection results are listed in the table. Figures 1-4 present the mean relative performance with different λ generated by the three methods

Table 1. Numerical results with $\lambda = 0.15$

	BFO	PSO	GA	BFO-LDC
x_1	0.1792	0.1845	0.1051	0.1829
x_2	0.2529	0.1636	0.1168	0.3878
x_3	0.3511	0.4848	0.5516	0.2066
x_4	0.2169	0.1671	0.2265	0.2227
Profit	0.00534	0.005628	0.005718	0.005279
Risk	0.125527	0.128607	0.141979	0.124968
Max	1.0003	1.5104	1.7837	0.9677
Min	0.9782	1.1627	1.0530	0.9664
Mean	0.9846	1.3206	1.4431	0.9669
Std	0.0065	0.0086	0.2258	3.60E-04

According to the tables and the figures, we could find that:

1) Base on the data of the maximum value, minimum value and standard deviations in the tables. The results generated by BFO-LDC are the most robustness (the smallest standard deviations) and most precise results (the smallest mean fitness value).At the same time, it can be concluded that the result obtained by BFO-LDC are better than that of others. BFO-LDC performance best. When the investor hates the risk

Table 2. Numerical results with different $\lambda = 0.85$

	BFO	PSO	GA	BFO-LDC
x_1	0.751	0.7544	0.6366	0.75029
x_2	0.213	0.2034	0.3051	0.24
x_3	0.0343	0.004	0.0032	0.00071
x_4	0.0017	0.0202	0.0551	0.009
Profit	0.004183	0.00397	0.00428	0.00413
Risk	0.092882	0.09159	0.09347	0.09277
Max	7.8803	10.0926	11.3107	7.7322
Min	7.7438	7.9064	8.057	7.7314
Mean	7.7807	9.1335	9.8877	7.7317
Std	0.038	0.4691	0.7987	2.01E-04

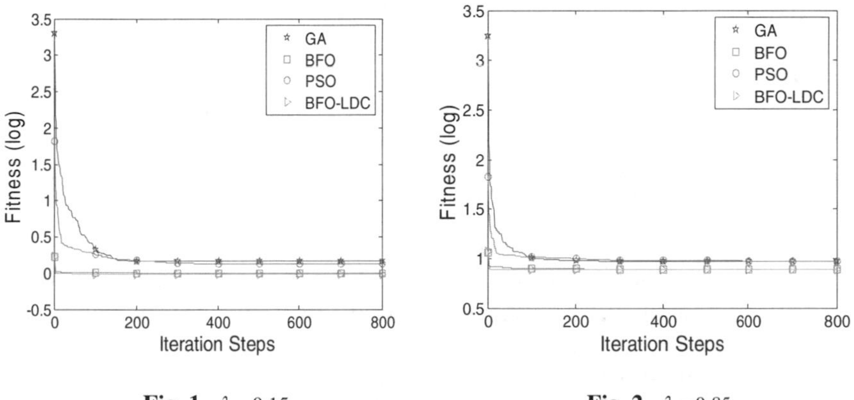

Fig. 1. $\lambda = 0.15$

Fig. 2. $\lambda = 0.85$

the least degree, that is $\lambda = 0.15$, BFO-LDC produced the highest revenue rate with the hightest risk rate. On the contrary, when $\lambda = 0.85$, it proposed the lowest risk rate with the lowest revenue rate.

2) Along with the different λ, the proportion of the four assets is different.When λ is bigger, asset 3 posses smaller proportion, and asset 1 occupy higher proportions. When $\lambda = 0.85$, the proportion of asset 1 become higher. The result show that high risk can be with the high profit, so the market can compensate for the liquidity risk.
3) Comparing the convergence graphs presented in Figs1-2, among other three algorithms, BFO-LDC is superior to others for all the test cases. BFO-LDC is highly competitive with BFO, usually surpassing its performance. The experiments show that the BFO-LDC is best choice to solve the portfolio optimization.

5 Conclusions

In this paper, we used the proposed BFO-LDC method and the other three swarm optimization algorithms to solve the improved portfolio optimization model. The obtained results indicate that the high performance of the proposed algorithm in searching for the optimal portfolios with high convergence rate and accuracy. The future research effort should focus on designing new biological mechanisms into basic BFO.

Acknowledgment

This work is supported by Shenzhen-Hong Kong Innovative Circle project (Grant no. SG200810220137A), The Natural Science Foundation of Guangdong Province (Grant no. 9451806001002294), 863 Project(No. 008AA04A105), and Project 801-000021 supported by SZU R/D Fund.

References

1. Goldberg, D.E.: Genetic Algorithms in Search, Optimization and Machine Learning. Addison-Wesley, New York (1989)
2. Wang, X.F., Elunluk, M.E.: The Application of Genetic Algorithm with Natural Networks to the Induction Machines Modeling. System Analysis Modeling Simulation 31, 93–105 (1998)
3. Eberchart, R., Kennedy, J.: A New Optimizer Using Particle Swarm Theory. In: 6th IEEE International Symposium on Micromachining and Human Science, pp. 39–43. IEEE Press, Piscataway (1995)
4. Passino, K.M.: Biomimicry of Bacterial Foraging for Distributed Optimization and Control. IEEE Control System Magazine 22, 52–67 (2002)
5. Markowitz, H.M.: Portfolio Selection: Efficient Diversification of Investments. Wiley, New York (1959)
6. Sharp, W.: A Theory of Capital Market Equilibrium under Conditions of Risk. Journal of Finance 19, 425–442 (1964)
7. Bangia, A., Diebold, F.X., Schuerman, T.: Liquidity on the Outside. Risk 12, 68–73 (1999)
8. Alexande, G., Baptista, J.: Economic Implications of Using a Mean-VAR Model for Portfolio Selection: A Comparison with Mean-Variance Analysis. Journal of Economic Dynamic and Control 26, 1159–1193 (2002)
9. Consigli, G.: Estimation and Mean-VAR Portfolio Selection in Markets Subject to Financial Instability. Journal of Banking Finance 26, 1355–1382 (2006)
10. Berkowitz, J.: Incorporating Liquidity Risk into Value-at-Risk Models. Working Paper, University of California, Irvine (2000)

A Combined System for Power Quality Improvement in Grid-Parallel Microgrid

Xiaozhi Gao*, Linchuan Li, and Wenyan Chen

Key Laboratory of Power System Simulation and Control of Ministry of Education, Tianjin University, Tianjin 300072, China
hebgirl1981@yahoo.com.cn

Abstract. The aim of this paper is to investigate the use of combined system constructed by Shunt Active Power Filter (SAPF) and Static Var Compensator (SVC) for power quality improvement in grid-parallel microgrid. Microgrid configuration is introduced first and appropriate control system is designed to ensure the microgrid operate well in grid-connected mode. In order to improve the power quality of the microgrid, this paper proposes a combined system, in which SAPF is adopted near the microsource to mitigate harmonic currents and SVC near the load to compensate reactive power so as to relieve the voltage variation. Simulation results show the effectiveness of the combined system.

Keywords: Microgrid, control system, SVC, SAPF, power quality.

1 Introduction

Microgrid, an intensive application of DG in the low voltage network, has gaining more and more attention. With appropriate control system, microgrid can safely operate in grid-parallel and stand-alone modes [1]. When connected to the utility, power quality issues of microgrid must be considered to ensure both the microgrid and the utility can operate in a stable and secure state. Currently the primary power quality issues of microgrid are the increase in harmonic content and the variation in amplitude of the supply voltage. Due to the increasing application of power-electronic loads and the unstable output of renewable energy sources such as photovoltaic and wind farms, harmonics in microgrid is becoming a serious problem. Meanwhile, the variation in amplitude of the supply voltage can cause abnormal operation of sensitive equipments[1][2] and it could be relieved by adopting reactive power compensation technology due to the "voltage-reactive power" droop characteristic.

Currently, Shunt Active Power filter (SAPF) has been researched and developed to restrain the harmonics .Reference[3] presents a three-phase four-wire grid-interfacing power quality compensator for microgrid applications to achieve an enhancement of both the quality of power within the microgrid and the quality of currents flowing between the microgrid and the utility system. Reference [4] proposes proposes the application of the strategy control approach used for the shunt active power filter previously proposed by the authors to the inverter interface of each single micro-sources

* This work was supported by grants from Key Project of Chinese National Programs for Fundamental Research and Development (973 Program) (No.2009CB219707).

K. Li et al. (Eds.): LSMS/ICSEE 2010, Part II, CCIS 98, pp. 23–29, 2010.

present in a Microgrid. The main advantage of the proposed strategy control approach lies on the fact that all sensitive loads connected to the Point of Common Coupling (PCC) are immunized from the power quality problems. Authors in [2] investigates the use of energy storage added to a shunt active filter configuration which can supply both reactive and real power for a short duration, in order to enhance its operation in maintaining power quality conditioning at a particular part of a grid system.

However, the researches do not consider the power quality problem under the situation that the microgrid operates with appropriate control strategy. Also the researches use APF to mitigate harmonics and compensate reactive power at the same time. Actually, both APF and SVC can achieve harmonic currents elimination and reactive power compensation. The two kinds of equipments have their own merits and demerits: with faster response speed than SVC, APF can be used to meet real-time requirement, but APF with a big reactive power capacity is much more expensive than SVC with the same capacity.

Therefore, based on economy and compensation performance, a combined system constructed by SVC and SAPF is designed to improve the power quality of microgrid operating with appropriate control strategy in this paper. And the paper is organized as followed: the microgrid configuration and control system are presented first, then the combined system is designed for the power quality improvement, the results obtained from the simulation verify the APF+SVC performance, at last the important conclusions of this study are summarized.

2 Microgrid Configuration and Control System

Fig.1 shows the microgrid configuration considered in this paper. The control system of the microgrid is designed to safely operate the system in grid-parallel and stand-alone modes. Here we adopt peer to peer control, in which each microsource is equivalent and has its particular control method, no communication between microsources is needed [1][5]. The control system for grid-parallel mode is designed as: the DG1 system adopts PQ control method and the DG2 system adopts V/f control method.

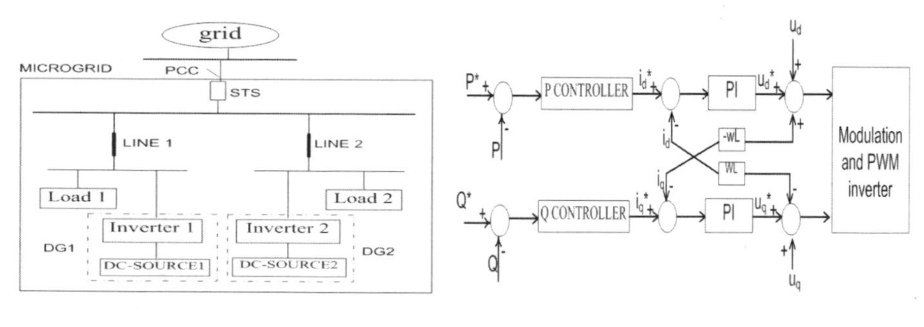

Fig. 1. Microgrid configuration

Fig. 2. PQ control scheme

PQ Control Method:

PQ control is that the inverter should be operated to meet a given real and reactive power set point [6]. The control scheme is as shown in Fig.2. In this scheme, P and Q are actual real power and reactive power output of DG system, respectively, P^* and Q^* are the dispatched real power and reactive power when in grid-connected mode respectively. And proportional-integral controller (PI) is adopted in this scheme.

V/f Control Method:

This control scheme presented in Fig.3 is adopted to keep the voltage and frequency stable [7]. This method can realize such two functions: force the capacitor voltages V_{abc} to track their sinusoidal reference waveforms V_{abc}^* and force the inductance current i_{abc} to track their reference waveforms i_{abc}^*.

And this control method is always used together with droop control method [8]. The droop control scheme is presented in Fig.4.

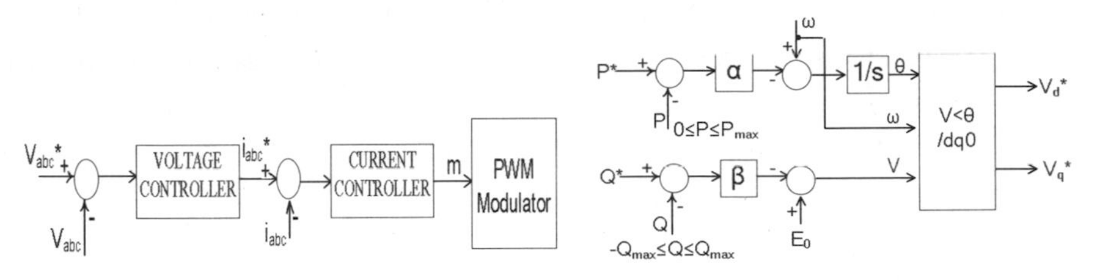

Fig. 3. V/f control scheme **Fig. 4.** Droop control scheme

In this paper, only V/f control method is applied and the droop control will be added to V/f control in our future work.

3 Combined System for Power Quality Improvement

With the rapid development of microgrid, new and stricter standards in respect to power quality are issued. In this paper, we focus on the research of harmonic currents elimination and reactive power compensation.

3.1 APF for Harmonic Elimination

Following with the conventional passive filters, Shunt Active Power filter (SAPF) has been researched and developed as a feasible solution to restrain the harmonics. The schematic diagram of SAPF is shown in Fig.5.

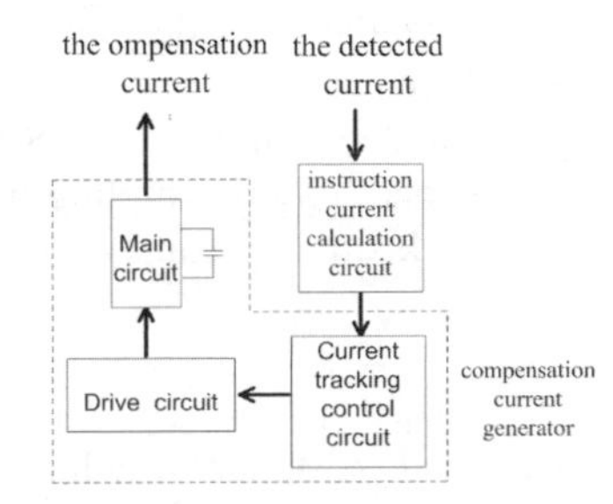

Fig. 5. The schematic diagram of SAPF

The current detection is the main parts of the SAPF. The current detection method is presented in Fig. 6 [9].

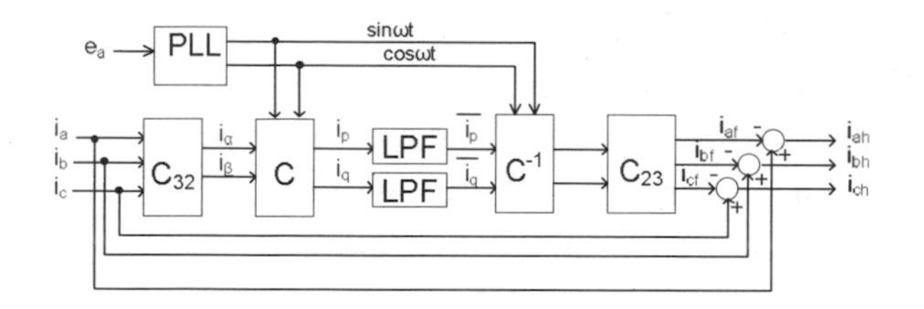

Fig. 6. Current detection method

The derivation for current detection is as follows: first, the current is transformed from stationary a-b-c frame to the rotating d-q-0 frame. Then fundamental current can be obtained, and $\bar{i}_p$ and $\bar{i}_q$ are the i_p and i_q filtered by LPF (Low Pass Filter) respectively, and i_{af} , i_{bf} and i_{cf} are the fundamental value of the current. Finally the compensation current (i_{ah} , i_{bh} and i_{ch}) can be obtained.

3.2 SVC for Reactive Power Compensation

The Static Var Compensator (SVC) is a shunt device using power electronics to control power flow and improve transient stability on power grids. The schematic diagram of SVC described in Matlab help file is shown in Fig.7.

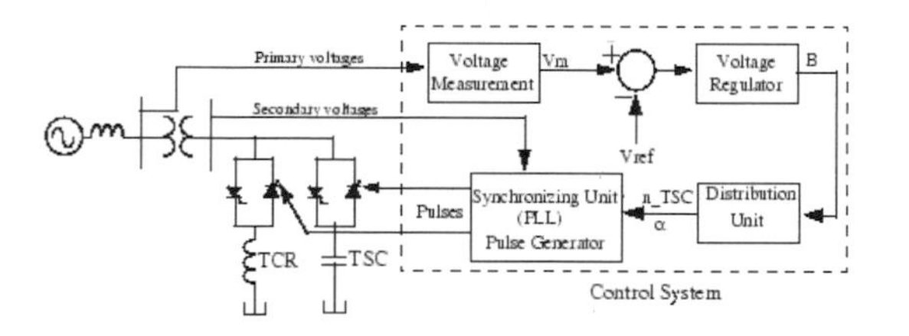

Fig. 7. The schematic diagram of SVC

For Simpowersystem help file in Matlab provides detailed information about SVC, the SVC part is simply introduced here.

The whole structure of the combined system is as: the SAPF is closed to the micro-source and the SVC is set near the load, so the coupling between APF and SVC can be avoided when SAPF and SVC work simultaneously [10].

4 Simulation Results

The proposed system (Fig.1) is verified in Matlab/Simulink simulations. The frequency of microgrid is 50Hz. The parameters used for simulations are given in Table 1.

Table 1. System parameters

DG1 (P Q)	2kw, 100var
Line 1	R=0.641/km X=0.101/km 100m
RL Load1	1kw 100var (adjustable)
Line 2	R=0.641/km X=0.101/km 50m
RL Load2	1kw 100var
C (DG1 DG2)	20uF
L (DG1 DG2)	60mH

4.1 Simulation of Microgrid without Combined System

From Fig.8 and Fig.9, it can be seen that both control methods achieve their goals: the reactive power and active power are kept constant in DG1 system, and the voltage and frequency are also kept constant in DG2 system.

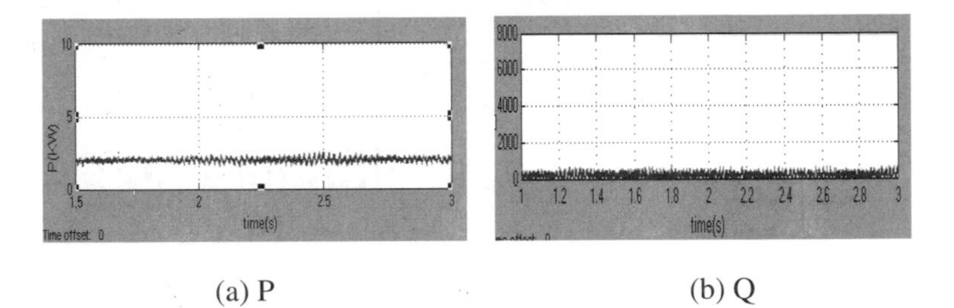

(a) P (b) Q

Fig. 8. Simulation results with P-Q control at DG1

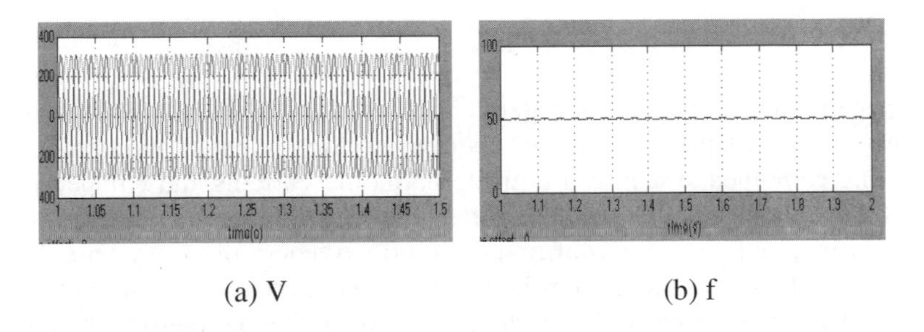

(a) V (b) f

Fig. 9. Simulation results with V/f control at DG2

4.2 Simulation of Microgrid with Combined System

With the control system mentioned above, the microgrid in grid-connected mode can operate in a desired steady state. However, we can see from Fig.10 if we adjust the reactive power of load1 from 100Var to 500 Var, the amplitude of voltage fluctuates. And the current of the microgrid always contains harmonic contents and in this situation THD reaches to 7.98% shown in Fig.11.

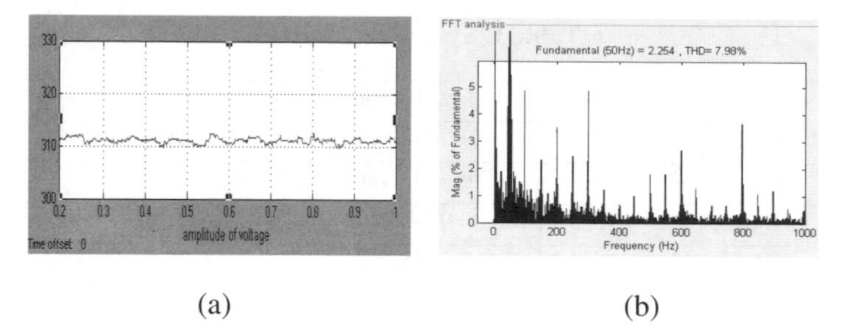

(a) (b)

Fig. 10. Voltage fluctuating and current distortion

After applying the combined system to DG1 system in the microgrid --the SAPF is closed to the microsource and the SVC is set near the load, it can be seen that the power quality has been enhanced from Fig.11, where the amplitude of voltage becomes steady and THD of current drops to 2.14%.

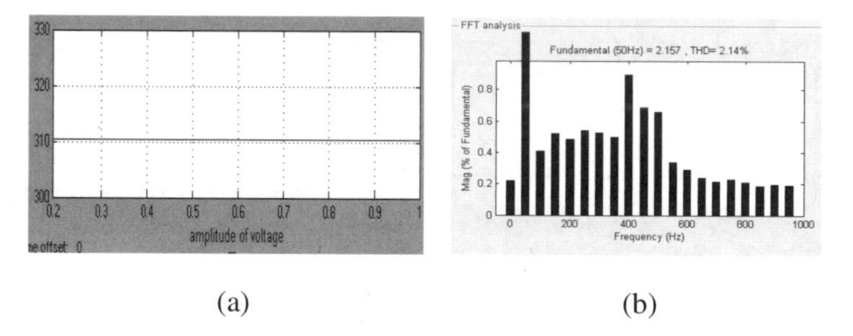

(a) (b)

Fig. 11. Voltage and current improvement

5 Conclusion

This paper proposed a combined system constructed by SAPF and SVC to improve the power quality of the grid-parallel microgrid which adopted appropriate control system. The combined system can mitigate harmonic currents and compensate reactive power simultaneously. The microgrid with the presented combined system has been tested in simulations for confirming the effectiveness of the system. The main advantages of the combined system lie on low cost, easy realization and good performance. And this combined system can be applied to the microgird with appropriate control strategy.

References

1. Lasseter, R.H.: CERTS microgrid. In: IEEE International Conference on System of Systems Engineering, San Antonio (2007)
2. Carastro, F., Sumner, M., Zanchetta, P.: An enhanced shunt active filter with energy storage for microgrids. In: 2008 IEEE Industry Applications Society Annual Meeting, Edmonton (2008)

3. Li, Y.W., Vilathgamuwa, D.M., Loh, P.C.: Microgrid power quality enhancement using a three-Phase four-wire grid-interfacing compensator. IEEE Transactions on Industry Applications 41(6), 1707–1719 (2005)
4. Menniti, D., Burgio, A., Pinnarelli, A., Sorrentino, N.: Grid-interfacing active power filters to improve the power quality in a microgrid. In: 13th International Conference on Harmonics and Quality of Power, Wollongong, NSW, Australia (2008)
5. Zhao, Y.S., Guo, L.: Dynamical Simulation of Laboratory Microgrid. In: 2009 Asia-Pacific Power and Energy Engineering Conference, Wuhan, China (2009)
6. Wang, C.S., Xiao, Z.X., Wabg, S.X.: Synthetical Control and Analysis of Microgrid. J. Automation of Electric Power Systems 32(7), 98–103 (2008)
7. Li, Y.W., Vilathgamuwa, D.M., Poh, C.L.: Design, analysis, and real-time testing of a controller for multibus microgrid system. J. IEEE Transactions on Power Electronics 19, 1195–1204 (2004)
8. Conti, S., Greco, A.M., Messina, N., Vagliasindi, U.: Generators control systems in intentionally islanded MV microgrids. In: SPEEDAM 2008 - International Symposium on Power Electronics, Electrical Drives, Automation and Motion, Ischia, pp. 399–405 (2008)
9. Yu, W.N., Zhu, J.X.: Simulation Research of Shunt Three Phase Active Power Filter. J. Journal of System Simulation. 19(20), 4624–4626+4638 (2007)
10. Deng, L.K., Jiang, X.J., Zhu, D.Q., Chen, J.L.: Stability Control of a Combined System of APF and SVC. J. Automation of Electric Power Systems 29(18), 29–32 (2005)

A Distance Sorting Based Multi-Objective Particle Swarm Optimizer and Its Applications

Zhongkai Li, Zhencai Zhu, Shanzeng Liu, and Zhongbin Wang

School of Mechatronics Engineering,
China University of Mining and Technology, Xuzhou, China
lizk@cumt.edu.cn

Abstract. Multi-objective particle swarm optimization (MOPSO) is an optimization technique inspired by bird flocking, which has been steadily gaining attention from the research community because of its high convergence speed. On the other hand, in the face of increasing complexity and dimensionality of today's application coupled with its tendency of premature convergence due to the high convergence speeds, there is a need to improve the efficiency and effectiveness of MOPSO. A novel crowding distance sorting based particle swarm optimizer is proposed (called DSMOPSO). It includes three major improvements: (I) With the elitism strategy, the evolution of the external population is achieved based on individuals' crowding distance sorting by descending order, to delete the redundant individuals in the crowded area; (II) The update of the global optimum is performed by selecting individuals with a relatively bigger crowding distance, which leading particles evolve to the disperse region; (III) A small ratio mutation is introduced to the inner swarm to enhance the global searching capability. Experiment results on the design of single-stage air compressor show that DSMOPSO handling problems with two and three objectives efficiently, and outperforms SPEA2 in the convergence and diversity of the Pareto front.

Keywords: multi-objective particle swarm optimization, elitism strategy, crowding distance sorting, air compressor design.

1 Introduction

Many engineering design problems are characterized by presence of multiple design objectives which often conflict with each other. This raises the question how to effectively search the feasible design region for optimal solutions, and simultaneously satisfy multiple constraints and scales of design variables. In the field of multi-objective optimization, a lot of evolutionary algorithms were proposed, such as NSGA, SPEA, NSGA-II and SPEA2 etc. Particle swarm optimization (PSO) [1] is a stochastic global optimization approach, which derived from the feeding simulation of fish school or bird flock. The main strength of PSO lies in its simplicity and fast convergence rate, so it is suitable to be expanded to solve the multi-objective optimization problems (MOP).

K. Li et al. (Eds.): LSMS/ICSEE 2010, Part II, CCIS 98, pp. 30–36, 2010.

Multi-objective particle swarm optimization (MOPSO) has received extensive research attention, and several MOPSO algorithms have been developed to solve a wide range of complex problems. Goh and Tan [2] adopted a competitive and cooperative co-evolutionary approach in MOPSO design, in which the problem was decomposed in the search space and the decision variables were evolved by different sub-swarms. Tripathi et al. [3] described a time variant multi-objective particle swarm optimization (TV-MOPSO), which is made adaptive by allowing its vital parameters (viz., inertia weight and acceleration coefficients) to change with iterations to explore the search space more efficiently. Abido [4] evolved a multi-objective version of PSO by redefinition of global and local best individuals, and used the algorithm in solving the environmental-economic dispatch problem. Jiang et al. [5] proposed a hybrid particle swarm optimization algorithm, which used the uniform design initialization, differential evolution operator and distance/volume fitness to improve the convergence and diversity of MOPSO. Cai et al. [6] employed the Pareto-dominance, fuzzifying, fitness sharing and turbulence factor, and proposed a multi-objective chaotic particle swarm optimization (MOCPSO). Li et al. [7] developed the new fitness assignment and random inertia weight strategy in the Pareto based MOPSO to avoid the algorithm getting trapped in local optima due to prematurity. Wang and Yang [8] proposed a MOPSO based on preference order ranking scheme and a novel updating formula for the particle's velocity, in order to solve the highly complex multi-objective problems.

The algorithms mentioned above improve the applications of MOPSO, but they still have the following shortcomings. (I) The diversity preservation for Pareto set and the update strategy for global best are complex, which leads to a high computational cost. (II) The MOPSOs tend to drop into a local extreme due to their weak global searching capability. (III) Some algorithms can only deal with MOP having two objectives, and the ability to solve problems with higher dimension is poor. With the elitism strategy and crowding distance operator, a crowding distance sorting based MOPSO (called DSMOPSO) is proposed. It can acquire the Pareto front closer to the true Pareto optimum in a smaller computational complexity, and the individuals satisfy the characteristic of diversity better. Efficiency and Effectiveness of the algorithm have been proved in the design for single-stage air compressor and the comparison with other algorithms.

2 Key Operators of DSMOPSO

Multi-objective evolutionary algorithms should guarantee the convergence and diversity characteristics of the Pareto front. For a PSO to be extended to MOPSO, the main challenge is to setup an appropriate diversity preservation strategy for the Pareto front and a swarm global best update mechanism. The crowding distance proposed by Deb in NSGA-II [9] can quickly judge the diversity among individuals under the same dominance level. By importing the elitism strategy and the crowding distance computation, DSMOPSO can preserve the diversity of Pareto front and update the global best based on particles' crowding distance sorting, in order to reduce the complex fitness computations.

2.1 Evolve of External Population

DSMOPSO uses the external population to preserve the nondominated individuals during the evolution process, and applies the crowding distance sorting to control the capacity of external population. This update strategy is shown in Fig. 1.

Supposing the external population A_t at iteration t includes m_1 individuals. M is the maximum capacity of the external population ($m_1 \leq M$). With the evolution of the internal swarm P, the newly produced n_1 nondominated individuals are copied into the external swarm to form A_t'. Individuals with the same objective values are regarded as repeated ones, and only one of them is preserved randomly. Now the number of nondominated individuals in A_t' is m_2 ($m_2 \leq m_1+n_1$). Then the crowding distance of all the individuals in A_t' are calculated and sorted in a descending order to form swarm A_t''. m_2 and M are compared. If $m_2 \leq M$ then A_t'' is denoted as a new external population A_{t+1}, in which M-m_2 individuals at the bottom of A_{t+1} are empty, as showed by Case 1 in Fig. 1. Otherwise, by calling the reduction process for the external population, the bottom m_2-M individuals with the minimum crowding distance will be deleted, and preserve the front M individuals in A_t''. So a reductive external population A_{t+1} will be formed as illustrated by Case 2 in Fig. 1.

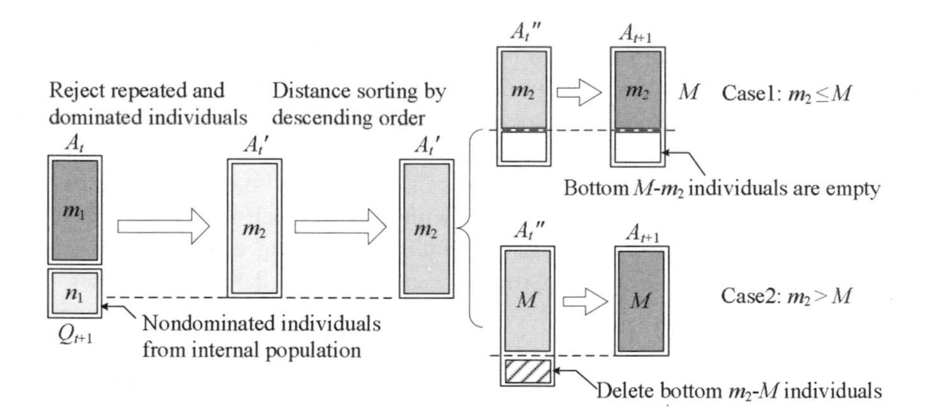

Fig. 1. Evolve strategy for external population

2.2 Update of Global Optimum

DSMOPSO needs to acquire a Pareto optimum with evenly distribution, so the selection of global best *gbest* is different from PSO, which only selecting particles with maximum or minimum objective. The particles in the disperse region should be selected in order to lead the swarm to evolve to the unattached region. After the updating of external population A, all individuals are sorted by crowding distance in a descending order. The update strategy for *gbest* is shown in Fig. 2, which includes the following two cases:

1. If the crowding distances of all individuals in A are INF (A only includes a small number of boundary individuals), a particle will be selected randomly as *gbest*, as shown in Fig.2(a).

2. If A includes individuals whose crowding distances are not INF as shown in Fig.2(b), a particle will be randomly selected as *gbest* whose crowding distance is much bigger but not INF. The calculation formula is:

$$gbest = A_k,\ k = \text{Irnd}(n, n + \text{Round}((m-n)\times 0.1)). \tag{1}$$

where n is the serial number of the first individual whose crowding distance is not INF in the population (n>1), m is the number of individuals in A ($m \geq n$). The *gbest* is randomly selected in the range n~n+Round((m-n)×0.1). The function Round denotes the rounding operation, and the function Irnd(n,k) returns a random integer between [n, k]. The function (m-n)×0.1 limits the selection range in individuals with bigger crowding distance. So the randomly selected *gbest* is an individual in the disperse region of the Pareto front.

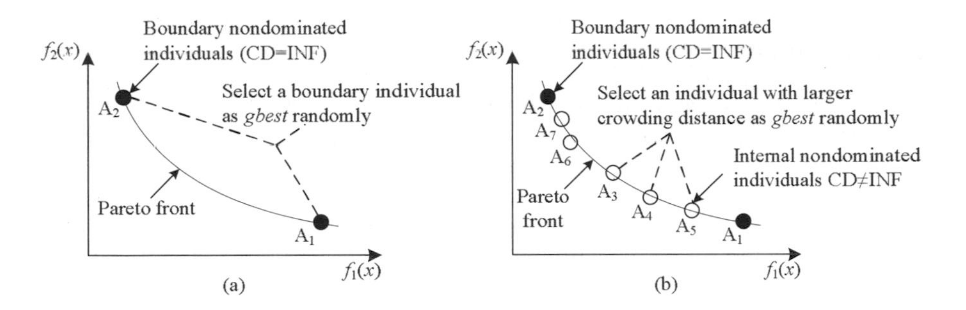

Fig. 2. Update strategy for global best

2.3 Swarm Mutation Mechanism

MOPSO preserves the characteristic of fast convergence from PSO. But if the convergence rate is too fast, it will also have some disadvantages in multi-objective optimization. Because of the fast convergence and limited searching space, MOPSO usually falls into the local Pareto optimal front, but not the global optimum. So we introduce a random mutation mechanism in the evolution process of the internal swarm. A small range of disturbance will be produced to enhance the global searching capability of the algorithm. The polynomial mutation [10] is employed to mutate the location of the particles, and the mutation probability p_m is usually set as a small number in the range [0, 1].

3 Detailed Computational Process

The computational framework of DSMOPSO is described as follows:

Step 1. Initialize the internal swarm. The particles in the internal swarm are randomly initialized in the predefined region. The initial velocities of particles are set zero, and the local best is the same as the variable values. The objectives are calculated, and the external population and the iteration are initialized as empty and zero, respectively.

STEP 2. The external population is updated by dominance between particles, and trimmed by the crowding distance sorting in a descending order.

STEP 3. Follow the global best update strategy to select the latest *gbest*.

STEP 4. Using PSO equations to update the velocity and location of internal particles, and the objective values are calculated. Then the new location and the local best of particles are compared with the constraint Pareto dominance criteria [9], to update each particle's local best.

STEP 5. Mutation operator is conducted in the internal swarm, during which the locations of particles are restricted between $x_{\min}$ and $x_{\max}$.

STEP 6. Repeat from STEP 2 for a predefined number of iterations. Then output the external population as the Pareto optimal set.

4 Experiment

In large scale air separation system with deep cooling method, the air compressor is a necessary equipment to supply the raw material and system energy. By making the compression ratio ε, the air flux rate Q and the isentropic efficiency η as optimization objectives, we construct optimization models with two and three objectives.

Setting the maximum compression ratio and air flux as the optimization objectives, DSMOPSO is used to solve MOP with two objectives and acquire the Pareto front containing 60 solutions as shown in Fig. 3. A fuzzy-based mechanism [11] is used to extract the best compromise solution in the optimal set. The variable values and their corresponding performances for the initial design, ε best, Q best and the compromised solution are illustrated in Table 1, respectively.

From Fig. 3 and Table 1, we get that (I) the ε best $\varepsilon^{\max}$=3.28 and Q best $Q^{\max}$ = 149.17Nm^3/min are both not superior to the initial design in view of both the two objectives; (II) the compromise solution ε=3.14 and Q=141.03Nm^3/min acquires the higher compression ratio and air flux rate than the initial design ε=3.10 and Q=140.27Nm^3/min, which improves the comprehensive performances of the air compressor.

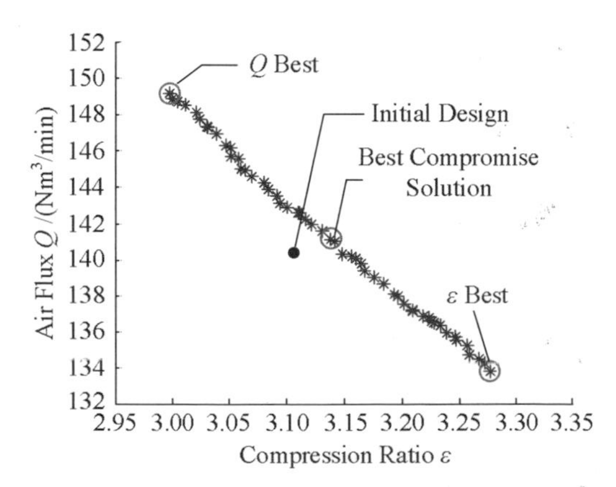

Fig. 3. Pareto fronts acquired by DSMOPSO

Table 1. Design schemes of initial and DSMOPSO with two objectives

	Initial Design	DSMOPSO		
		ε Best	Q Best	Compromise Best
ε	3.10	3.28	3.00	3.14
Q	140.27	133.79	149.17	141.03

The Pareto front solving the model with three objectives is shown in Fig. 4, during which the Pareto optimal solutions have a uniform distribution. The trends of the compression ratio and air flux rate are corresponded with the boundary and distribution in Fig. 3. So in the premise of satisfying the rated power and isentropic efficiency requirements, the efficiency of the compressor is higher with the reducing of the compression ratio and the increasing of the air flux rate.

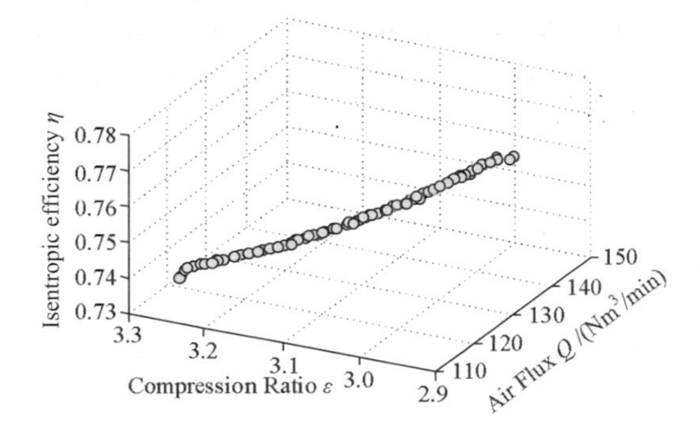

Fig. 4. Pareto Front by DSMOPSO for three objectives model

The improved strength Pareto evolutionary algorithm (SPEA2 [12]) has been used in many engineering applications. In order to compare the computational performance between DSMOPSO and SPEA2, the Pareto fronts for two objectives model by the algorithms is shown in Fig. 5. It illustrates that DSMOPSO achieves a better convergence and diversity than SPEA2. In another computational time experiment, DSMOPSO based on the particles flying scheme has a faster convergence speed than SPEA2 with crossover and mutation operators. These prove that DSMOPSO has a more powerful capability in solving multi-objective optimization problems.

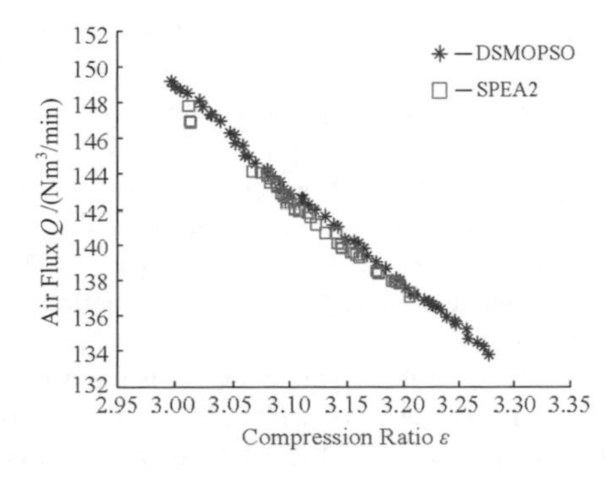

Fig. 5. Comparison of Pareto fronts between DSMOPSO and SPEA2

5 Conclusion

A multi-objective particle swarm optimizer based on crowing distance sorting is proposed. Its characteristics can be summarized as: the solutions are compared by constrained Pareto dominance, a crowding distance sorting is used to update the global optimum and a small probability mutation is introduced enhance the global searching capability. Through the experiments for single-stage air compressor design with two and three objectives, DSMOPSO acquire Pareto fronts with better convergence and diversity characteristics than SPEA2, and the computational time for MOP is shortened significantly. Future work is planned to introduce an adaptive parameters changing strategy to further improve the global searching capability of the algorithm.

Acknowledgments. This paper is supported by the National High-Tech. R & D Program of China (NO. 2009AA04Z415), the Postdoctoral Science Foundation of China (NO. 20100471007), the Talents Introduction Foundation and the Career Sailing Foundation of China University of Mining and Technology.

References

1. Kennedy, J., Eberhart, R.: Particle swarm optimization. In: Proceedings of IEEE International Conference on Neutral Networks, pp. 1942–1948 (1995)
2. Goh, C.K., Tan, K.C., Liu, D.S., Chiam, S.C.: A competitive and cooperative co-evolutionary approach to multi-objective particle swarm optimization algorithm design. European Journal of Operational Research 202(1), 42–54 (2010)
3. Tripathi, P.K., Bandyopadhyay, S.B., Pal, S.K.: Multi-objective particle swarm optimization with time variant inertia and acceleration coefficients. Information Sciences 177(22), 5033–5049 (2007)
4. Abido, M.A.: Multiobjective particle swarm optimization for environmental/economic dispatch problem. Electric Power Systems Research 79(7), 1105–1113 (2009)
5. Jiang, S.W., Cai, Z.H.: A novel hybrid particle swarm optimization for multi-objective problems. In: Deng, H., Wang, L., Wang, F.L., Lei, J. (eds.) AICI 2009. LNCS, vol. 5855, pp. 28–37. Springer, Heidelberg (2009)
6. Cai, J.J., Ma, X.Q., Li, Q., Li, L.X., Peng, H.P.: A multi-objective chaotic particle swarm optimization for environmental/economic dispatch. Energy Conversion and Management 50(5), 1318–1325 (2009)
7. Li, Y., Jing, P.P., Hu, D.F.: Optimal reactive power dispatch using particle swarms optimization algorithm based Pareto optimal set. In: Yu, W., He, H., Zhang, N. (eds.) ISNN 2009. LNCS, vol. 5553, pp. 152–161. Springer, Heidelberg (2009)
8. Wang, Y.J., Yang, Y.P.: Particle swarm optimization with preference order ranking for multi-objective optimization. Information Sciences 179(12), 1944–1959 (2009)
9. Deb, K., Pratab, A., Agarwal, S., Mey Arivan, T.: A fast and elitist multi-objective genetic algorithm: NSGA-II. IEEE Trans. on Evolutionary Computation 6(2), 182–197 (2002)
10. Deb, K., Agrawal, R.B.: Simulated binary crossover for continuous search space. Complex Systems 9(4), 115–148 (1995)
11. Abido, M., Bakhashwain, J.: Optimal VAR dispatch using a multiobjective evolutionary algorithm. Electrical Power and Energy Systems 27(2), 13–20 (2005)
12. Zitzler, E., Laumanns, M., Thiele, L.: SPEA2: Improving the strength Pareto evolutionary algorithm. Swiss Federal Institute of Technology. Tech. Rep. (2001)

A Discrete Harmony Search Algorithm

Ling Wang, Yin Xu, Yunfei Mao, and Minrui Fei

Shanghai Key Laboratory of Power Station Automation Technology,
School of Mechatronics and Automation, Shanghai University,
Shanghai, 200072
{wangling,panda_chris}@shu.edu.cn

Abstract. Harmony search (HS), inspired by the music improvisation process, is a new meta-heuristic optimization method and has been used to tackle various optimization problems in discrete and continuous space successfully. However, the standard HS algorithm is not suitable for settling discrete binary problems. To extend HS to solve the binary-coded problems effectively, a novel discrete binary harmony search (DBHS) algorithm is proposed in this paper. A new pitch adjustment rule is developed to enhance the optimization ability of DBHS. Then parameter studies are performed to investigate the properties of DBHS, and the recommended parameter values are given. The results of numerical experiments demonstrate that the proposed DBHS is valid and outperforms the discrete binary particle swarm optimization algorithm and the standard HS.

Keywords: harmony search, binary code, meta-heuristic.

1 Introduction

Harmony Search (HS) algorithm is a recent meta-heuristic algorithm firstly developed by Geem et al. [1] in 2001. It imitates the musician seeking to find pleasing harmony determined by an aesthetic standard, just as the optimization process seeks to find a global optimal solution determined by an objective function.

Due to its excellent characteristics such as easy implementation and good optimization ability, HS has drawn more and more attention and dozens of variants have been proposed to enhance the optimization ability. Pan [2] proposed a harmony search algorithm with ensemble of parameter sets which can self-adaptively choose the best control parameters during the evolution process. Mahdavi [3] developed an adaptive pitch adjustment rate strategy to improve HS. Li [4] proposed a hybrid PSO-HS algorithm where HS was used to deals with the variable constraints. Li and Wang [5] combined the HS with Differential Evolution algorithm and proposed two hybrid algorithms and tested their performance with a set of benchmark functions. Li and Li [6] presented a hybrid HS algorithm combined with particle swarm optimization to solve high dimensional optimization problems and achieve better optimization results. Jang [7] developed a hybrid Simplex Algorithm-Harmony Search algorithm where Simplex Algorithm was used to improve the accuracy and convergence speed. Omran and Mehrdad [8] introduced a new

K. Li et al. (Eds.): LSMS/ICSEE 2010, Part II, CCIS 98, pp. 37–43, 2010.

global-best harmony search inspired the concept of Particle Swarm Optimization (PSO) algorithm and validated its efficiency on the numerical problem and integer programming problem. In summary, hybrid HS algorithm has been the hotspot, there are other related hybrid algorithm research works such as HS combined with genetic algorithm (GA) [9] and Clonal Selection Algorithm [10]. So far, HS has been successfully applied to various fields, such as slope stability analysis [11], groundwater management model optimization [12], multiple dam system scheduling [13], and energy dispatch problem [14]. The application results promise that HS is a powerful search and optimization technique that may yield better solutions to these problems compared with Genetic Algorithm and PSO algorithm.

As far as we know, all of the previous works on HS concentrated in solving optimization problems in discrete or continuous space expect Greblicki [15] analyzed the properties of HS on the one dimensional binary knapsack problem which is a binary-coded problem. According to his result, HS gains a poor performance in the problem. To make up for it, a discrete binary harmony search (DBHS) algorithm is proposed for solving the binary-coded problems in this paper.

2 Discrete Binary Harmony Search Algorithm

Although the standard HS can be used to tackle the binary-coded problems, the performance is not satisfied, even poor [15]. To extend HS and tackle binary-valued problems effectively, we proposed a novel discrete binary HS algorithm by developing a new pitch adjustment rule.

2.1 Initialization of Harmony Memory

In DBHS, an individual is formed by the binary-string, so HM is initialized as (1):

$$H = \left[h_{1j} \ h_{2j} \ldots \ h_{ij} \ h_{HMS,M} \right] \ h_{ij} \in \{0,1\}, i \in HMS, j \in \{1,2,\ldots,M\} \tag{1}$$

where HMS is the size of HM and M is the length of binary-string, i.e., the dimensionality of the solution.

2.2 Harmony Memory Consideration Rule

The harmony memory consideration rate is of great importance for improvising, it indicates whether the element of new candidate is generated from the HM or randomization. Imitating the standard HS, the HMCR is the probability of picking up a value from HM in DBHS, while the (1-HMCR) is the rate of randomly choosing a feasible value not limited to HM, that is, it is re-initialized stochastically to be "0" or "1".

Here we propose a refined harmony memory consideration rule called individual selection strategy. In this strategy, all the elements of a solution are only selected from one HM vector. The individual strategy operation is defined in (2-3)

$$x_j = \begin{cases} h_{tj},\ t \in \{1,2,...,HMS\} & r_1 < HMCR \\ R & else \end{cases} \quad (2)$$

$$R = \begin{cases} 0 & r_2 < 0.5 \\ 1 & else \end{cases} \quad (3)$$

where x_j is the *j-th* element of the new candidate; r_1 and r_2 are two independent random number between 0 and 1; t is a random integer.

2.3 Pitch Adjustment Rule

Pitch adjustment operator chooses an adjacent value from the HS memory with probability of PAR in standard HS. For binary optimization problems, it only has two values, i.e., "0" or "1". Thus Pitch adjustment operator of the standard HS is a NOT gate for binary-coded problems as shown in (4-5)

$$x_j = \begin{cases} \overline{x_j} & r < PAR \\ x_j & else \end{cases} \quad (4)$$

$$\overline{x_j} = \begin{cases} 1 & if\ h_{ij} = 0 \\ 0 & if\ h_{ij} = 1 \end{cases} \quad (5)$$

As the pitch adjustment operator is used to find locally improved solutions, it is not strange that the optimization ability of standard HS for binary problems is poor. To make up of it, we propose a new pitch adjustment rule which choose the value from its structural neighborhood rather an adjacent value in HS memory. For easy implementation, the neighbor for each HS vector is defined the global optimal HS vector in HS memory. So the pitch adjustment rule in DBHS is re-defined as (6):

$$x_j = \begin{cases} h_{bj} & r < PAR \\ x_j & else \end{cases} \quad (6)$$

where r is a random number between 0 and 1; h_{bj} denotes the corresponding element value of the global optimal HS vector.

Till now, the new candidate has been generated in DBHS. In summary, the whole process of DBHS can be described as follows.

Step1: Initialize the harmony memory.
Step2: Initialize the parameters such as HMCR, PAR.
Step3: Improvise new harmonies by conducting harmony memory consideration, pitch adjustment and randomization according to (2-3, 6).
Step4: Update the harmony memory and the global best HS vector.
Step5: Repeat Steps 3 and 4 till some termination criterion is satisfied.

3 Parameter Study on DBHS

Like other meta-heuristic algorithms, the optimization ability of DBHS depends on the value of its control parameters, so parameter study is necessary. Four criterions, i.e., the best found value (BEST), the mean best value (MEAN), the rate of finding the global optima (SR) and the fitness calculation number (FCN), are adopted to judge the optimization performance of DBHS in our work.

3.1 The Number of New Harmony Memory Generating

The standard HS only improvises one new candidate once time and then updates HS. The parallel strategy has been reported to have a better performance for some optimization problems, that is, multiple candidates are generated and used to update the HM during each iteration. To check its effects on DBHS, here the performances of DBHS with the different number of new generating candidates (NGC) were studied. The parameters of DBHS were set as the values recommended in [15], i.e., HMS=30, HMCR=0.74 and PAR=0.1. For a fair comparison, the total number of fitness calculating is set as a constant, i.e., 90000.

Table 1. The effects of the number of new generating candidates

	NGC	1	10	15	20	30
Goldstein& Price	BEST	3.0	3.0	3.0	**3.0**	3.0
	MEAN	3+1.91E-08	3+1.61E-08	3+1.50E-08	**3+1.45E-08**	3+1.66E-08
	SR	24%	36%	40%	**42%**	34%
	FCN	71286	62802	46155	**58320**	64950
Schwefel 2.26	BEST	-837.96577	-837.96577	-837.96577	**-837.96577**	-837.96577
	MEAN	-837.62955	-837.96570	-837.96571	**-837.96571**	-837.96570
	SR	40%	48%	60%	**64%**	54%
	FCN	62053	27209	33030	**38640**	67590

From Table 1, we can find that parallel strategy is beneficial to the optimization ability of DBHS, but a small number of NGC is harmful. DBHS gained a better performance only when the number of NGC is bigger than 15, but a too big number, for instance, 30, also spoiled the performance of DBHS. According to the results in the Table 1, DBHS achieved the best optimization ability when NGC=20. So NGC=20 is recommended and used as the default value in the following sections.

3.2 Analysis of HMCR and PAR

Studying the HMCR and PAR parameters is important as they help the algorithm find globally and locally improved solutions. Apparently, there is a coupling between HMCR and PAR on optimization ability of DBHS. To study their influence on DBHS exactly, HMCR was tuned from 0.3 to 0.9 with step 0.2 and PAR was set within {0.1 0.3 0.5 0.7 0.9} simultaneously. The test results are drawn in Fig.1-2.

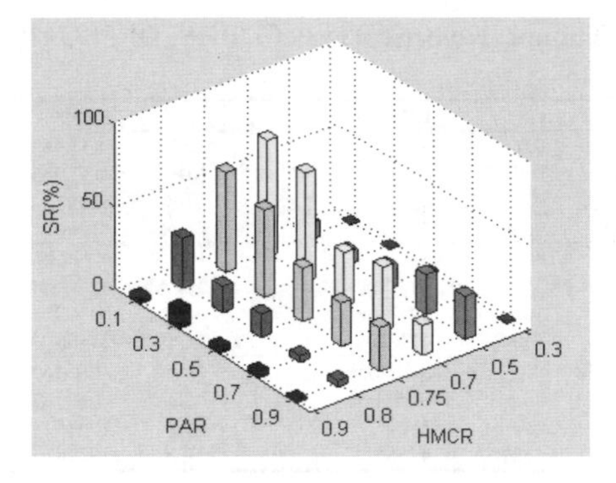

Fig. 1. The influence of PAR and HMCR on Rosenbrock

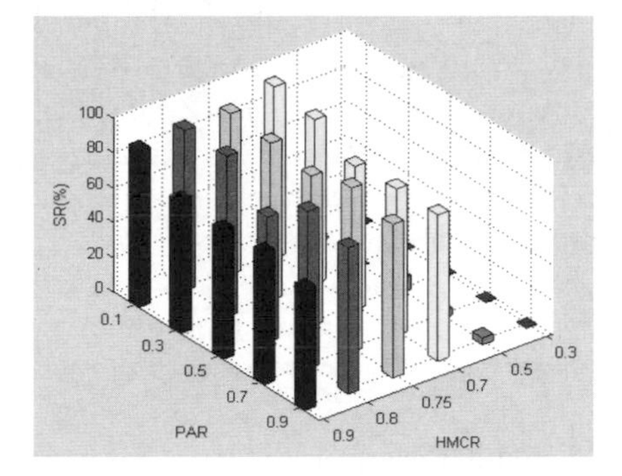

Fig. 2. The influence of PAR and HMCR on Schaffer F6

From Fig.1-2, it is shown that DBHS will be severely spoiled with a too big or small value of HMCR, and DBHS gains a better performance when the value of HMCR is between 0.7 and 0.8. Overall, DBHS achieved the best performance when HMCR=0.7 and PAR=0.1which are adopted as the default parameter values in DBHS.

4 Numerical Experiment

For a comparison, DBHS, the discrete binary Particle Swarm Optimization (DBPSO) [16] and the standard HS were used to solve 6 numerical optimization problems. The number of population size of all algorithms was set to 30. The parameters of DBPSO were used the values in [16], i.e., c_1=2, c_2=2, w=0.8. As the optimization results of the standard HS with HMS=30, NGC=1, HMCR=0.82, PAR=0.5 suggested in [15] are poor, we conduct another parameter study for HS and the optimal parameter values, i.e., HMCR=0.95, PAR=0.001 were used. The parameters of DBHS were HMS=30, NGC=20, HMCR=0.7, PAR=0.1. The results are shown in Table 2, for the limitation of the page, only the figure of the convergence of Rosenbrock is given as Fig.3.

Table 2. Numerical test of DBHS, DBPSO,HS

		DBHS	HS	DBPSO			DBHS	HS	DBPSO
Rosenbrock	BEST	**0.0**	0.0	0.0	Schwefel 2.26	BEST	**-837.965774**	-837.965774	-837.965774
	MEAN	**8.58E-10**	0.029915	1.22E-8		MEAN	**-837.965728**	-837.629500	-837.965681
	SR	**24%**	0%	6%		SR	**64%**	42%	36%
	FCN	**29460**	90000	76320		FCN	**41260**	61998	34770
Goldstein& Price	BEST	**3.0**	3.0	3.0	Branin	BEST	**0.3978873**	0.3978873	0.3978873
	MEAN	**3+1.55E-08**	3+2.11E-08	3+1.89E-08		MEAN	**0.3978983**	0.4090033	0.3979143
	SR	**34%**	16%	24%		SR	**94%**	20%	86%
	FCN	**82540**	76546	47130		FCN	**24600**	78110	42159
Rastrigin	BEST	**80.70658**	80.70658	80.70658	Camel-6	BEST	**-1.031628**	-1.031628	-1.031628
	MEAN	**80.70450**	80.63075	80.70036		MEAN	**-1.031628**	-1.031102	-1.031623
	SR	**74%**	4%	56%		SR	**100%**	58%	66%
	FCN	**45822**	88245	47556		FCN	**39278**	48774	48030

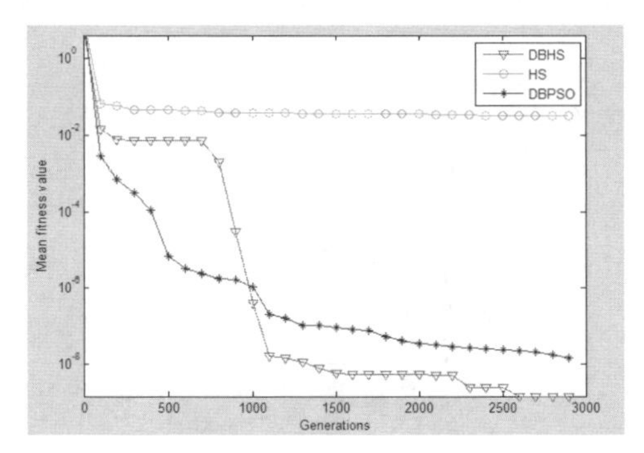

Fig. 3. Convergence of Rosenbrock

According to the results, it is obvious that the optimization ability of the standard HS for binary-coded problems is not satisfied. However, the proposed DBHS achieved the best performances and outperforms the standard HS and DBPSO on 6 functions.

5 Conclusion

In this paper, a novel discrete binary harmony search algorithm is proposed to extend HS to tackle the binary-coded optimization problems effectively. Based on the analysis on the drawback of HS for binary-valued problems, a new pitch adjustment rule is developed in DBHS. Then parameter studies are performed and the recommend parameter values are given. Numerical experiments are conducted and the comparison with DBPSO and the standard HS demonstrates that the proposed DBHS is effective and efficient and it is superior to DBPSO and the standard HS in terms of accuracy and convergence speed.

Acknowledgement

This work is supported by ChenGuang Plan (2008CG48), the Projects of Shanghai Science and Technology Community (10ZR1411800, 08160705900 & 08160512100), and Shanghai University "11th Five-Year Plan" 211 Construction Project, Mechatronics Engineering Innovation Group project from Shanghai Education Commission.

References

1. Geem, Z.W., Kim, J.H., Loganathan, G.V.: A new heuristic optimization algorithm: harmony search. J. Simulations 76, 60–68 (2001)
2. Pan, Q.K., Suganthan, P.N., Fatih Tasgetiren, M.: A Harmony Search Algorithm with Ensemble of Parameter Sets. In: IEEE Congress on Evolutionary Computation, pp. 1815–1820 (2009)
3. Mahdavi, M., Fesanghary, M., Damangir, E.: An improved harmony search algorithm for solving optimization problems. J. Applied Mathematics and Computation 188, 1567–1579 (2007)
4. Li, L.J., Huang, Z.B., Liu, F., et al.: A heuristic particle swarm optimizer for optimization of pin connected structures. J. Computers & Structures 85, 340–349 (2007)
5. Li, L.P., Wang, L.: Hybrid Algorithms Based on Harmony Search and Differential Evolution for Global Optimization. In: 2009 World Summit on Genetic and Evolutionary Computation, pp. 271–278 (2009)
6. Li, H.Q., Li, L.: A Novel Hybrid Particle Swarm Optimization Algorithm Combined with Harmony Search for High Dimensional Optimization Problems. In: 2007 International Conference on Intelligent Pervasive Computing, pp. 94–97 (2007)
7. Jang, W.S., Kang, H.I., Lee, B.H.: Hybrid Simplex-Harmony Search Method for Optimization Problems. In: 2008 IEEE Congress on Evolutionary Computation, pp. 4157–4163 (2008)
8. Omran, M., Mehrdad, M.: Global-best harmony search. J. Applied Mathematics and Computation 198, 643–656 (2008)
9. Li, H.Q., Li, L.: A novel hybrid real-valued genetic algorithm for optimization problems. In: 2007 International Conference on Computational Intelligence and Security, pp. 91–95 (2007)
10. Wang, X., Gao, X.Z., Ovaska, S.J.: A hybrid optimization method for fuzzy classification systems. In: 8th International Conference on Hybrid Intelligent Systems, pp. 264–271 (2008)
11. Li, L., Yu, G.M., Lu, S.B.: An Improved Harmony Search Algorithm for the Location of Critical Slip Surfaces in Slope Stability Analysis. In: 5th International Conference on Intelligent Computing, pp. 215–222 (2009)
12. Ayvaz, M.T.: Application of Harmony Search algorithm to the solution of groundwater management models. J. Advances in Water Resources 32, 916–924 (2009)
13. Geem, Z.W.: Optimal scheduling of multiple dam system using harmony search algorithm. In: Sandoval, F., Prieto, A.G., Cabestany, J., Graña, M. (eds.) IWANN 2007. LNCS, vol. 4507, pp. 316–323. Springer, Heidelberg (2007)
14. Ngonkham, S., Buasri, P.: Harmony search algorithm to improve cost reduction in power generation system integrating large scale wind energy conversion system. In: 1st World Non-Grid-Connected Wind Power and Energy Conference, pp. 95–99 (2009)
15. Greblicki, J., Kotowski, J.: Analysis of the Properties of the Harmony Search Algorithm Carried Out on the One Dimensional Binary Knapsack Problem. In: Moreno-Díaz, R., Pichler, F., Quesada-Arencibia, A. (eds.) Computer Aided Systems Theory - EUROCAST 2009. LNCS, vol. 5717, pp. 697–704. Springer, Heidelberg (2009)
16. Kennedy, J., Eberhurt, R.: A discrete binary version of the particle swarm algorithm. In: Proc. 1997 Conf. Systems, Man, Cybernetics, Piscataway, NJ (1997)

CFBB PID Controller Tuning with Probability based Binary Particle Swarm Optimization Algorithm

Muhammad Ilyas Menhas [1,2], Ling Wang[2], Hui Pan[2], and Minrui Fei[2]

[1] Ali Ahmed Shah University College of Engineering and Technology Mirpur Azad Kashmir, University of Azad Jammu and Kashmir, Pakistan
[2] Shanghai Key Laboratory of Power Station Automation Technology, School of Mechatronics and Automation Shanghai University, Shanghai, 200072
ilyasminhas75@yahoo.com

Abstract. The high combustion efficiency, extensive fuel flexibility and environment friendly characteristics have made circulating fluidized bed boiler (CFBB) an alternate choice for coal fired thermal power plants for clean energy production. But CFBB is a highly nonlinear and complex combustion system because of coupling characteristics and time delays. PID controller tuning of such a complex system with traditional tuning methods cannot meet required control performance. In this paper, a new variant of binary particle swarm optimization algorithm (PSO), called probability based binary PSO is presented to tune the parameters of CFBB. The simulation results show that PBPSO can effectively optimize the controller parameters and achieve s a better control performance than those based on that of a standard discrete binary PSO and a modified binary PSO.

Keywords: Circulating fluidized bed boiler, PID controller tuning, binary particle swarm optimization.

1 Introduction

The circulating fluidized bed boiler (CFBB) is a slow time-varying, nonlinear, multivariable complex combustion system with significantly large time delays and strong coupling characteristics. However, the system is widely used in power plants because of its high combustion efficiency, wide range of fuel flexibility and environment friendly characteristics [1].It is difficult to get an accurate mathematical model of the system due to the complex combustion process and its special structure [2].Therefore, the control of such a complex object is a challenging task as there are strong coupling relationships between inputs (coal, primary air ,secondary air, drawing wind, recycle material) and outputs(bed temperature, main steam pressure, the furnace negative pressure, oxygen content).The most prominent coupling relationship is between bed temperature and main steam pressure [3].

At present, the Proportional –Integral-Derivative (PID) controllers are used in process control. It is due to their robust performances, simple structure and easy understanding of algorithm for field engineers [4].It is necessary to tune the controller to

K. Li et al. (Eds.): LSMS/ICSEE 2010, Part II, CCIS 98, pp. 44–51, 2010.

find optimal PID parameters after a process system is upgraded or a new system is installed [5].The most famous classical auto tuning methods include Ziegler-Nichols rules, Cohen coon method and relay feedback methods [6].An MIMO system comprising nonlinear and coupled terms is much more complex than a SISO system. PID parameters tuning for such system often involve trial and error methods which not only consume a considerable amount of time and labor but also it is hard to find optimal or near optimal parameters for complex system with strong nonlinearities. In order to overcome such difficulties and complexities in tuning controllers many intelligent and heuristic methods such as Artificial Neural networks, Fuzzy Logic Particle Swarm Optimization (PSO), Ant Colony Optimization (ACO), Differential Evolution (DE) etc. and a lot of hybrid approaches have been developed in literature and applied in control in addition to finding global optimal for continuous functions and other real world optimization problems.

2 Probability Based Discrete Binary Particle Swarm Optimization Algorithm

2.1 Basic Particle Swarm Optimization Algorithm

The basic particle swarm optimization algorithm as was proposed by J. Kennedy and Eberhart [7] consists of a group of particles called swarm. In this swarm each particle is represented by its position $x_{ij} = (x_{11}, x_{12} ..., x_{nd})$ where subscripts *i* represent a specific particle in *n* particles and *j* represents its coordinates in a *d* dimensional search space. Each particle moves with velocity in the aforesaid *d* dimensional search space. All particle positions and velocities are initialized randomly from normal population U *(0, 1)* in the beginning, thereafter, an iterative process starts, and particles begin their movements in the search domain. The fitness of each particle is evaluated at every position according to an objective function, each particle remembers its best position during previous movements, then their positions and velocities for further search are updated according to the following equations (1) and (2)

$$v_{ij_{new}} = w * v_{ij_{old}} + c_1 * r_1(p_{ij} - x_{ij}) + c_2 * r_2(g_{1j} - x_{ij}) \tag{1}$$

$$x_{ij_{new}} = x_{ij_{old}} + v_{ij_{new}} \tag{2}$$

2.2 Discrete Binary Particle Swarm Optimization (DBPSO)

2.2.1 Basic Discrete Binary PSO (DBPSO)

In 1997 Kennedy and Eberhart further extended the PSO algorithm to the discrete space called DBPSO[8]. In discrete binary particle swarm algorithm (DBPSO) particle positions and velocities are binary vector of appropriate length. DBPSO preserves the original velocity Eq. (1) and converts the velocity of each particle into a number named S between 0 and 1 using a sigmoid function. Finally, a comparison is done between S and a random number [0, 1] to determine each bit either 0 or 1 for the particle position vector as in eq.(3) and eq. (4).

$$S(v_{id}) = 1/(1+e^{-v_{id}}) \tag{3}$$

$$x_{id} = \begin{cases} 1 & \text{if rand()} \leq S(v_{id}) \\ 0 & \text{otherwise} \end{cases} \tag{4}$$

Kennedy and Eberhart used DBPSO algorithm to optimize certain functions. In fact, the optimization ability of basic DBPSO was not satisfactory. However, the algorithm possesses the mutation ability[11],[12]. Therefore, DBPSO has some probability of escaping from local optimal .

2.3 Probability based Binary Particle Swarm Optimization Algorithm (PBPSO)

Considering the limitations of DBPSO, we have proposed a probability based binary particle swarm optimization algorithm (PBPSO). In fact, it is a hybrid approach that combines the idea of probability optimization and particle swarm optimization algorithm. This hybrid approach has significantly improved the performance of DBPSO in terms of both the optimization ability and computing burden for the processor. The myth lies in introducing a new linear estimator of pseudo probability for the actual binary solution and preservation of originality of particle swarm optimization algorithm. As PBPSO preserves both the velocity and position update rules, therefore it does not alter the original information sharing mechanism of PSO. The DBPSO only preserves the velocity update rule of PSO and considers velocity to estimate the pseudo probability for the actual binary bit of particle's position using a nonlinear sigmoid function. In contrast, the proposed PBPSO algorithm employs position update rule of PSO to determine the pseudo probability for the actual binary bit using a linear estimator. The new paradigms are very simple to realize, so the modification has not increased any complexity. The proposed PBPSO algorithm is explained below.

2.3.1 Initialization

It includes setting of initial state of particle's positions, their velocities, number of dimensions, number of binary bits to encode each particle position and velocity, population size, stopping criteria, search boundaries, and algorithm's parameters such as inertia weight and acceleration coefficients. In PBPSO, setting of population size, number of binary bits for each component of real vectors for position of particles and the stopping criteria are similar to that of DBPS .However, all components of binary vector for velocity and pseudo probability $x_{r_{ij}}$ are set to zero. The initial position vector for the particle's position is determined using a linear estimator in the following equation (9) to determine actual binary bits for the initial candidate solutions.

Let $x_{b_{ij}}$, $x_{r_{ij}}$ v_{ij} be $[m \times n * codel]$ binary candidate solutions, velocities and pseudo probability for actual binary bits for position vectors respectively. Where, m , n and $codel$ are the population size ,number of dimensions of each vector and number of binary bits to code each dimension. For simplicity let n be one, the initial binary vector for particle positions, velocities and pseudo probability $x_{r_{ij}}$ are

$$x_{b_{ij}} = \begin{bmatrix} x_{b_{(1,1)}} & x_{b_{(1,2)}} \cdots & x_{b_{(1,codel)}} \\ \vdots & \ddots & \vdots \\ x_{b_{(m,1)}} & x_{b_{(m,2)}} \cdots & x_{b_{(m,codel)}} \end{bmatrix} \tag{5}$$

$$x_{r_{ij}} = \begin{bmatrix} x_{r_{(1,1)}} & x_{r_{(1,2)}} \cdots & x_{r_{(1,codel)}} \\ \vdots & \ddots & \vdots \\ x_{r_{(m,1)}} & x_{r_{(m,2)}} \cdots & x_{r_{(m,codel)}} \end{bmatrix} \tag{6}$$

$$v_{ij} = \begin{bmatrix} v_{(1,1)} & v_{(1,2)} \cdots & v_{(1,codel)} \\ \vdots & \ddots & \vdots \\ v_{(m,1)} & v_{(m,2)} \cdots & v_{(m,codel)} \end{bmatrix} = \begin{bmatrix} 0 & 0 \cdots & 0 \\ \vdots & \ddots & \vdots \\ 0 & 0 \cdots & 0 \end{bmatrix} \tag{7}$$

The values for each $x_{b_{ij}}$ are determine on the outcome of probability p_{ij} according to following eq. (8) and (9)

$$p_{ij} = \frac{x_{r_{ij}} - x_{\min}}{x_{\max} - x_{\min}} . \tag{8}$$

As the initial value of $x_{r_{ij}}$ is zeros and $x_{\min} = -x_{\max}$ therefore the estimator for the pseudo probability gets an estimate of $p_{ij} = 0.5$ to determine initial actual binary bits for the position $x_{b_{ij}}$ vectors. Here $x_{\max}$ and $x_{\min}$ are predefined upper and lower bounds of pseudo probability $x_{r_{ij}} \in [x_{\min}, x_{\max}]$.Now a random number $r \in [0,1]$ is generated and compared with $p_{ij} \in [0,1]$ to determine actual bits for $x_{b_{ij}}$,the binary position vectors.

$$x_{b_{ij}} = \begin{cases} 1 & if \ r \le p_{ij} \\ 0 & if \ r > p_{ij} \end{cases} . \tag{9}$$

2.3.2 Updating

Now, the velocity and pseudo probability $x_{r_{ij}}$ are updated using the velocity update and position update rules described in eq. (10) and eq. (11) as

$$v_{ij_{new}} = w * v_{ij_{old}} + c_1 * r_1(p_{b_{ij}} - x_{b_{ij}}) + c_2 * r_2(g_{b_{1j}} - x_{b_{ij}}) \quad (10)$$

$$x_{r_{ijnew}} = x_{r_{ijold}} + v_{ij_{new}} \quad (11)$$

3 PBPSO Implementation for PID Controller Tuning of CFBB Using Matlab

3.1 Mathematical Model of Circulating Fluidized Bed Boiler (CFBB)

An approximate transfer function matrix of the circulating fluidized bed boiler (CFBB) model [10] that captures most of essential features is given below.

$$\begin{bmatrix} T_b \\ P_o \end{bmatrix} = \begin{bmatrix} P_{11}(s) & P_{12}(s) \\ P_{21}(s) & P_{22}(s) \end{bmatrix} \begin{bmatrix} C_m \\ A_p \end{bmatrix} = \begin{bmatrix} \dfrac{6.5e^{-30s}}{(1520875s^3+39675s^2+3445s+1)} & \dfrac{5e^{-80s}}{(50625s^2+450s+1)} \\ \dfrac{-4e^{-40s}}{(130s+1)} & \dfrac{7.5e^{-45s}}{(22500s^2+300s+1)} \end{bmatrix} \begin{bmatrix} C_m \\ A_p \end{bmatrix} \quad (12)$$

Where, T_b, P_0 are bed temperature and main steam pressure; C_m, A_p are amount of coal and primary air respectively.

3.2 Experiments and Simulation Results

3.2.1 PBPSO Parameter Setting

As PSO algorithms require a careful selection of certain parameters such as acceleration coefficients, inertia weight, population size and total number of iterations therefore, values for the acceleration coefficients C1 and C2 are set to 2.The inertia weight W, population size n and total number of iterations N are set to 0.7,60 and 200 respectively as commonly used in literature.

After setting algorithmic parameters each variable of PID controller was set as a 16 bit binary string. Therefore, a total of 96 bits were used to encode each candidate solution comprising six variables for two PID controllers. These candidate solutions were decoded to their corresponding decimal equivalents using standard binary to decimal conversion rules. The resultant decimal values were passed to both PID controllers simultaneously to evaluate fitness of each candidate solution to minimize the following objective function.

$$f = \int_0^t t \cdot |y_{d1} - y_{Out1}| dt + \int_0^t t \cdot |y_{d2} - y_{Out2}| dt + \int_0^t t \cdot |y_{d3} - y_{Out3}| dt + \int_0^t t \cdot |y_{d4} - y_{Out4}| dt \quad (13)$$

where $y_{d1} = y_{d4} = 1, y_{d2} = y_{d3} = 0$

There were 60 candidate solutions during each iteration and this process was repeated for 200 iterations to determine the global optimal for controller parameters as required by objective function defined in equation (13).

3.2.2 Experimental Results

The experiments were carried out using Pentium ®4 CPU 3.06GHz, 512 MB of RAM. The data obtained from experiments is presented in table1.

Table 1. Controller parameters obtained by PBPSO algorithm

	Kp_1	Ki_1	Kd_1	Kp_2	Ki_2	Kd_2	*Fitness*
PBPSO	0.1149	$4.5787e^{-4}$	4.9978	0.4073	$9.1564e^{-4}$	9.8590	$2.3921e^{+5}$

The transient response of the system in response to a unit step signal can be viewed from figure3 (a) and figure3 (b) respectively.

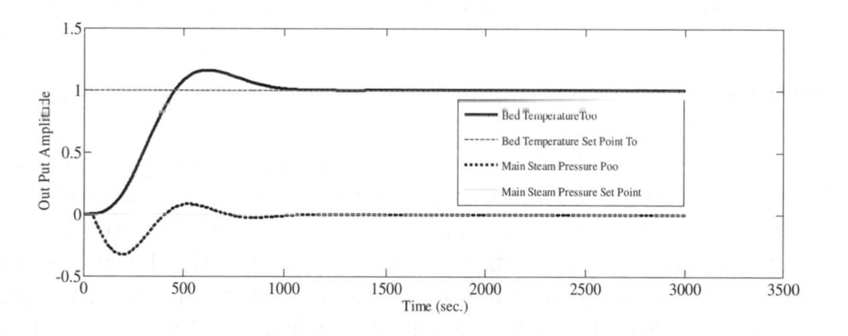

Fig. 1. Transient performance of bed temperature control loop with PBPSO

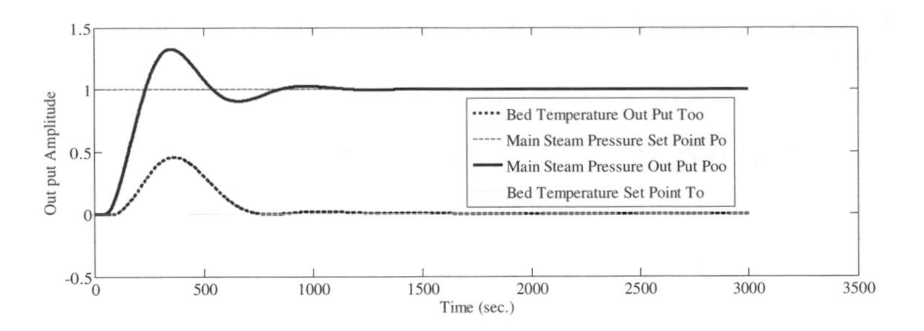

Fig. 2. Transient performance of main steam pressure control loop with PBPSO

4 Comparison Results

A comparative study of the proposed PBPSO algorithm was done with two former versions of discrete binary particle swarm optimization algorithm under identical experimental conditions. The comparison results are given in table 2. An overall fitness comparison of PBPSO, DBPSO[7], and MBPSO[9] shows that PBPSO performs 29.5% better than DBPSO and 4% better than MBPSO algorithms. The proposed PBPSO algorithm can effectively optimize the CFBB PID controllers with least over shoot, quicker rise time and settling time in comparison to DBPSO and MBPSO.

Table 2. Comparison results

	Kp_1	Ki_1	Kd_1	Kp_2	Ki_2	Kd_2	*Fitness*
DBPSO	0.1563	$6.1046e^{-4}$	9.5320	0.1953	$4.5787e^{-4}$	3.5184	$3.3928e^{+5}$
MBPSO	0.1563	$4.5877e^{-4}$	9.9998	0.2654	$6.1136e^{-4}$	9.9997	$2.4806e^{+5}$
PBPSO	0.1149	$4.5787e^{-4}$	4.9978	0.4073	$9.1564e^{-4}$	9.8590	$2.3921e^{+5}$

5 Conclusions

In this paper the discrete binary PSO (DBPSO) is modified to a new binary PSO (PBPSO) by introducing a new linear estimator to estimate the pseudo probability for actual bit of the particle position, and the estimator determines the probability of actual bit using particle's position update rule of the PSO. The new linear estimator has significantly reduced computational time in comparison to the nonlinear sigmoid function used in the DBPSO employing velocity update rule of the PSO to determine bits for particle position. The modification has improved optimization performance of the DBPSO and has reduced computational time. The new binary PSO (PBPSO) is further used to optimize the CFBB controller parameters. The CFBB is a complex, nonlinear, interacting MIMO system with large time delays. The traditional controller tuning methods cannot meet required control performance of the CFBB. The simulation results show that the proposed PBPSO can effectively optimize the controller parameters of the CFBB in comparison to DBPSO and MBPSO.

Acknowledgements

This research is supported by Mechatronics Engineering Innovation Group project from Shanghai Education Commission and the Projects of Shanghai Science and Technology Community (08160705900, 08160512100 &10ZR1411800).

References

1. Lu, C.M.: Equipment and operation of circulating fluidized bed boiler, pp. 101–158. China Electric Power Press, Beijing (2003)
2. Ma, S.X., Yang, X.Y.: Study on dynamic characteristics of the combustion system of circu-lating fluidized bed boilers. Proceedings of the CSEE 26(9), 1–6 (2006)
3. Ma, S.X., Xue, Y.L.: Multi variable control of circulating fluidized bed boilers com busti on system. Journal of Power Engineering 27(4), 528–532 (2007)
4. Shinskey, F.G.: Process Control Systems: Application, Design, and Tuning. McGraw-Hill, New York (1988)
5. Yoon, M.H., Shin, C.H.: Design of online auto-tuning PID controller for power plant process control. System and communication research laboratory Korea electric power research institute 103-16 Munji-dong yusung-ku, Taejon, korea, pp. 305–380 (1997)
6. Ziegler, J.G., Icholos, N.B.: Optimum setting for automatic controllers. Trans. ASME 65, 433–444 (1943)
7. Kennedy, J., Eberhart, R.C.: Particle Swarm Optimization. In: Proc. IEEE Int. Conf. Neural Networks, Perth, Australia, vol. 4, pp. 1942–1948 (1995)
8. Kennedy, J., Eberhart, R.C.: A discrete binary version of the particle swarm algorithm. In: 1997 IEEE International Conference on Systems, Man, and Cybernetics. Computational Cybernetics and Simulation, vol. 5, pp. 4104–4108 (1997)
9. Shen, Q., Jiang, J.H.: Modified Particle Swarm optimization algorithm for variable selection in MLR and PLS modeling: QSAR studies of antagonism of angoitensin II antagonists. European Journal of Pharmaceutical Sciences 22, 145–152 (2004)
10. Hu, Y.M., Zhang, G.Z., Chen, Y.F.: Neural network decoupling control strategy and its applications in combustion system of fluidized bed boiler. J. Electric Power Automation Equipment 23, 7–10 (2003)
11. Wang, L., Yu, J.S.: Fault feature selection based on modified binary PSO with mutation and its application in chemical process fault diagnosis. In: Wang, L., Chen, K., Ong, Y. S. (eds.) ICNC 2005. LNCS, vol. 3612, pp. 832–840. Springer, Heidelberg (2005)
12. Zhen, L.L., Wang, L., Huang, Z.Y.: Probability-based Binary Particle Swarm Optimization Algorithm and Its Application to WFGD Control. In: 2008 International Conference on Computer Science and Software Engineering (2008)

A Novel Cultural Algorithm and Its Application to the Constrained Optimization in Ammonia Synthesis

Wei Xu, Lingbo Zhang, and Xingsheng Gu*

Research Institute of Automation, East China University of Science and Technology, 200237 Shanghai, China
xsgu@ecust.edu.cn

Abstract. A novel cultural differential evolution algorithm with multiple populations (MCDE) is proposed. The single individual in each population is affected by the situational and normative knowledge from belief space simultaneously. The populations communicate with each other following a rule of knowledge exchange, which helps to enhance the search rate of evolution. The concept of culture fusion is introduced to develop an adaptive mechanism of preserving the population diversity. The mechanism ensures that populations are diverse along the whole evolution and excellent candidate solutions are not rejected. The performance of MCDE algorithm is validated by typical constrained optimization problems. Finally, MCDE is applied to maximizing the net value of ammonia in an ammonia synthesis loop. The results indicate that the proposed algorithm has the potential to be used in other problems.

Keywords: Cultural algorithm, multi-population, diversity preservation, constrained optimization, ammonia synthesis.

1 Introduction

Culture is a system of symbolically encoded conceptual phenomenon that is socially and historically transmitted within and between populations [1]. Cultural algorithms (CA), which is a class of computational models from observing the cultural evolution process in nature, were developed by Robert G. Reynolds [2] in 1994. It is a dual inheritance system that characterizes evolution in human culture at both the macro-evolutionary level and the micro-evolutionary level [3]. Some population-based evolutionary algorithms (EA) [4][5][6], which support the mechanism of the cultural change, have been adopted as the population space. Because of the introduction of the culture, CA has been successfully applied in many fields [7][8][9]. However, basic cultural algorithm also has the problems of prematurity and low accuracy. In this paper, we propose a novel multi-population cultural algorithm, where differential evolution is employed as the population space. The communication between populations follows a rule of knowledge exchange. To escape from local optima, the populations are diversified without being messed up. The proposed algorithm is validated through some constrained optimization problems. The algorithm is applied to a real-world

* Corresponding author.

K. Li et al. (Eds.): LSMS/ICSEE 2010, Part II, CCIS 98, pp. 52–58, 2010.

constrained problem in an ammonia synthesis production, whose objective is to maximize the net value of ammonia.

The rest of the paper is arranged as follows. In Section 2, the basic concepts of cultural algorithm and differential evolution are described. Section 3 presents the proposed algorithm, and then evaluates the algorithm using some constrained optimization problem. In Section 4, the algorithm is applied to an ammonia synthesis system. Section 5 concludes the article.

2 Cultural Algorithm and Differential Evolution

Cultural algorithm consists of two components: the population space and the belief space. The population space, in which a set of individuals is employed as candidate solutions to the problem, can be adopted by any population-based algorithms. The belief space stores the experience which is acquired by the individuals along the evolutionary process. To interact with each other, a communication protocol is established which provides the way that information can be exchanged between these two spaces. In cultural algorithm, individuals are evaluated by the performance function to represent the problem solving experience. And then, some individuals from the population space are selected by means of the acceptance function and contribute to the update of the knowledge in the belief space. The knowledge is used to guide the inheritance of individuals through the influence function. The selection function is in charge of selecting the offspring for the next generation.

As a new population-based evolutionary algorithm, differential evolution (DE) [10] was developed by Storn and Price in 1995. For a population consisting of N individuals in the D-dimensional problem space, mutation and crossover operations are executed to generate N trial individuals, each of which is given by

$$y_{id}^{t} = \begin{cases} x_{r_3 d}^{t} + F \times (x_{r_1 d}^{t} - x_{r_2 d}^{t}) & \text{if} \quad rand \le C_{\mathrm{R}} \\ x_{id}^{t} & \text{otherwise} \end{cases}, \tag{1}$$

where y_{id}^{t} is the dth component of the ith trial individual at iteration t; r_1, r_2, r_3 are random indices of individuals, which are different from i. $rand$ is a random number distributed uniformly in [0, 1]. C_{R} and F are given by the user in [0, 1] and [0, 2], respectively. After all N trial individuals are acquired, each of them is compared with the corresponding target individual, and the better one would be the offspring for the next generation. All these operations don't stop until terminal conditions are reached.

3 Proposed Cultural Algorithm

3.1 Design of MCDE Algorithm

Belief Space. The belief space includes two knowledge sources in the paper, i.e., situational knowledge and normative knowledge. Situational knowledge stores typical individuals from population space, which provides the guidance for other individuals. In MCDE, the mutation operation of each individual is influenced by situational knowledge as follows:

$$s_{id}^{t} = E_{d}^{t} + F \times (x_{r_1 d}^{t} - x_{r_2 d}^{t}), \tag{2}$$

where E_d^t is the dth component of a typical individual in the situational knowledge at iteration t. An improvement on the constitution of situational knowledge is introduced, which is that the situational knowledge includes a center individual besides the optimum one found so far. E is any one of them. The center individual is the arithmetic mean vector of all individuals, which presents the concentrated message of the whole population and has the potential to obtain good solutions. The center individual would be recalculated at the next generation.

A set of the range information for the decision variables where good individuals locate is referred to as normative knowledge. The following expression describes the influence of normative knowledge on the mutation operation of each individual.

$$n_{id}^{t} = \begin{cases} x_{id}^{t} + F \times rand \times (u_d^t - l_d^t) & \text{if } \; x_{id}^{t} < l_d^t \\ x_{id}^{t} - F \times rand \times (u_d^t - l_d^t) & \text{if } \; x_{id}^{t} > u_d^t \\ x_{id}^{t} + F \times randn \times (u_d^t - l_d^t) & \text{otherwise} \end{cases}, \tag{3}$$

where u_d^t, l_d^t are upper and lower bound of dth decision variable at iteration t, respectively; *rand* is a uniformly random number in [0, 1]; *randn* is a random number following a Gaussian distribution with zero mean and constant variance 1. When the accepted individuals locate outside the range, the update of the normative knowledge implements the expansion of the range.

Acceptance Function. Acceptance function controls the amount of good individuals which impact on the update of belief space. The amount is 20% of the population size in this paper.

Influence Function. In MCDE algorithm, situational knowledge and normative knowledge are involved to influence the mutation operation of each individual separately, and then two mutated populations are generated.

Knowledge Exchange. The communication among populations is a promising way to strengthen the search. Two populations, POP_1 and POP_2, are adopted in this paper. Ten percent of individuals in each population are picked out randomly after each given generation. And then, comparisons between the two groups of selected individuals are implemented correspondingly. The individual with better fitness value in a population replaces the corresponding one in another population. The knowledge exchange realizes the information sharing of populations, thus greatly promotes the search efficiency.

Diversity Preservation. Knowledge exchange increases the evolution rate, but it also might be easy for the populations to get stuck to local optima since population diversity declines. To solve this specific issue, we introduce an adaptive mechanism based on the concept of culture fusion so as to enhance the performance.

In this paper, culture fusion between the populations is thought to be the loss of diversity. The fusion state would be judged at every generation. Let's define

$$radius_j = \max\left\{\sqrt{\sum_{d=1}^{D}(x_{id}^j - cen_d^j)^2}, (i=1,...,N)\right\}, j \in \{1,2\}, \tag{4}$$

$$discen = \sqrt{\sum_{d=1}^{D}(cen_d^1 - cen_d^2)^2}, \tag{5}$$

where x_{id}^j is the dth decision variable of the ith individual from the jth population, where j=1, 2; cen is the center individual; $radius_j$ is the radius of the jth population; $discen$ is the Euclidean distance between cen^1 and cen^2. Moreover, we define the following factors, τ_1 and τ_2.

$$\tau_1 = \frac{discen}{\max(radius_1, radius_2)}, \tau_2 = \frac{discen}{\min(radius_1, radius_2)}. \tag{6}$$

If the two populations overlap closely enough, i.e., τ_1 or τ_2 is less than a small given threshold, and the theoretical global optimum has not been found, it is thought that the fusion between POP_1 and POP_2 occurs. In this case, POP_1 and POP_2 are integrated and N top good-performing individuals are chose to construct a new population for the next generation first. And then, the population with shorter radius is reinitialized in the bounds. The purpose of the above operations is to preserve the populations' diversity and help them escape from the local optima while the achievements made before won't be ruined.

3.2 Performance Evaluation

Eleven constrained optimization problems [11] are used to validate the performance of MCDE algorithm. The penalty function method [12] is used to construct the fitness function for evaluating individuals. 30 trials are performed independently for each problem. The parameters in MCDE algorithm are set as the following: the size of each population is 50; the maximum generation is 1000; F=1.0; C_R=0.6; the interval between knowledge exchanges is 100 generations; the thresholds, which τ_1 and τ_2 have to be less than so that the diversifying mechanism proceeds, are 0.1 and 0.3. All these parameters are obtained through many experiments, and the above setting is a compromise for all the problems. Our proposed MCDE algorithm is compared with two other approaches, i.e. the homomorphous mappings (HM) [11] and the stochastic ranking method (SR) [13], which are well known to handle constrained problems. The results obtained by each approach are showed in Table 1. The best results in 30 trials reveal that MCDE can reach the global optima in seven problems (g01, g03, g04, g05, g06, g08, g011) and get very close to those in the remaining. MCDE also presents better performance in the comparison of mean results with another two methods. The behaviors of MCDE indicate that MCDE is a competitive approach to solve the constrained problems.

Table 1. The comparison of best and mean results of HM, SR and MCDE

	Optimum	Best result			Mean results		
		HM	SR	MCDE	HM	SR	MCDE
*g*01	-15	-14.7864	-15.000	-15.0000	-14.7082	-15.000	-15.0000
*g*02	0.803619	0.79953	0.803515	0.803364	0.79671	0.781975	0.799597
*g*03	1	0.9997	1.000	1.0000	0.9989	1.000	1.0000
*g*04	-30665.539	-30664.5	-30665.539	-30665.539	-30655.3	-30665.539	-30665.539
*g*05	5126.4981	——	5126.497	5126.4981	——	5128.881	5126.4981
*g*06	-6961.8138	-6952.1	-6961.814	-6961.8138	-6342.6	-6875.940	-6961.8138
*g*07	24.306209	24.620	24.307	24.311654	24.826	24.374	24.333340
*g*08	0.095825	0.095825	0.095825	0.095825	0.089157	0.095825	0.095825
*g*09	680.630057	680.91	680.630	680.630121	681.16	680.656	680.630370
*g*10	7049.25	7147.9	7054.316	7049.262	8163.6	7559.192	7057.695
*g*11	0.75	0.75	0.750	0.750000	0.75	0.750	0.750000

4 Application of MCDE in Ammonia Synthesis

Ammonia synthesis system is an important chemical process. Ammonia is the chemical material for the production of fertilizers, such as nitrogenous fertilizer, phosphate fertilizer, and urea. The net value of ammonia, which is the difference between the outlet and inlet ammonia concentration of ammonia converter, concerns the product quality of ammonia synthesis. The high net value of ammonia results in strong production capability and high yield of ammonia. In other words, it reduces energy consumption. For an ammonia synthesis loop with four-stage quench-type axial-radial reactor, an optimization model maximizing the net value of ammonia is developed. The variables are five operational parameters, which are defined as follows: x_1 is the hydrogen concentration in recycle gas(mole percent); x_2 is the hot-spot temperature of axial stage (℃); x_3, x_4, x_5 are the hot-spot temperature of 1st, 2nd, and 3rd radial stages (℃). The objective function is given by:

$$\max \Delta(NH_3) = \Phi(x_1, x_2, x_3, x_4, x_5) \,, \tag{7}$$

where $\Delta(NH_3)$ is the net value of ammonia; $\Phi(\bullet)$ is the relationship description between $\Delta(NH_3)$ and the variables, characterized based on BP neural network.

Based on the experience of operators, the inequality constraints are given by:

$$x_3 - x_2 < 0,\ x_4 - x_3 < 0,\ x_4 - x_5 < 0,\ x_5 - x_2 < 0. \tag{8}$$

The variables must not exceed the upper and lower limits:

$$55 \le x_1 \le 65,\ 470 \le x_2 \le 500,\ 475 \le x_3 \le 495,\ 460 \le x_4 \le 485,\ 470 \le x_5 \le 490. \tag{9}$$

Our proposed MCDE algorithm is applied to this constrained optimization problem. The maximum net value of ammonia is 10.88%, and the optimal variables are: x_1^*=61.81, x_2^*=500.00, x_3^*=480.37, x_4^*=480.36, x_5^*=490.00. The convergence curve of the MCDE is illustrated in Fig.1. Before 98th generation, the optimal solution is infeasible and its corresponding value of objective function makes no sense. Individuals are gradually affected by both of situational and normative knowledge together, and two populations communicate with each other in a diversified environment along the evolution. All of these encourage individuals to move towards feasible regions rapidly. At 98th generation, the global feasible optimum is found. The behavior implies that MCDE algorithm can deal with the constraints well and is qualified to solve other constrained optimization problems.

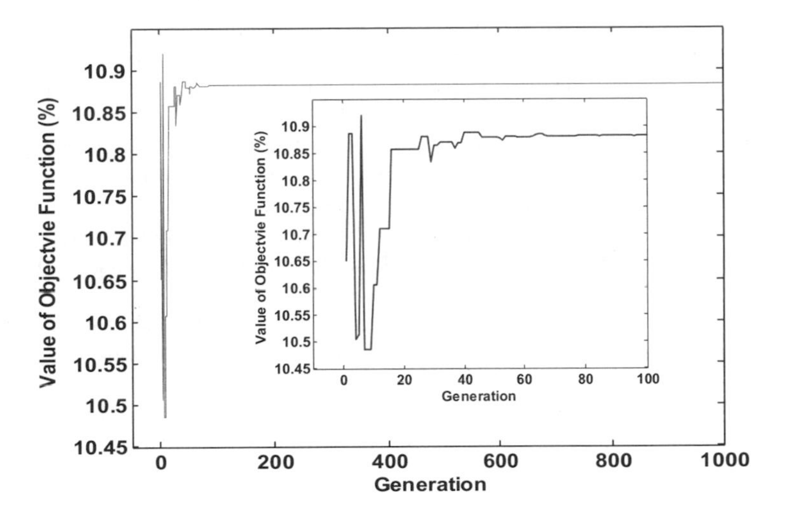

Fig. 1. The convergence curve of MCDE

5 Conclusions

In this paper, a multi-population cultural differential algorithm (MCDE) was proposed for constrained optimization problems. In each population, individuals are influenced by two knowledge sources together, i.e., situational knowledge and normative knowledge from the belief space. The situational knowledge stores the center individual as well as current optimum. Populations are linked through a new scheme, which states the way that they exchange the knowledge to enhance the convergence rate. In order to avoid trapping in local optima, a strategy for maintaining populations' diversity is introduced. Some benchmark constrained problems were used to validate the performance of MCDE algorithm. Furthermore, MCDE algorithm was applied to a practical optimization problem in ammonia synthesis. The results indicated that the proposed MCDE algorithm is capable of being further applied to other real-world constrained optimization problems.

Acknowledgments. We are very grateful to the editors and anonymous reviewers for their valuable comments and suggestions to help improve our paper. This work is supported by National High Technology Research and Development Program of China (863 Program) (No. 2009AA04Z141), Shanghai Commission of Science and Technology (Grant no. 08JC1408200), Shanghai Leading Academic Discipline Project (Grant no. B504).

References

1. Durham, W.: Co-evolution: Genes, Culture, and Human Diversity. Stanford University Press, Stanford (1994)
2. Reynolds, R.G.: An Introduction to Cultural Algorithm. In: Proceedings of the 3rd Annual Conference on Evolutionary Programming, pp. 131–139. World Scientific, Singapore (1994)
3. Reynolds, R.G., Peng, B., Brewster, J.J.: Cultural Swarms: Knowledge-driven Problem Solving in Social Systems. In: IEEE International Conference on Systems, Man, and Cybernetics, pp. 3589–3594. IEEE Press, New York (2003)
4. Gao, F., Cui, G., Liu, H.: Integration of Genetic Algorithm and Cultural Algorithms for Constrained Optimization. In: King, I., Wang, J., Chan, L.-W., Wang, D. (eds.) ICONIP 2006. LNCS, vol. 4234, pp. 817–825. Springer, Heidelberg (2006)
5. Lin, C., Chen, C., Lin, C.: A Hybrid of Cooperative Particle Swarm Optimization and Cultural Algorithm for Neural Fuzzy Networks and Its Prediction Applications. IEEE Trans. Syst. Man. Cy. C 39, 55–68 (2009)
6. Ricardo, L.B., Carlos, A.C.C.: A Cultural Algorithm with Differential Evolution to Solve Constrained Optimization Problems. In: Lemaître, C., Reyes, C.A., González, J.A. (eds.) IBERAMIA 2004. LNCS (LNAI), vol. 3315, pp. 881–890. Springer, Heidelberg (2004)
7. Jin, X., Reynolds, R.G.: Date Mining using Cultural Algorithms and Regional Schemata. In: 14th IEEE International Conference on Tools with Artificial Intelligence, pp. 33–44. IEEE Press, New York (2002)
8. Huang, H., Gu, X.: Neural Network based on Cultural Algorithms and Its Application on Modeling. Control and Decision 23, 477–480 (2008) (in Chinese)
9. Yuan, X., Nie, H., He, L., Li, C., Zhang, Y.: A Cultural Algorithm for Scheduling of Hydro Producer in the Power Market. In: Second International Conference on Genetic and Evolutionary Computing, pp. 364–367. IEEE Press, New York (2008)
10. Storn, R., Price, K.: Differential Evolution–a Simple and Efficient Adaptive Scheme for Global Optimization over Continuous Spaces. Technical report, International Computer Science Institute 8, 22–25 (1995)
11. Koziel, S., Michalewicz, Z.: Evolutionary Algorithms, Homomorphous Mappings, and Constrained Parameter Optimization. Evol. Comput. 7, 19–44 (1999)
12. Zangwill, W.I.: Nonlinear Programming via Penalty Functions. Management Science 13, 344–358 (1967)
13. Runarsson, T.P., Yao, X.: Stochastic Ranking for Constrained Evolutionary Optimization. IEEE Trans. Evol. Comput. 4, 284–294 (2000)

Pareto Ant Colony Algorithm for Building Life Cycle Energy Consumption Optimization

Yan Yuan[1], Jingling Yuan[2], Hongfu Du[2], and Li Li[1]

[1] School of Urban Design,Wuhan University,
Luojia Hill, Wuhan, China 430072
Shyy98@163.com
[2] Computer Science and Technology School, Wuhan University of Technology
122 Luoshi Road, Wuhan, 430070, China
yjl@whut.edu.cn

Abstract. This article aims at realizing optimal building energy consumption in its whole life cycle, and develops building life cycle energy consumption model (BLCECM), as well as optimizes the model by Ant Colony Algorithm (ACA). Aiming at the complexity and multi-objective principle of building life cycle energy consumption, this research tries to modify Pareto Ant Colony Algorithm (PACA), making it fit the needs of finding solution to least energy consumption in a building's whole life cycle. In the initial stage of ant colony constructing solution, each objective weighing is defined randomly, which improves the optimal determination mechanism of Pareto solution, perfects the renovation principle of pheromone, and finally realize the goal of optimization. This research is a innovative application of ACA in building energy-saving area, and it provides definite as well as practical calculation method for building energy consumption optimization in terms of a whole life cycle.

Keywords: ant colony algorithm, building life cycle, building energy consumption, multi-objective optimization.

1 Introduction

Sustainable development is a significant issue for future human society, in which energy consumption of building occupies a great proportion. In order to save energy consumed by building, we have to evaluate and optimize building energy consumption. In order to achieve comprehensive evaluation and optimization of building energy consumption, more and more attention is drawn to the evaluation and optimization of whole life cycle of building energy consumption.

In order to realize optimization of whole life cycle of building energy consumption, and to set up a scientific as well as effective way of optimization method, this research involves building life cycle energy consumption model（BLCECM）, and optimizes it by Pareto Ant Colony Algorithm (PACA). PACA is an intellectual optimization method based on ACA. ACA has excellent global searching capability and the nature of concurrent computation, and it performs especially well in finding solution to optimization grouping problem.

K. Li et al. (Eds.): LSMS/ICSEE 2010, Part II, CCIS 98, pp. 59–65, 2010.

2 Description of Model

To establish the Building life cycle energy consumption model (BLCECM), building can be to transform four components, which are door&window, wall, roof, and floor. Possible materials of components include steel, cement, skeletal material, timber, glass, tile, aluminium, lime, etc. Each component has variety of styles and material composition, and thus have different standards for energy consumption. The optimization goal is to achieve the minimum life cycle energy consumption, in the premise to satisfy the constraints, the limited of total construction cost.

2.1 Objective Functions

$$f = \sum_{k=1}^{4} f_k$$

Where,

f building life cycle energy consumption;

f_k the whole energy consumption of k component k=1,2,3,4, to express door&window, wall, roof, and floor.

$$f_k = \sum_{j_k=1}^{n_k} (f'_{j_k} + f''_{j_k})$$

f'_{j_k}, among k component the whole energy consumption of j_k style component in the stage of preparing, construction and dismantling;

f''_{j_k}, among k component the operation energy consumption of j_k style component;

n_k, the number of the kinds in k component.

In which, the f'_{j_k} can be expressed as:

$$f'_{j_k} = (\sum_{i_k=1}^{m_k} p_{i_k} q_{i_k}^{(j_k)} Q_{j_k}^{(k)}) + (\sum_{i_k=1}^{m_k} r_{i_k} q_{i_k}^{(j_k)} Q_{j_k}^{(k)}) + c_{j_k} Q_{j_k}^{(k)} + b_{j_k} Q_{j_k}^{(k)}$$

m_k, the number of the Material variety in k component;

p_{i_k}, unit productive energy consumption of the i_k material in k component;

r_{i_k}, unit transporting energy consumption of the i_k material in k component;

$q_{i_k}^{(j_k)}$, the amount of i_k material in j_k style component unit amount;

c_{j_k}, unit constructing energy consumption of j_k in k component;

b_{j_k}, unit dismantling energy consumption of j_k in k component ;

$Q_{j_k}^{(k)}$, the amout of j_k style component in k component.

The f_{j_k}'' can be expressed as:

k=1 (door&window)

$$f_{j_k}'' = (R_c t_c - R_h t_h) Q_{j_k}^{(k)} S_{j_k} D_{j_k} + Q_{j_k}^{(k)} K_{j_k} (T_{outc} - T_c) t_c + Q_{j_k}^{(k)} K_{j_k} (T_{outh} - T_h) t_h$$

R_c、R_h, the solar radiation intensity of air-conditioning&heating-period;

t_c、t_h, the time of air-conditioning&heating-period in operation;

$Q_{j_k}^{(k)}$, the amount of j_k style component in k component (door&window);

S_{j_k}, the sun-shading coefficient of j_k style component;

D_{j_k}, the solar absorption coefficient of j_k style component;

K_{j_k}, the heat transfer coefficient of j_k style component;

T_{outc}、T_{outh}, the average temperature outdoor in air-conditioning& heating-period;

t_c、t_h, the set temperature indoor in air-conditioning&heating-period.

k=2,3 (wall, roof)

$$f_{j_k}'' = Q_{j_k}^{(k)} K_{j_k} (T_{outc} - T_c) t_c + Q_{j_k}^{(k)} K_{j_k} (T_{outh} - T_h) t_h$$

K_{j_k}, the heat transfer coefficient of j_k component in k component.

k=4 (floor)

$$f_{j_k}'' = Q_{j_k}^{(4)} K_{j_k} (T_{soilc} - T_c) t_c + Q_{j_k}^{(k)} K_{j_k} (T_h - T_{soilh}) t_h$$

T_{soilc}、T_{soilh}, the Soil temperature Under floor in air-conditioning&heating-period.

2.2 Constraints

$$J = \sum_{k=1}^{4} \sum_{j_k=1}^{n_k} J_{j_k} Q_{j_k}^{(k)} \le J \max$$

J, total construction cost;

J_{j_k}, the unit construction cost of j_k style component in k component;

$J_{\max}$, acceptable maximum construction cost.

3 Pareto Ant Colony Optimization of BLCECM

3.1 Introduce of PACA

Pareto ant colony algorithm (PACA) is an intellectual optimization method based on species group. It possesses high concurrency, especially in finding solution for multi-objective problem. In PACA, K objects must correspond to K pheromones, which is expressed by pheromone quantity τ_{ij}^{K} . In the early stage, mark weighing quantity which correspond to objective function as

$$W=\{ w_1, w_2, \ldots, w_n \}^T$$

While

$$\sum_{g=1}^{n} w_g = 1, \quad 0 \leqslant w_g \leqslant 1$$

Renovating pheromones which correspond to various objects can make every object optimize toward best value. In consequence, the randomicity of weighing quantity helps coordinate the various components to a simultaneously best condition.

The mode shifting rule of PACA goes as follows :

$$P_{ij}^{k}(t) = \begin{cases} \dfrac{[\sum_{k=1}^{K} w_g \tau_m^k(i,j)]^{\partial} \eta_j^{\beta}}{\sum_{s \in J_m(i)} \{[\sum_{k=1}^{K} w_g \tau_m^k(i,s)]^{\partial} \eta_s^{\beta}\}}, & j \in J_m(i) \\ 0 & , otherwise \end{cases} \tag{1}$$

$$j = \begin{cases} \arg \max_{s \in J_m(i)} \{[\sum_{k=1}^{K} w_g \tau_m^k(i,j)]^{\partial} \eta_j^{\beta}\}, q \le q_0 \\ P_{ij}^{k}(t) & , otherwise \end{cases} \tag{2}$$

Where, q is a random number evenly scattered in an interval of [0,1], and q_0 is a parameter ($0 \le q_0 \le 1$). $P_{ij}^{k}(t)$ represents ants' shift probability to next route in step i.

$J_m(i)$ represents all possible routes for ant m in step i.

$\sum_{k=1}^{K} w_g \tau_m^k(i, j)$ represents the weight sum of pheromone vectors in route j.

The global updating rule of PACA goes as follows:

$$\tau^k(i,j) = (1-\rho_0)\tau^k(i,j) + \rho_0 \Delta\tau^k(i,j) \tag{3}$$

while

$$\Delta\tau^k(i,j) = \begin{cases} 1/L_{gb} \\ 0 \end{cases}$$

Where, (i,j) is the global best route while L_{gb} is the shortest route. ρ_0 is a parameter in interval [0,1], representing volatilisation of pheromone.

After all ants finish a loop cycle, add pheromone update to the worst ant's route. If (i,j)is one side of worst ant's route, while it is not a side of best ant's route, then adjust the pheromone of that side as follows:

$$\tau^k(i,j) = (1-\rho_0)\tau^k(i,j) - \varepsilon * f_{k-worst} / f_{k-best} \quad (4)$$

Where, ε is a parameter in interval [0,1], and $f_{k-worst}$ represents the energy consumption of worst ant in current loop cycle under object number k. In addition, f_{k-best} represents the energy consumption of best ant in current loop cycle under object number k.

The partial updating rule of PACA goes as follows:

$$\tau^k(i,j) = (1-\rho)\tau^k(i,j) + \rho\Delta\tau^k(i,j) \quad (5)$$

Where, ρ is a parameter in interval [0,1], representing volatilisation of pheromone.

3.2 Implementation of PACA in BLCECM

The multi-objective optimization problem discussed in this article is a kind of optimization grouping problem. Hence, this article tries to switch the problem to the traveling salesman's problem (TSP), then using ACA to search Pareto optimal solutions.

In the beginning stages of each ant's constructing solutions, defining the total energy consumption weighing of doors and windows w_1 ($0 \le w_1 \le 1$), walls w_2 ($0 \le w_2 \le 1$), roofs w_3 ($0 \le w_3 \le 1$), and floors w_4 ($0 \le w_4 \le 1$), and $w_1 + w_2 + w_3 + w_4 = 1$.

For the updating of global pheromones, only update the pheromone in route that is opted by minimum value of current combined object:

$$\Delta\tau^1(i,j) = global1/\min\{f\},\ \Delta\tau^2(i,j) = global2/\min\{f\} \quad (6)$$

$$\Delta\tau^3(i,j) = global3/\min\{f\},\ \Delta\tau^4(i,j) = global4/\min\{f\}$$

Where, global1, global2, global3, global4 are all constants..

For the updating of partial pheromones, $\Delta\tau^k(i,j)$ selection rule goes as follows:

$$\Delta\tau^1(i,j) = local1/f_1,\ \Delta\tau^2(i,j) = local2/f_2, \quad (7)$$

$$\Delta\tau^3(i,j) = local3/f_3,\quad \Delta\tau^4(i,j) = local4/f_4,$$

Where, local1, local2, local3, local4 are all constants, f_1 , f_2 , f_3 , f_4 are all objective function value.

3.3 Flow Chart

The flow chart is as Fig.1.

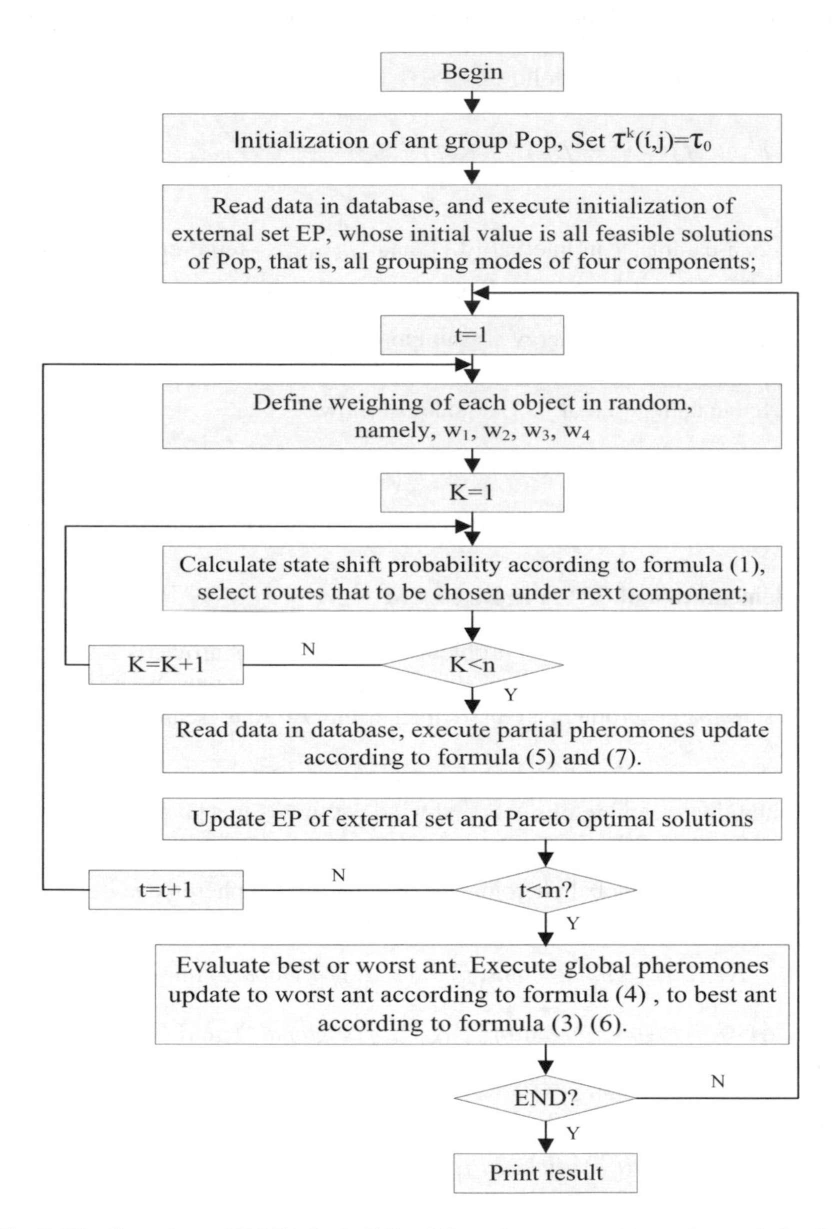

Fig. 1. The flow chart of PACA for building life cycle energy consumption optimization

4 Conclusions

In a society which aims at sustainable development, building energy saving design plays a important part in human environment and resources. In the beginning of building plan, suitable tool should be used to help architects find better multi-objective optimization model. That's the area in which our research attempts. This article integrates the characteristic of finding solution to multi-object problems, and tries to improve PACA on relatively foundation, making it suitable to find the least energy consumption in energy-saving building design considering a whole life cycle.

Method constructed in this article is different from previous ones, which basically converted multi-objective problems to solo ones. Instead, it explore solutions to multi-objective problems directly, which result in great decrease in calculation work, and increase in efficiency. Value of each object's weighing coefficient is randomly defined, to endow every object with same position, thus making various objects achieve ideal condition. By using best and worst ant to carry on global update, this method enhances optimal solutions more greatly while reducing worst solution. In that case, it makes multiple objects achieve optimal ideal condition simultaneously as much as possible.

As an interim achievement, the current of this study focuses on building envelope only. The energy consumption model is simplified as a primary result. Hence, still a lot more work should be done in the future, to achieve deeper research results.

References

1. Verbeeck, G., Hens, H.: Life cycle inventory of buildings: A calculation method. Building and Environment 45(4), 1037–1041 (2010)
2. The Weidt Group: Integrated cost-estimation methodology to support high-performance building design. Energy Efficiency 2(1), 69–85 (2009)
3. Gu, D., Zhu, Y., Gu, L.: Life cycle assessment for China building environment impacts. Journal of Tsinghua University 46(12), 1953–1956 (2006)
4. Coloni, D.M., Maniezzo, V., et al.: Distributed Optimization by Ant Colonies. In: Proceedings of European Conference on Artificial Life, Paris, France, pp. 134–142 (1991)
5. Dorigo, M., Maniezzo, V., Coloni, A.: The Ant System: Optimization by a Colony of Cooperating Agents. IEEE Transactions on Systems, Man,and Cybernetics—Part B 26(1), 29–41 (1996)
6. Dorigo, M., Gambardella, L.M.: Ant colony system:a cooperative learning approach to the traveling salesman problem. IEEE Trans. on Evolutionary Computation 1(1), 53–66 (1997)

Particle Swarm Optimization Based Clustering: A Comparison of Different Cluster Validity Indices

Ruochen Liu, Xiaojuan Sun, and Licheng Jiao

Key Laboratory of Intelligent Perception and Image Understanding of Ministry of Education of China, Institute of Intelligent Information Processing, Xidian University, Xi'an, 710071

Abstract. Most of clustering algorithms based on natural computation aim to find the proper partition of data to be processed by optimizing certain criteria, so–called as cluster validity index, which must be effective and can reflect a similarity measure among objects properly. Up to now, four typical cluster validity indices such as Euclid distance-based PBM index, the kernel function induced CS measure, Point Symmetry (PS) distance-based index, Manifold Distance (MD) induced index have been proposed. But, there is not a detailed comparison among these indexes. In this paper, we design a particle swarm optimization based clustering algorithm, in which, four different cluster validity index above mentioned are used as the fitness of a particle respectively. By applying the proposed algorithm to a number of artificial synthesized data and UCI data, the performance of different validity indices are compared in terms of clustering accuracy and robustness at length.

Keywords: particle swarm optimization, clustering, cluster validity, PBM index, CS measure, point symmetry distance, manifold distance.

1 Introduction

Clustering is an important unsupervised classification technique where a set of patterns, usually vectors in a multi-dimensional space, are grouped into clusters in such a way that patterns in different clusters are dissimilar in the same sense and in the same clusters are similar in some sense.

Many types of clustering algorithms have been developed. Whatever the clustering method may be, one has to determine the numbers of clusters and the validity of the clusters formed. The measure of validity of the clusters should be such that it will be able to impose an ordering of the clusters in terms of its goodness. Most of existed works focus on proposing a certain cluster validity index and finding the optimal partition for some special data. For example, Sanghamitra, et. al. proposed PBM index[1] for clustering complex spherical data. They also proposed a point symmetry distance [2] based index to process data with center symmetry. Bo, et al. [3] designed a density-sensitive similarity measure or a manifold distance to define cluster validity index for data with manifold characters. Swagatam, et. al. used a kernel function-induced index[4] to cluster some special data. Each approach has its own merits and disadvantages. Milligan, et. al. [5] provided a comparison of several validity indices for data sets containing distinct non-overlapping clusters while using only hierarchical

K. Li et al. (Eds.): LSMS/ICSEE 2010, Part II, CCIS 98, pp. 66–72, 2010.

clustering algorithms. Meila, et. al. made a comparison of some clustering methods and initialization strategies in [6]. Bandyopadhyay, et. al. [7] provided a comparison of several validity indices for not assuming any underlying distribution of the data sets while using nonparametric genetic clustering algorithms.

In this paper, a Particle Swarm Optimization based Clustering technique is designed, and we aim to evaluate the performance of four validity indices, namely, PBM index, CS measure, PS-based index, Manifold distance induced index applied in the proposed PSO based clustering technique.

The rest of the paper is organized as follow: Section 2 briefly outline the standard PSO. Section 3 describe the PSO based Clustering technique and four validity indices. Experimental results are presented and discussed in Section 4. Finally the paper is concluded in Section 5.

2 Particle Swarm Optimization

Particle swarm optimization (PSO) is an optimization algorithm inspired by the social interaction behavior of birds flocking [8]. In PSO, a population of conceptual 'particle' is initialized with random positions X_i and velocities V_i, and a function f, is evaluated, using the particle's positional coordinates as input values. In an n-dimensional search space, $X_i = (X_{i1}, X_{i2}, \cdots, X_{in})$ and $V_i = (V_{i1}, V_{i2}, \cdots, V_{in})$, positions and velocities are adjusted, and the function is evaluated with the new coordinates at each time-step. The basic update equations for the d th dimension of the i th particle in PSO may be given as

$$\left.\begin{aligned} V_{id}(t+1) &= \omega \bullet V_{id}(t) + C_1 \bullet \Phi_1 \bullet (P_{lid} - X_{id}(t)) + C_2 \bullet \Phi_2 \bullet (P_{gd} - X_{id}(t)) \\ X_{id}(t+1) &= X_{id}(t) + V_{id}(t+1) \end{aligned}\right\} \tag{1}$$

where ω is called the inertia weight that controls the impact of previous velocity of particle on its current one, $\Phi_1, \Phi_2 \sim U(0,1)$, C_1 and C_2 are positive constant parameters called acceleration coefficients which control the maximum step size between successive iterations, P_l is the local best solution found so far by the ith particle, while P_g represents the positional coordinates of the fittest particle found so far in the entire community or in some neighborhood of the current particle. Once the iterations are terminated, most of the particles are expected to converge to a small radius surrounding the global optima of the search space.

As a population-based iterative algorithm like Genetic algorithm, PSO need not complicated evolutionary operators such as crossover and mutation and requires less computational bookkeeping and generally fewer lines of code [9]. Owing to these advantages, PSO will be employed to construct a clustering algorithm in this paper.

3 PSO Based Clustering Technique

The procedure for PSO based clustering algorithm is as follows:

PSO based clustering algorithm
Begin
1. $t=0$.
2. Initialize population $P(t)$.
3. Assign all points to clusters as the nearest neighbor, and compute the fitness function values of $P(t)$.
4. If $t<t_{\max}$
5. $t=t+1$.
6. Update the global best and local best positions.
7. Update the positions and velocities of the population according to Eq. (1).
8. Go to step 3.
9. End if
10. Output the best and stop.
End

Particle Representation and Population Initialization: In the proposed algorithm, for n data with d-dimension, a particle is a vector of real numbers of dimension $k\times d$ representing the k cluster centers. The k cluster centers encoded in each particle are initialized to k randomly chosen points from the data sets. This process is repeated for each particle in the population. A single particle can be shown as

$$P_i(t)=\{\underbrace{a_{11}a_{12}\cdots a_{1d}}_{c_1}\cdots\underbrace{a_{i1}\cdots a_{id}}_{c_i}\cdots\underbrace{a_{k1}a_{k2}\cdots a_{kd}}_{c_k}\} \tag{2}$$

Where $c_1,c_2,\cdots,c_k$ are the k cluster centers. Thereafter five iterations of the K-means algorithm is executed with the set of centers encoded in each particle. The resultant centers are used to replace the centers in the corresponding particles.

Clustering Validity Indices: In this paper, we adopted four different validity indices for computing the fitness of the particle in the PSO based clustering algorithm, namely Euclid distance-based PBM index, the kernel function induced CS measure, Point Symmetry (PS) distance-based index and Manifold distance (MD) induced index . PBM index was proposed by Pakhira, et. al. in 2004[1]. The CS measure was proposed by Chou and Su in 2004[10] and then Swagatam, et. al. reformulated the CS measure by using a kernelized distance metric in 2008[4]. A point symmetry distance was proposed by Su, et. al. in 2001[11]. But it will fail for data sets where clusters themselves are symmetrical with respects to some intermediate points. Bandyopadhyay, et. al. proposed a new definition of the PS-based distance in 2007[2]. Bo, et al. designed a density-sensitive similarity measure or a manifold distance to define cluster validity index for data with manifold characters in 2007 [3]. Due to limited space, detailed description about these validity indices can be found in reference.

4 Experimental Results and Discussions

We denote the PSO based clustering algorithms using four validity indices as PBMPSO, CSKPSO, PSIPSO, and MDPSO respectively. In order to evaluate four validity indices, we apply four algorithms to 31 benchmark clustering problems.

All parameters of the proposed algorithm are determined experimentally. And it is implemented with the following parameters: the population size $P = 40$; the acceleration coefficients $c_1 = 2.0$, $c_2 = 2.0$; the inertia weight $\omega = 0.75$; the maximum number of generation $t_{max} = 100$; the termination criterion parameter $\varepsilon = 1e-5$.

4.1 Comparisons on the Clustering Accuracy

In this section, we choose 31 data to compare the performance of the four validity indices. These data can be classed into five groups, the first group include AD_5_2, AD_9_2, AD_10_2, AD_11_2, AD_12_2, AD_13_2, AD_14_2, AD_15_2, AD_20_2 and synthetic3[4][7], which are all sphere data. The second group includes Breast-cancer, iris, Liver-disorder, Lung-Cancer, New-thyroids, wine and glass, which are all UCI data [2] with unknown structure. The third group including data1, data2, data3, data5, data6 and data9 are proposed by Bandyopadhyay and used for data with center symmetric [2]. The fourth include special data and the fifth include many data which possess manifold structure [3]. The detailed description of all data sets can be found in reference.

We performed 20 independent runs on each problem. The mean and standard deviation of the clustering accuracy of data sets are shown in Table 1. The best solution in each case has been shown in bold.

For group one data sets, it is easy to see from Table 1, the average accuracy of CSKPSO algorithm is the best for these ten data sets. Because these data sets which are all sphere data and internally symmetrical, the accuracy of PSIPSO is higher than PBMPSO. And the performance of MDPSO is found to be poor for these data sets. For group two data sets, the average accuracy of PBMPSO algorithm is the best for these seven UCI data sets. The accuracy of PSIPSO is higher than MDPSO, while the CSKPSO performs the worst. For group three data sets, the accuracy of PSIPSO is higher than the other three algorithms, while the performance of MDPSO is second. The performance of CSKPSO is found to be poor for these six data sets. For group four data sets, the accuracy of the MDPSO is the best for long1 data, while the PBMPSO is the second one. The performance of CSKPSO is found to be poor for this data. For sizes5 data, as seen from Table 1, the accuracy of the MDPSO is the best for this data, while the PSIPSO is the second one and the CSKPSO performs the worst. For group five data sets which are manifold structure data sets. As can be seen from Table 1, the accuracy of MDPSO algorithm for these data sets all are 100%. It is evident that the MDPSO performs the best than the other three algorithms.

4.2 Robustness Analysis

In order to compare the robustness of these methods, we follow the criteria used in Ref. [12]. In detail, the relative performance of the algorithm m on a particular data set is represented by the ratio b_m of the mean value of its adjusted rand index (R_m) to the highest mean value of the adjusted rand index among all the compared methods:

$$b_m = \frac{R_m}{\max_k R_k} \tag{3}$$

Table 1. Mean and standard deviation of the clustering accuracy of data sets by each clustering algorithm over 20 independent runs

Datasets	Algorithms			
	PBMPSO	CSKPSO	PSIPSO	MDPSO
AD_5_2	0.9408(0.0037)	**0.9904**(0.0100)	0.972(0.0135)	0.926(0.0060)
AD_9_2	0.9991(0.0021)	**1**(0)	0.9973(0.0023)	0.9904(0.0021)
AD_10_2	0.9944(0.0018)	**0.9992**(0.0017)	0.9952(0.0037)	0.9884(0.0053)
AD_11_2	0.9942(0.0011)	**1**(0)	**1**(0)	0.993(0.0011)
AD_12_2	0.9965(0.0005)	0.9997(0.0011)	**1**(0)	0.9911(0.0011)
AD_13_2	0.9926(0.0019)	**1**(0)	0.9992(0.0008)	0.9863(0.0005)
AD_14_2	0.9931(0.0011)	**0.9996**(0.0010)	0.9991(0.0007)	0.9874(0.0011)
AD_15_2	0.9977(0.0009)	**0.9995**(0.0011)	0.9987(0)	0.9925(0.0011)
AD_20_2	**1**(0)	**1**(0)	**1**(0)	**1**(0)
synthetic3	**1**(0)	**1**(0)	**1**(0)	**1**(0)
Breast-cancer	**0.9605** (0)	0.9477(0.0100)	0.9078(0)	0.937(0)
iris	0.8933(0)	0.8787(0.1284)	**0.9367**(0.0105)	0.9000(0.0070)
Liver-disorder	**0.4406**(0.0072)	0.418(0.0018)	0.4093(0.0032)	0.4380(0.0109)
Lung-Cancer	**0.5656**(0.0925)	0.3625(0.0535)	0.5281(0.1346)	0.5250(0.1257)
New-thyroids	0.8558(0.0135)	0.6270(0.0583)	**0.8891**(0.0160)	0.7609(0.1106)
wine	0.7096(0.0027)	0.7371(0.2843)	0.7427(0.0181)	**0.9270**(0)
glass	**0.407**(0.0321)	0.3879(0.0851)	0.3444(0.0517)	0.4033(0.0998)
data1	0.773(0.0028)	0.7(0.0020)	**0.9940**(0.0046)	0.8002(0.0204)
data2	0.8965(0.0064)	0.5048(0.2833)	**0.9888**(0.0067)	0.9370(0.0011)
data3	0.9628(0.0122)	0.8558(0.057)	0.9995(0.0016)	**1**(0)
data5	0.7395(0.0093)	0.7465(0.0238)	**0.9945**(0.0044)	0.7345(0.0150)
data6	0.74(0.0071)	0.6320(0.0284)	**0.9208**(0.1277)	0.7455(0.0511)
data9	0.8367(0.1187)	0.5673(0.1782)	0.957(0.0395)	**1**(0)
long1	0.5073(0.0149)	0.4764(0.0153)	0.4921(0.0335)	**1**(0)
sizes5	0.8295(0.1143)	0.4155(0.0857)	0.9880(0.0102)	**0.9890**(0.0017)
Line blobs	0.6395(0.1472)	0.6936(0.1209)	0.6640(0.1218)	**1**(0)
spiral	0.5993(0.0177)	0.5089(0.0742)	0.4987(0.0023)	**1**(0)
sticks	0.6014(0.2258)	0.4096(0.0846)	0.7203(0.1513)	**1**(0)
synthetic1	0.5222(0.0093)	0.4198(0.0092)	0.5442(0.1006)	**1**(0)
eyes	0.462(0.0028)	0.4893(0.0123)	0.8963(0.2313)	**1**(0)
synthetic5	0.837(0.0061)	0.5003(0.0110)	0.5142(0.0557)	**1**(0)

The adjusted Rand index proposed by Hubert and Arabie [13] assumes the generalized hyper-geometric distribution as the model of randomness, i.e., the partitions U and V are picked at random so that the numbers of points in the classes are fixed. Let n_{ij} be the number of points in both class u_i and class v_j .Let $n_{i\cdot}$ and $n_{\cdot j}$ be the numbers of points in class u_i and class v_j , respectively.

Then the adjusted Rand index is given as

$$R(U,V)=\frac{\sum_{i,j}\binom{n_{ij}}{2}-\left[\sum_{i}\binom{n_{i\cdot}}{2}\cdot\sum_{j}\binom{n_{\cdot j}}{2}\right]\Big/\binom{n}{2}}{\frac{1}{2}\left[\sum_{i}\binom{n_{i\cdot}}{2}+\sum_{j}\binom{n_{\cdot j}}{2}\right]-\left[\sum_{i}\binom{n_{i\cdot}}{2}\cdot\sum_{j}\binom{n_{\cdot j}}{2}\right]\Big/\binom{n}{2}} \tag{4}$$

The adjusted Rand index returns values in the interval $[0,1]$ and is to be maximized. From Eq. (3), the best method m^* on that data set has $b_{m^*}=1$, and all the other methods have $b_m \le 1$. The large the value of b_m, the better the performance of the method is in relation to the best performance on that data set. Thus the sum of b_m over all datasets provides a good measure of the robustness of the method m. A larger value of the sum indicates good robustness.

The distribution of b_m of each method over the 31 data sets is shown in Fig. 1. For each method, the 31 values of b_m are stacked, and the sum is given on the top of the stack. Fig. 1 reveals that MDPSO has the highest sum value. Thus MDPSO is the most robust method among the compared methods.

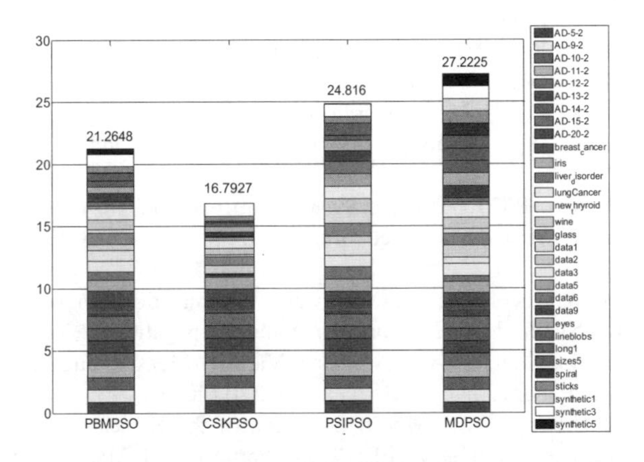

Fig. 1. Robustness of the algorithm compared for 31 data sets

5 Conclusions

In this paper, we have proposed a PSO clustering algorithm using four validity indices to perform unsupervised clustering problems. The experiment results on a large number of artificial synthesized data and UCI data showed that the MDPSO algorithm outperformed the PBMPSO, CSKPSO, and PSIPSO in terms of clustering accuracy and robustness. Our future work is to find a more proper or faster method of computing the manifold distance.

Acknowledgments. This work was supported by the National Natural Science Foundation of China (Grant No. 60703108, 60703198), Research Fund for the Doctoral Program of Higher Education of China (No. 20070701022); the China Postdoctoral Science Foundation Special funded project (No. 200801426), the Fundamental Research Funds for the Central Universities (No. JY10000902040), and the Fund for Foreign Scholars in University Research and Teaching Programs (the 111 Project) (No. B07048).

References

1. Pakhira, M.K., Bandyopadhyay, S., Maulik, U.: Validity Index for Crisp and Fuzzy Clusters. Pattern Recognition 37, 487–501 (2004)
2. Sanghamitra, B., Sriparna, S.: GAPS: A clustering method using a new point symmetry-based distance measure. Pattern Recognition 40, 3430–3451 (2007)
3. Ling, W., Liefeng, B., Licheng, J.: Density-Sensitive Semi-Supervised Spectral Clustering. Journal of Software 18(18), 2412–2422 (2007)
4. Swagatam, D., Ajith, A., Amit, K.: Automatic Kernel Clustering with a Multi-Elitist Partical Swarm Optimization Algorithm. Pattern Recognition Letters 29, 688–699 (2008)
5. Milligan, G.W., Cooper, C.: An Examination of Procedures for Determining the Number of Clusters in a Data Set. Psychometrika 50(2), 159–179 (1985)
6. Meila, M., Heckerman, D.: An Experimental Comparison of Several Clustering and Initialization Methods. In: Proc. 14th Conf. Uncertainty in Artificial Intelligence, pp. 386–395. Morgan Kanfmann, Canada (1998)
7. Bandyopadhyay, S., Maulik, U.: Nonparametric Genetic Clustering: Comparison of Validity Indices. IEEE Trans. Syst. Man Cybern. Part C: Application and Reviews 31(1), 120–125 (2001)
8. Kennedy, J., Berhart, R.C.E.: Particle Swarm Optimization. In: Proc. The IEEE International Joint Conference on Neural Network, vol. 4, pp. 1942–1948. IEEE Press, New York (1995)
9. Boeringer, D.W., Werner, D.H.: Particle Swarm Optimization Versus Genetic Algorithms for Phased Array Synthesis. IEEE Trans. Antennas Propagation 52(3), 771–779 (2004)
10. Chou, C.H., Su, M.C., Lai, E.: A New Cluster Validity Measure and Its Application to Image Compression. Pattern Anal. Appl. 7(2), 205–220 (2004)
11. Muchun, S., Chienhsing, C.: A Modified version of the K-means Algorithm with a Distance Based on Cluster Symmetry. IEEE Trans. Pattern Anal. Mach. Intell. 23(6) (2001)
12. Geng, X., Zhan, D.C., Zhou, Z.H.: Supervised Nonlinear Dimen-sionality Reduction for Visualization and Classification. IEEE Trans. Syst. Man, Cybern., Part B: Cybern. 35(6), 1098–1107 (2005)
13. Hubert, L., Arabie, P.: Comparing partitions. Journal of Classification 2(1), 193–218 (1985)

A Frequency Domain Approach to PID Controllers Design in Boiler-Turbine Units

Hui Pan[1], Minrui Fei[1,*], Ling Wang[1], and Kang Li[2]

[1] Shanghai Key Laboratory of Power Station Automation Technology, Shanghai University, Shanghai, 20072, China

[2] School of Electronics, Electrical Engineering and Computer Science, Queen's University Belfast, United Kingdom

panhui001@163.com, mrfei@staff.shu.edu.cn, wangling@shu.edu.cn, k.li@ee.qub.ac.uk

Abstract. This paper proposes a frequency domain approach—direct Nyquist array (DNA) method, to the design of PID controllers for multivariable boiler-turbine units based on gain and phase margins. The main objective is to propose an integrated method for the design and auto-tuning of simple yet robust PID controllers that can be more easily implemented for the boiler-turbine units in modern power plants. For this, the model of the original multi-input multi-output (MIMO) system is first transformed into a diagonal or diagonal dominance matrix after the system is appropriately compensated. Then, various PID controller design methods for single-input single-output (SISO) systems can be easily extended to decoupled or quasi-decoupled MIMO systems. In particular, the proposed method allows the user to specify the robustness and other key performances of the system through the gain and phase margin specifications. Simulation results illustrate the efficacy of the proposed method, showing that the designed controller for a boiler-turbine unit has a reduced number of elements by a half and produces much better dynamic performances than the one designed by Tan's method.

Keywords: PID control, gain and phase margins, boiler-turbine unit, direct Nyquist array method.

1 Introduction

A boiler-turbine unit is a typical configuration that is widely used in modern power plants. The configuration employs a single boiler to feed steam into a single turbine to generate the power [1].

For the boiler-turbine control system, the primary task is to adjust the output power to meet system demand while minimizing pressure and temperature fluctuation. The system known to undertake this main control task in a boiler-turbine unit in coal-fired power plants is the coordinated control system (CCS). It is a complex time-varying multivariable system with large uncertainty and strong coupling, and practically, control parameters in the CCS are often chosen by trial-and-error method, and the CCS is

* Corresponding author.

K. Li et al. (Eds.): LSMS/ICSEE 2010, Part II, CCIS 98, pp. 73–81, 2010.

usually put in a manual state. It becomes a key issue to improve the control performance in coal-fired power plants. Although significant progress has been made on the design of more advanced CCS in the past decades [1-5], most of the new developments, especially inspired from intelligent computing techniques, such as fuzzy logic, neural networks, intelligent computing, are still restricted to simulation study or laboratory tests. Other tuning methods, e.g [5], however, are strongly dependent on the accuracy of the process models.

Rosenbrock extended the Nyquist stability and design concepts from single-input single-output (SISO) systems to multi-input multi-output (MIMO) systems [6]. The direct Nyquist array (DNA) method was proposed in [7]. It shows that after a MIMO system is compensated, four equations similar as in [8] can be obtained. To solve the equations, numerical algorithms such as the Newton-Raphson method is required. This process however can be quite time consuming, especially for auto-tuning of PID controllers on-line. Further, the method uses the relay method for the parameter identification by simplifying the system as a first-order plus dead-time (FOPDT) model.

In [8], simple procedure is proposed to tune the PID parameters to meet user specified gain margin and phase margin. This is very useful in identifying the parameters of PID controllers for real-time applications. It is possible to apply the PID control design method for SISO systems proposed in [8] to MIMO systems by converting the open-loop transfer function of the MIMO system into a diagonal or diagonal dominance matrix after using linear approximation of the arcsin function in section 2. Further, the area-based method proposed in [9] is used to identify the parameters of the FOPDT model, which is robust and simple. These methods are integrated within one framework in this paper, aiming to design much simpler yet robust PID controllers for boiler-turbine units under different load demands.

2 Direct Nyquist Array (DNA) Design of PID Controllers

A multivariable feedback control system is shown in Fig. 1, where $P(s)$ is the multivariable process, while D and $C(s)$ are a compensator and controllers, respectively.

Define

$$Q(s) = P(s) \cdot D \cdot C(s) = \begin{bmatrix} q_{11}(s) & \cdots & q_{1n}(s) \\ \vdots & \ddots & \vdots \\ q_{n1}(s) & \cdots & q_{nn}(s) \end{bmatrix}. \tag{1}$$

Usually, the direct Nyquist array (DNA) design method consists of two stages. In the first stage, a compensator is designed to decouple the transfer function of a multivariable process into a diagonal or diagonal dominance matrix. The second stage is to design diagonal multiloop controllers to shape the Gershgorin bands. Closed-loop performance and stability can be guaranteed by shaping the Gershgorin bands such that they do not enclose the point $(-1+j0)$ and encircle it at an appropriate number of times, in accordance with the generalized Nyquist stability theorem [7].

Theorem (Nyquist Stability Theorem [7]) Let the Gershgorin bands centered on the diagonal elements $q_{ii}(s)(i=1,2,\cdots,n.)$ of $Q(s)(Q(s)\in C^{n\times n})$ which exclude the point $(-1+j0)$, and let the ith Gershgorin band encircle the $(-1+j0)$ point N_i times anticlockwise. Then, the closed-loop system is stable if and only if

$$\sum_{i=1}^{n} N_i = p_0 \cdot \tag{2}$$

where p_0 is the number of unstable poles of $Q(s)$.

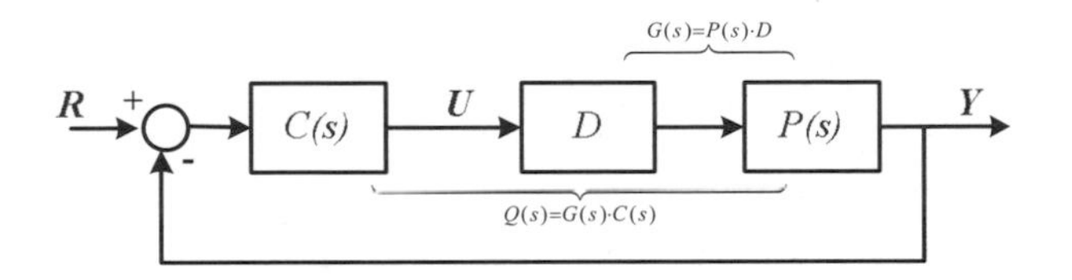

Fig. 1. Multivariable feedback system

Since most industrial processes are minimum phase systems and open-loop stable [10], $p_0=0$ is assumed in this paper. The procedure in DNA design is summarized as follows:

1) Decoupling

Consider an n-input n-output open-loop stable multivariable plant with a transfer function matrix $P(s)$, A constant compensator D is designed to decouple $G(s)$.

Then, there are two possible cases:

a) Steady-state decoupling: This can be achieved by selecting $D=\left[\lim_{s\to 0} P(s)\right]^{-1}$;
b) Approximate decoupling at the bandwidth frequency ωb: This is usually obtained by choosing $D=P_r^{-1}$, where P_r is a real approximation of $P(j\omega_b)$. P_r may be obtained, for example, using the ALIGN algorithm [11]. The bandwidth frequencyωb is defined as the frequency where the singular value of the sensitivity function crosses $1/\sqrt{2}$ from below.

2) Shaping the Gershgorin bands

The second stage is to design n SISO PID controllers to shape the Gershgorin bands. The controller transfer function matrix C(s) is a diagonal matrix, which can be PI or PID controllers. The Gershgorin bands are often shaped by trial and error manually. Using the technique as follows, however, they can be shaped automatically.

3) PID controllers design

The technique of shaping the Nyquist curve for SISO systems through gain and phase margins design [8] can be extended to the DNA design as follows.

An approximate analytical solution can be obtained if we make the following approximation for the arcsin function (see Fig. 2) :

$$2\arcsin(x/2) \approx x \qquad (0 \le x < 1). \tag{3}$$

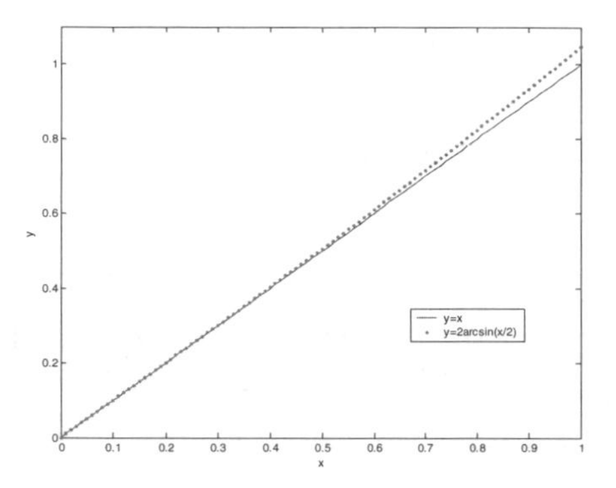

Fig. 2. Approximation of arcsin function

The aim of the controller design based on gain and phase margins is that gain margin $A_{mi}^{'}$ and $\phi_{mi}^{'}$ have sufficient margins (see [7]). A rule of thumb is suggested in [7] as $2 \le A_{mi}^{'} \le 5$ and $30^\circ \le \phi_{mi}^{'} \le 60^\circ$. However, for the sake of simplicity, in the proposed method, it is assumed that A_{mi} and ϕ_{mi} at the center of Gershgorin bands are specified 'a priori'. Using the approximation above, the relationship between them is as follows:

$$\begin{aligned} A_{mi}^{'} &\le A_{mi} < (1+\theta_{\omega_p})A_{mi}^{'}, \\ \phi_{mi}^{'} &\le \phi_{mi} < (1+\theta_{\omega_g})\phi_{mi}^{'}. \end{aligned} \tag{4}$$

Where $\theta_{\omega_p}, \theta_{\omega_g} \in [0,1)$. Often, after the process is compensated, $\theta_{\omega_p}, \theta_{\omega_g} \ll 1$.Then, a rule of thumb here is given as $A_{mi} = 5$ and $\phi_{mi} = 75^\circ$.

3 Simulation

Consider a 300MW coal-fired once-through boiler-turbine unit, the schematic diagram of the control structure of a boiler-turbine unit with a compensator is shown in Fig. 3. The transfer function of the plant [3] is as follows:

$$P(s) = \begin{bmatrix} P_{11}(s) & P_{12}(s) \\ P_{21}(s) & P_{22}(s) \end{bmatrix}. \tag{5}$$

Table 1 gives a linearized transfer function $P(s)$ when the boiler-turbine unit is being operated at 100% load or at at 70% load.

The controllers design is based on the full load. The Nyquist array of the process, $P(s)$, is shown in Fig. 4. It is easy to verify that this process is not row diagonal dominant. Therefore, a steady-state decoupler,

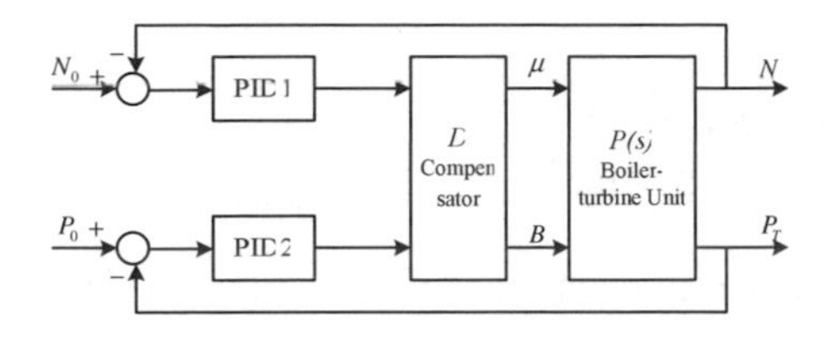

Fig. 3. Control structure of a boiler-turbine unit with a compensator

Table 1. Linearized transfer function $P(s)$ of a 300MW boiler-turbine unit

	At full (100%) load	At 70% load
$P_{11}(s)$	$\frac{4.665s(99s+1)}{(58^2s^2+50s+1)(4.1s+1)}$	$\frac{1.483s(150s+1)}{(63^2s^2+40s+1)(2.7s+1)}$
$P_{12}(s)$	$\frac{2.069(311s+1)}{(149s+1)^2(22.4s+1)}$	$\frac{2.116(457s+1)}{(221s+1)^2(21.8s+1)}$
$P_{21}(s)$	$\frac{-1.42(2.8s+1)}{70s+1}$	$\frac{-0.828(0.97s+1)}{97s+1}$
$P_{22}(s)$	$\frac{1.265(205s+1)}{(128s+1)^2(11.7s+1)}$	$\frac{1.649(275s+1)}{(168s+1)^2(11.5s+1)}$

$$D=\left[\lim_{s\to 0}P(s)\right]^{-1}=\begin{bmatrix}0.43 & -0.7\\ 0.48 & 0\end{bmatrix} \tag{6}$$

is used to achieve diagonal dominance.

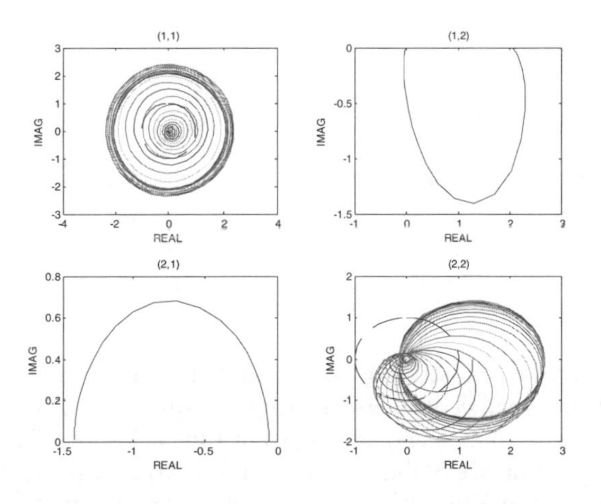

Fig. 4. Nyquist array of $P(s)$ (at full load)

$$G(s) = P(s) \cdot D = \begin{bmatrix} \frac{1.01}{1+3.15s} e^{-21.43s} & * \\ * & \frac{1}{1+64.5s} e^{-3.26s} \end{bmatrix}. \quad (7)$$

The Nyquist array of the compensated process plant, $G(s)$, is shown in Fig. 5. Adopting the robust model-based methods [9], namely, the area-based methods, $G(s)$ is identified as

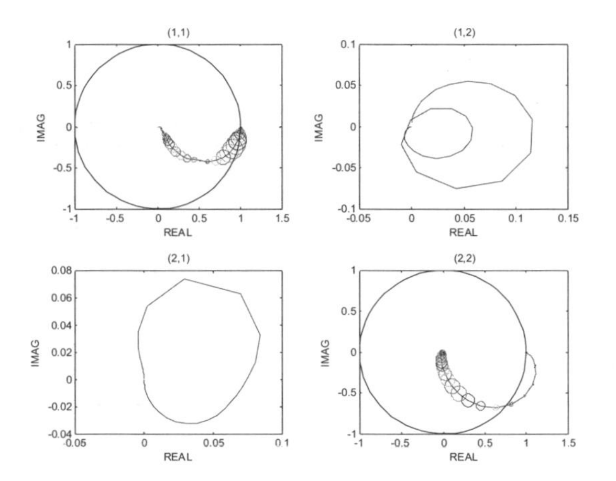

Fig. 5. Nyquist array of $G(s)$ (at full load point)

The results of PID controllers can be given as

$$C(s) = \begin{bmatrix} 0.046\left(1+\frac{1}{3.158s}\right) & 0 \\ 0 & 6.27\left(1+\frac{1}{114s}\right) \end{bmatrix}. \quad (8)$$

The PID controllers proposed in [12] (using Tan's method) are shown as

$$C_{Tan's}(s) = \begin{bmatrix} 0.43\left(1+\frac{1}{85.4s}\right) & 2.25s \\ 3.28 & -\frac{0.05}{s} \end{bmatrix}. \quad (9)$$

The step responses for the closed-loop system at full load are shown in Fig. 6. To test the robustness of the tuned controllers, the step responses at 70% load with the same PID controllers were examined and are shown in Fig. 7. Further, Table 2 summarizes the key transient performance of the system with the two controllers, including the rise time, settling time, and peak overshoot, at different load requirements. From Fig. 6, Fig. 7 and Table 2, it is shown clearly that the proposed controller achieves far better overall transient response than the one designed using Tan's method.

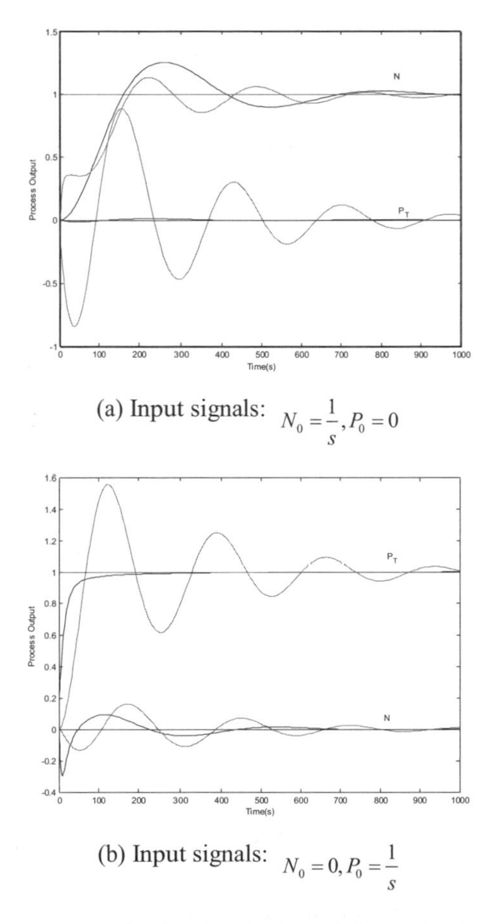

(a) Input signals: $N_0 = \frac{1}{s}, P_0 = 0$

(b) Input signals: $N_0 = 0, P_0 = \frac{1}{s}$

Fig. 6. Closed-loop step responses for the plant model at full load (solid: proposed method; dash: Tan's method)

Table 2. Performance comparison of the proposed method with Tan's method

Load (%)	Step signal input terminal	Method	t_r (s)	t_s (s)	Overshoot (%)
	N_0	proposed	162	632	25.35
100		Tan's	175	822	13.2
	P_0	proposed	67	66	—
		Tan's	162	188	55.34
	N_0	proposed	195	812	23.59
70		Tan's	237	995	25.49
	P_0	proposed	8	10	1.93
		Tan's	88	739	24.64

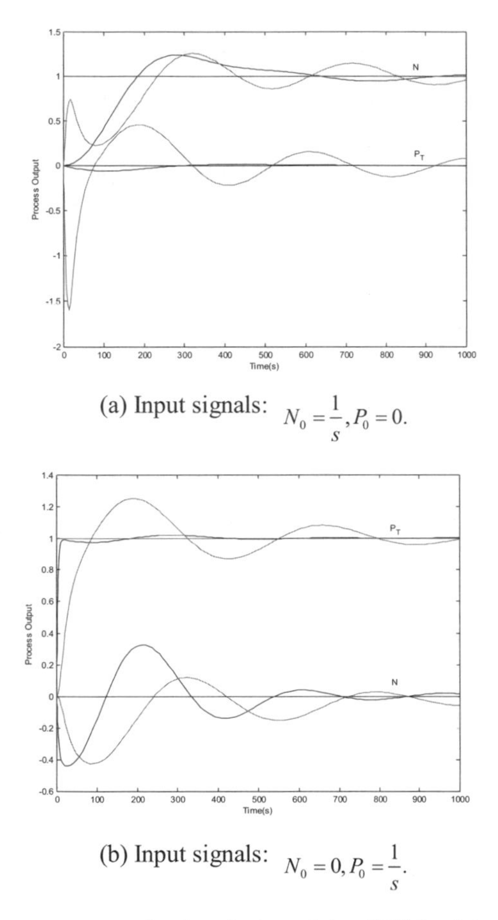

(a) Input signals: $N_0 = \frac{1}{s}, P_0 = 0.$

(b) Input signals: $N_0 = 0, P_0 = \frac{1}{s}.$

Fig. 7. Closed-loop step responses for the plant model at 70% load (solid: proposed method; dash: Tan's method)

4 Conclusions

This paper presents an integrated design method of PID controllers based on gain and phase margins and employs the DNA design method for a multivariable boiler-turbine unit. The main objective has been to develop simpler and yet more effective controllers against another ready-to-wear method. Results show that the controller designed using the proposed method has a reduced number of control elements by a half and outperformed Tan's method at different load demands. It therefore has the potential for real applications to design controllers that are more easily implemented, tuned and maintained, and are more robust and energy efficient.

Acknowledgments. This work is supported by Key Project of Science and Technology Commission of Shanghai Municipality under Grant 08160512100 and 08160705900, and supported by Mechatronics Engineering Innovation Group project from Shanghai Education Commission.

References

1. Tan, W., Fang, F., Tian, L., Fu, C., Liu, J.: Linear control of a boiler-turbine unit: Analysis and design. ISA Transactions 47, 189–197 (2008)
2. Li, S.Y., Liu, H.B., Cai, W.J., Soh, Y.C., Xie, L.H.: A new coordinated control strategy for boiler-turbine system of coal-fired power plant. IEEE Transactions on Control Systems Technology 13, 943–954 (2005)
3. Hu, K.D., Wang, Z.Q., Qian, Z.H.: A Frequency Domain Method for Design of Coordinated Control System of a Unit Power Plant. Journal of Southeast University 19, 69–77 (1989) (in Chinese)
4. Moon, U.C., Lee, K.Y.: A boiler-turbine system control using a fuzzy auto-regressive moving average (FARMA) model. IEEE Transactions on Energy Conversion 18, 142–148 (2003)
5. Tan, W., Liu, J.Z., Fang, F., Chen, Y.Q.: Tuning of PID controllers for boiler-turbine units. Isa Transactions 43, 571–583 (2004)
6. Rosenbrock, H.H.: Design of multivariable control systems using the inverse Nyquist array. IEE proceedings: Control Theory and Applications 116, 1929–1936 (1969)
7. Ho, W.K., Lee, T.H., Xu, W., Zhou, J.R.R., Tay, E.B.: The direct Nyquist array design of PID controllers. IEEE Transactions on Industrial Electronics 47, 175–185 (2000)
8. Ho, W.K., Hang, C.C., Cao, L.S.: Tuning of PID controllers based on gain and phase margin specifications. Automatica 31, 497–502 (1995)
9. Bi, Q., Cai, W.J., Lee, E.L., Wang, Q.G., Hang, C.C., Zhang, Y.: Robust identification of first-order plus dead-time model from step response. Control Engineering Practice 7, 71–77 (1999)
10. Chen, Seborg, D.E.: Multiloop PI/PID controller design based on Gershgorin bands. IEE Proceedings-Control Theory and Applications 149, 68–73 (2002)
11. Edmunds, J., Kouvaritakis, B.: Extensions of the frame alignment technique and their use in the characteristic locus design method. International Journal of Control 29, 787–796 (1979)

An Estimation of Distribution Algorithm Based on Clayton Copula and Empirical Margins*

L.F. Wang, Y.C. Wang, J.C. Zeng, and Y. Hong

College of Electrical and Information Engineering, Lanzhou University of Technology, Lanzhou, 730050, China
Complex System and Computational Intelligence Laboratory, Taiyuan University of Science and Technology, Taiyuan, 030024, China
wlf1001@163.com

Abstract. Estimation of Distribution Algorithms (EDAs) are new evolutionary algorithms which based on the estimation and sampling the distribution model of the selected population in each generation. The way of copula used in EDAs is introduced in this paper. The joint distribution of the selected population is separated into the univariate marginal distribution and a function called copula to represent the dependence structure. And the new individuals are obtained by sampling from copula and then calculating the inverse of the univariate marginal distribution function. The empirical distribution and Clayton copula are used to implement the proposed copula Estimation of Distribution Algorithm (copula EDA). The experimental results show that the proposed algorithm is equivalent to some conventional continuous EDAs in performance.

Keywords: Estimation of distribution algorithms (EDAs), copula theory, the joint distribution, the marginal distribution, Clayton copula.

1 Introduction

Estimation of Distribution Algorithms (EDAs) [1] are a class of evolutionary algorithms(EAs) in which the main operators are the estimation of a probability distribution from the selected solutions and the subsequent sampling from the estimated distribution. EDAs can be understood and further developed without the background of Genetic Algorithms (GAs) though they were introduced as an extension of GAs. It has been shown in previous work that EDAs can outperform traditional evolutionary algorithms on a number of difficult benchmark problems [2].

Many kinds of EDAs [3,4] were studied by scholars which uses different approaches to estimating the probability distributions of the selected population. But the computational costs to estimating the distribution are large in many algorithms [5]. This paper introduces how the copula theory can be used in an EDA. The benefit of the copula EDA is that only marginal distributions need to be estimated from the population of solutions, thus promising efficiency savings.

* This work was supported in part by the Youth Research Fund of ShanXi Province (No. 2010021017-2).

K. Li et al. (Eds.): LSMS/ICSEE 2010, Part II, CCIS 98, pp. 82–88, 2010.

Copula theory [6] is a recently developed mathematical theory for multivariate probability analysis. It separates joint probability distribution function into the product of univariate margins and a copula which represents the dependency structure of random variables. Copulas are functions that join the multivariate distribution functions to their one-dimensional marginal distribution functions. On the grounds of it, given a joint distribution with margins, there exists a copula uniquely determined; conversely, the joint distribution function is determined for a given copula and margins. The study about EDAs utilizing two dimensional copula can be found in [7]and [8].

This paper is organized as follows. It is discussed in section 2 how copula theory can be used in EDA. Section 3 talks about a sampling algorithm for Clayton copula proposed by Marshall and its application in the copula EDA. The experimental results are shown in section 4. The paper ends with concluding remarks in section 5.

2 The Framework of Copula EDA

Consider a continuous random vector $(x_1, x_2, \ldots, x_n)$ with the joint cumulative distribution function H and margins $F_1, F_2, \ldots, F_n$. The copula representation of H is given by $H(x_1, \ldots, x_n) = C(F_1(x_1), \ldots, F_n(x_n))$ [6], where C is a unique cumulative distribution function having uniform margins on (0,1). The copula function is the joint distribution of the probability integral transforms. Therefore, the copula function is also known as the "dependence function." The copula function contains all the information regarding the cross-dependence structures of $(x_1, x_2, \ldots, x_n)$.

By virtue of copula theory, if the joint distribution of random vector $(x_1, x_2, \ldots, x_n)$ is denoted by a copula C and n marginal distribution functions $F_1, \ldots, F_n$, *i.e.*, the joint is $C(F_1(x_1), \ldots, F_n(x_n))$, we need only generate a vector $(u_1, u_2, \ldots, u_n)$ of observation of uniform (0,1) random variables $(U_1, U_2, \ldots, U_n)$ whose joint distribution function is C, and then transform the variables $u_1, u_2, \ldots, u_n$ to $x_1, x_2, \ldots, x_n$ via $x_i = F_i^{-1}(u_i)$. There are many algorithms about sampling $(u_1, u_2, \ldots, u_n)$ from copula[9][10]. In this paper, we focus on the following generating method proposed by Marshall and Olkin [9].

Algorithm 1: sampling algorithm from copula

1) Sample $v \sim F = \mathcal{LS}^{-1}\left[\phi^{-1}\right]$, where $\mathcal{LS}^{-1}\left[\phi^{-1}\right]$ is the inverse Laplace-Stieltjes transform of the generator ϕ^{-1}.

2) Sample *i.i.d.* $x_i \sim U[0,1]$, $i = 1, \ldots, n$.

3) Return $(u_1, \ldots, u_n)$, where $u_i = \phi^{-1}\left((-\log x_i)/v\right), i = 1, \ldots, n$.

EDAs are algorithm implemented by the iterative run containing three main steps. The first step is to select a subpopulation, which contains the distribution information of the better-performed population, according to a certain selection strategy. Many selection strategy used in GAs can be used in EDAs too. The second step is to estimate the distribution model of the selected population. The distribution model is denoted in different ways, for example, the Baysian network or multivariate Gaussian distribution. The joint distribution function is a good way to exactly reflect the relationship of random variables and the distribution characters of each random variable. But the work to directly estimate the joint is very difficult and time consuming. Copula theory provides the theory basis to simplifying the work as to estimate the marginal distribution functions and a copula. Obviously, estimating one-dimensional distribution is easier than estimating the multivariate joint distribution. The Gaussian distribution, empirical distribution and other distributions can be used to estimate the margins. There are many known copulas and many studies about them. We can use one of the known copulas to denote the dependence structure of the selected solution. Then the joint distribution can be obtained according to copula theory. Thus the second step is finished. The third step is to sample from the estimated distribution. On the grounds of copula theory, the sampling work can be done by doing the following two works. First, sample a vector in $[0,1]^n$ from the copula. Second, get the value of each random variable according to the inverse function of its marginal distribution function. In other words, if the optimization problem is

$$\min f(X) = f(x_1,x_2,\ldots,x_n),\ x_i \in [a_i,b_i]\ \ (i=1,2,\ldots,n) \tag{1}$$

Denote the corresponding random vector as $(X_1,X_2,\ldots,X_n)$ and the estimated margins as $F_1,F_2,\ldots,F_n$. C is the copula. Sample a vector $(u_1,u_2,\ldots,u_n)$ from C firstly. Obviously $u_i \in [0,1]$, $i=1,2,\cdots n$. Calculate $x_i=F_i^{-1}(u_i)$ $(i=1,2,\cdots n)$ secondly. And then the $(x_1,x_2,\ldots,x_n)$ is the sampling result of $(X_1,X_2,\ldots,X_n)$ whose joint distribution is $C(F_1,F_2,\ldots,F_n)$.

Concluding the above discussion, the framework of the copula EDA is expressed as in Fig.1.

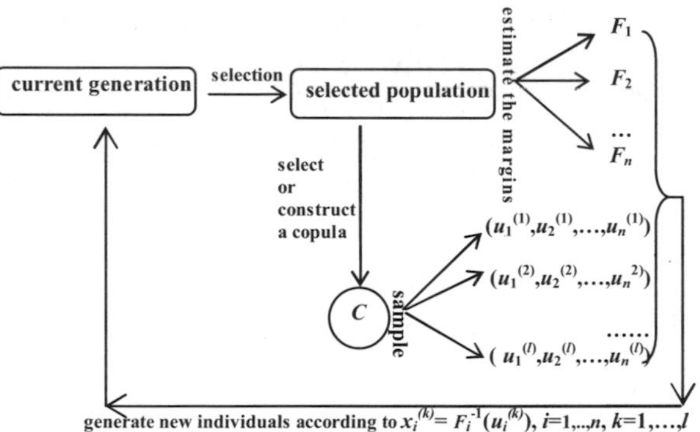

Fig. 1. The flow chart of copula EDA

As shown in Fig.1, the copula EDA is an iterative run starting with a randomly generated population of solutions. The better solutions are selected from the current population of solutions. Let $\mathbf{x}=\{\boldsymbol{x}^j=(x_1^j,x_2^j,\ldots x_n^j),\ j=1,2,\ldots,s\}$ denote the selected population. Therefore, the set $\{\ x_i^j,\ j=1,2,\ldots,s\ \}$ can be considered as the sample of the random variable X_i, $i=1,2,\ldots,n$. The following two operators can be implemented in parallel. One is to estimate the margins F_i for each random variable $X_i(i=1,2,\ldots,n)$ according to the samples $\{\ x_i^j,\ j=1,2,\ldots,s\ \}$. The other is to select or to construct a copula C according to $\mathbf{x}$, and then sample from C. supposing the sampled vectors are $\{\boldsymbol{u}^{(k)}=(\ u_1^{(k)},u_2^{(k)},\ldots,u_n^{(k)}),\ k=1,2,\ldots,l\ \}$, the new individuals $\{\boldsymbol{x}^{(k)}=(\ x_1^{(k)},x_2^{(k)},\ldots,x_n^{(k)}),\ k=1,2,\ldots,l\ \}$ are obtained by calculating $x_i^{(k)}= F_i^{-1}(u_i^{(k)})$, $i=1,..,n$, $k=1,\ldots,l$. copula EDA replaces some old individuals of the original population with the new generated individuals and progresses the new evolution.

3 Copula-EDA Based on Clayton Copula

According to the frame work of copula EDA, we use Clayton copula to estimate the dependence structure of variables and use the empirical distribution to estimate the marginal distribution of each variable. Because the generator of Clayton copula is $\phi(t) = (t^{-\theta} - 1)/\theta$, the inverse function is $\phi^{-1}(t) = (1+\theta t)^{-1/\theta}$ and the inverse Laplace-Stieltjes transform of the generator ϕ^{-1} is $\mathcal{LS}(\phi^{-1}) = F(v) = \frac{(1/\theta)^{1/\theta}}{\Gamma(1/\theta)} e^{-v/\theta} \cdot v^{1/\theta-1}$.

For the selected population **x**, the samples of the ith variable X_i are $\{x_i^1, x_i^2, \ldots, x_i^s\}$. the sorted list of these s samples is $x_i^{<1>} \le x_i^{<2>} \le \ldots \le x_i^{<s>}$. Denote the search space of X_i is $[x_i^{<0>}, x_i^{<s+1>})$, then the empirical distribution function of X_i is

$$F_i(x) = j/s, (x_i^{<j>} \le x < x_i^{<j+1>}), \tag{2}$$

and the inverse function of $F_i(x)$ used in this paper is

$$x = F_i^{-1}(u) = \begin{cases} rand[x_i^{<j>}, x_i^{<j+1>}) & \text{if } \lfloor u \times s \rfloor = j \text{ and } x_i^{<j>} \ne x_i^{<j+1>} \\ rand(x_i^{<j>}; \delta) & \text{if } \lfloor u \times s \rfloor = j \text{ and } x_i^{<j>} = x_i^{<j+1>} \end{cases}, \tag{3}$$

where, $rand[x_i^{<j>}, x_i^{<j+1>})$ denote the random number in the interval $[x_i^{<j>}, x_i^{<j+1>})$, and $rand(x_i^{<j>}; \delta)$ denote the random number in the δ-neighbor of $x_i^{<j>}$.

Concluding the analysis presented above, the pseudo code for copula EDA is shown in Algorithm 3.

Algorithm 2: Pseudo code for copula EDA using Clayton copula and empirical margins
1) Initialize randomly the population P_0 with m individuals and set g←0.
2) Select a subpopulation S_t of size s from P_g according to certain select-strategy.
3) Estimate the unvariate marginal distribution function F_i for each dimension from S_t.
4) For k=1 to l do

4.1) Sample $v \sim F(v) = \frac{(1/\theta)^{1/\theta}}{\Gamma(1/\theta)} e^{-v/\theta} \cdot v^{1/\theta-1}$.

4.2)Sample *i.i.d.* $v_i \sim U[0,1]$, i=1,…,n, get u_i from $u_i = \phi^{-1}((-\log v_i)/v) = (1 - \frac{\theta \log v_i}{v})^{-1/\theta}$.

4.3) Get new individual $(x_1^{(k)}, x_2^{(k)}, \ldots, x_n^{(k)})$ by calculating

$$x_i^{(k)} = F_i^{-1}(u_i) = \begin{cases} rand[x_i^{<j>}, x_i^{<j+1>}) & \text{if } \lfloor u_i \times s \rfloor = j \text{ and } x_i^{<j>} \ne x_i^{<j+1>} \\ rand(x_i^{<j>}; \delta) & \text{if } \lfloor u_i \times s \rfloor = j \text{ and } x_i^{<j>} = x_i^{<j+1>} \end{cases}. \tag{4}$$

5) Replace the old individuals of P_g with the new individuals, set g←g+1.
6) If stopping criterion is not reached go to step (2).

4 Experiments

The three benchmark functions in Table 1 are used to test the performance of the proposed algorithm in this paper. They are also used in reference [11] and the parameters are set as the same as in [11], *i.e.* the dimensions are 10, the population size are 2000, 2000 and 750 respectively for each function, and the maximal evaluation times are 300000.

Table 1. The description of the test functions

Function :	$F_1(x) = -\{10^{-5} + \sum_{i=1}^{n} \mid y_i \mid\}^{-1}$, where, $y_1 = x_1, y_i = y_{i-1} + x_i$
Search space :	$-0.16 \le x_i \le 0.16, i = 1,...,10$
Minimum value :	$F(0) = 100000$
Function :	$F_2(x) = \sum_{i=1}^{n} [(x_1 - x_i^2)^2 + (x_i - 1)^2]$
Search space :	$-10 \le x_i \le 10, i = 1,...,10$
Minimum value :	$F(0) = 0$
Function :	$F_3(x) = 1 + \sum_{i=1}^{n} \frac{x_i^2}{4000} - \Pi_{i=1}^{n} \cos(\frac{x_i}{\sqrt{i}})$
Search space :	$-600 \le x_i \le 600, i = 1,...,10$
Minimum value :	$F(0) = 0$

The mutation operator is also used in the experiments. The selection rate is 0.2 and the mutation rate is 0.05. To increase the diversity of the population, 1/3 individuals of the selected population are selected by truncation selector and the other 2/3 are selected by roulette selector. The mutation operator is performed by randomly producing individuals in the neighbor areas of the selected population. The parameter θ of Clayton copula is set to 1. The experimental results are shown in Table 2. The total run time is 50. The StdVar, min and max in Table 2 represent respectively the standard variance, the minimal value and the maximal value of the 50 search results.

The copula EDA performs well in function F_2 but it fails in function F_3, and its performance in function F_1 is only a little better than $UMDA_c^G$, $MIMIC_c^G$ and ES. From the standard variance of the search results, we know that the performance of the proposed algorithm is not robust. The copula EDA encounter the same problem as other EDAs that is the search result is affected by the initial population. The parameter θ of Clayton copula influence the shape of the copula, it is need to be studied in details. Clayton copula is one of the Archimedean copulas, it may not reflect correctly the relationship of random variables for all problems. The diversity of population is limited because of the property of the empirical distribution which is used to estimate the marginal distribution.

Table 2. The performances of copula EDA and some conventional evolutionary algorithms

Algorithm	mean	StdVar	min	max
		F1		
$UMDA_c^G$	-53460			
$MIMIC_c^G$	-58775			
$EGNA_{ee}$	-100000			
$EGNA_{BGe}$	-100000			
ES	-5910			
copula EDA	-81944	19609	-93394	-31875
		F2		
$UMDA_c^G$	0.13754			
$MIMIC_c^G$	0.13397			
$EGNA_{ee}$	0.09914			
$EGNA_{BGe}$	0.0250			
ES	0			
copula EDA	8.7786e-005	1.4235e-004	3.7552e-007	3.7416e-004
		F3		
$UMDA_c^G$	0.011076			
$MIMIC_c^G$	0.007794			
$EGNA_{cc}$	0.008175			
$EGNA_{BGe}$	0.012605			
ES	0.034477			
copula EDA	0.038551	0.0239	0	0.074655

5 Conclusions

EDAs are evolutionary algorithms by iteratively estimating the distribution model of the selected population and sample new individuals from the model. The precision of the model and the fitness between the distribution of the ancestors and those of the offspring affects the convergence of the algorithm. The joint distribution is an entire and precise way to describe the distribution of random vector. By using of copula theory, the joint distribution is simplified to a copula and the univariate marginal distribution. Therefore, the complexity of EDA will decrease by using of copula. We propose the frame work of copula EDA and implement it by the exchangeable Clayton copula and the empirical marginal distributions. The sampling algorithm from copula is used the algorithm proposed by Marshall and Olkin. The experimental results show that the performance of the proposed algorithm is equivalent to the conventional evolutionary algorithms in performance. The following topics need to be studied in the future.

1) The choose criteria for copulas. For the given selected population, the criteria decide which copula is the most proper function to reflect the dependence of the random variables.

2) The choice of the parameter of copulas. The copula with different parameter values correspond to different dependence structure which influence the precision of the joint distribution reflecting the selected population.

3) The marginal distribution in the copula EDA.

References

1. Mühlenbein, H., Paaß, G.: From combination of genes to the estimation of distributions: Binary parameters. In: Ebeling, W., Rechenberg, I., Voigt, H.-M., Schwefel, H.-P. (eds.) PPSN 1996. LNCS, vol. 1141, pp. 178–187. Springer, Heidelberg (1996)
2. Larranaga, P., Lozano, J. (eds.): Estimation of distribution algorithms. A new tool for evolutionary computation. Kluwer Academic Publishers, Dordrecht (2002)
3. Pelikan, M.: Hierarchical Bayesian Optimization algorithm: Towards a new generation of evolutionary algorithms. Springer, New York (2005)
4. Shakya, S.: DEUM: A Framework for an Estimation of Distribution Algorithm based on Markov Random Fields. PhD thesis, The Robert Gordon University, Aberdeen, UK (April 2006)
5. Duque, T.S., Goldberg, D.E., Sastry, K.: Enhancing the efficiency of the ECGA. In: Rudolph, G., et al. (eds.) PPSN 2008. LNCS, vol. 5199, pp. 165–174. Springer, Heidelberg (2008)
6. Nelsen, R.B. (ed.): An introduction to copulas, 2nd edn. Springer, New York (2006)
7. Wang, L.F., Zeng, J.C., Hong, Y.: Estimation of Distribution Algorithm Based on Copula Theory. In: Proceedings of the IEEE Congress on Evolutionary Computation (CEC 2009), Trondheim, Norway, May 18-21, pp. 1057–1063 (2009)
8. Salinas-Gutierrez, R., Hernandez-Aguirre, A., Villa-Diharce, E.R.: Using Copulas in Estimation of Distribution Algorithms. In: Hernandez Aguirre, A., et al. (eds.) MICAI 2009. LNCS (LNAI), vol. 5845, pp. 658–668. Springer, Heidelberg (2009)
9. Marshall, A.W., Olkin, I.: Families of Multivariate Distributions. Journal of the American statistical association 83, 834–841 (1988)
10. Whelen, N.: Sampling from Archimedean copulas. Quantitative Finance 4(3), 339–352 (2004)
11. Larranaga, P., Etxeberria, R., Lozano, J.A., Pena, J.M.: Optimization in continuous domains by learning and simulation of Gaussian networks. In: Proceedings of the GECCO 2000 Workshop in Optimization by Building and Using Probabilistic Models, pp. 201–204. Morgan Kaufmann, San Francisco (2000)

Clonal Selection Classification Algorithm for High-Dimensional Data

Ruochen Liu, Ping Zhang, and Licheng Jiao

Key Laboratory of Intelligent Perception and Image Understanding of Ministry of Education of China, Institute of Intelligent Information Processing, Xidian University, Xi'an, 710071

Abstract. Many important problems involve classifying high-dimensional data sets, which is very difficult because learning methods suffer from the curse of dimensionality. In this paper, Clonal Selection Classification Algorithm is proposed for high-dimensional data. First, an automatic non-parameter uncorrelated discriminant analysis (UDA) is adopted for dimensionality reduction (DR). Due to the favorable global search and local search, Clonal Selection Algorithm (CSA) is used to design classifier. The proposed method has been extensively compared with nearest neighbor (NN) based on Principal Component Analysis and linear discrimination analysis (PCA+LDA), nearest neighbor (NN) based on UDA (UDA+NN) and FCM based on UDA (FCM+UDA) when classifying six UCI data sets and a SAR image classification problems. The results of experiment indicate the superiority of the proposed algorithm over the three other classification algorithms in term of classification accuracy and stability.

Keywords: dimensionality reduction, UDA, Clonal Selection Algorithm (CSA), SAR image classification.

1 Introduction

Many machine leaning and data mining problems involve analyzing data in very high-dimensional spaces [1]. So, it is necessary to reduce the dimensionality of the data without losing important information. The idea here is that data can be approximated reasonably well even if only a relatively small number of dimensions are kept.

Due to its simplicity and easy use, the most popular dimensionality reduction (DR) approach is the principal component analysis (PCA) [2], in which, features are searched by mapping the data into the range of the total scatter matrix. To discriminate the established features, linear discrimination analysis (LDA) [3], has been widely used. LDA computes the optimal projection, which maximizes the ratio of between class scatter against within-class scatter (Fisher criterion) [4]. In this paper, we use an automatic non-parameter uncorrelated discriminant analysis (UDA) [5] algorithm, In this method, there is no parameter in the whole process and an entire automatic strategy is established.

Clonal Selection Algorithm (CSA) is a kind of artificial immune algorithm, which was proposed by De Castro [6]. CSA actually is a global optimization method inspired by biological clone selection principle to solve real world problems. It has been successfully applied into several challenging domains, such as multimodal optimization and pattern recognition [7, 8].

K. Li et al. (Eds.): LSMS/ICSEE 2010, Part II, CCIS 98, pp. 89–95, 2010.

Based on the CSA, we propose a new classification algorithm, Clonal Selection Classification Algorithm for the high-dimensional data in this paper. We first use UDA to do DR, and then CSA is applied to execute classifier. We have selected PCA+LDA, UDA+NN, and UDA+FCM for comparison to evaluate the performance of the proposed algorithm.

The rest of the paper is organized as follow: Section 2 briefly outlines uncorrelated discriminant analysis and the CSA technique. In Section 3, we present the proposed algorithm. Experimental results are presented and discussed in Section 4. Finally the paper is concluded in Section 5.

2 Background

2.1 Uncorrelated Discriminant Analysis

Assume that $\omega_1, \omega_2, \cdots, \omega_c$ are c known pattern classes [5]. Given a data matrix $X = [x_1, x_2, \cdots, x_N] \in R^{d \times N}$, where x_i denotes the i-th training data point in the d-dimensional space. Suppose that m_i and n_i $(i = 1, \cdots, c)$ are the mean vector and sample number of class ω_i respectively, and m is the total mean vector. The between-class scatter matrix S_b, the within-class scatter matrix S_w and the total scatter matrix S_t are determined by the following formulas:

$$S_w = \frac{1}{N} \sum_{i=1}^{c} \sum_{j \in \omega_i} (x_j - m_i)(x_j - m_i)^T \tag{1}$$

$$S_b = \frac{1}{N} \sum_{i=1}^{c} n_i (m_i - m)(m_i - m)^T \tag{2}$$

$$S_t = \frac{1}{N} \sum_{i=1}^{n} (x_i - m_i)(x_i - m_i)^T \tag{3}$$

It is easy to verify that $S_t = S_b + S_w$. Define matrices H_t.

$$H_t = \frac{1}{\sqrt{N}} [x_1 - m, x_2 - m \cdots, x_N - m] \tag{4}$$

The scatter matrices S_t can be expressed as $S_t = H_t H_t^T$. Then compute the left singular matrix U of H_t by SVD [5]. By the transformation U, the sum within-class and between-class scatter matrices in the lower-dimensional space become

$$S_t^L = U^T S_t U, S_b^L = U^T S_b U, S_w^L = U^T S_w U \tag{5}$$

The UDA algorithm searches the statistical uncorrelated discriminant vectors in the subspace Φ. We first describe a discriminant subspace and its properties. Then we select discriminant vectors from the discriminant subspace which satisfy statistical uncorrelation property.

Since not all discriminant vectors are useful in pattern classification, we introduce a discriminant subspace

$$\Phi = \left\{ \varphi \mid \varphi^T S_b^L \varphi - \varphi^T S_w^L \varphi > 0 \right\} \tag{6}$$

The next discriminant vector φ must maximize the criterion J on subspace Φ, and simultaneously satisfy the statistical uncorrelated property. In general, the m-th discrimination vector $\varphi_m \left(m \geq 1 \right)$ satisfies:

$$\varphi_{m+1} = \arg \max_{\varphi^T \varphi = 1} \left(\varphi^T S_b^L \varphi - \varphi^T S_w^L \varphi \right) \tag{7}$$

$$\text{Subject to } \begin{cases} \varphi \in \Phi, \\ \varphi_l^T S_t^L \varphi = 0, l = 1, \cdots, m \end{cases}$$

Compute the discriminant vector according to the theory referred before. Finally, generate the linear transformation matrix Ψ and let the transformation matrix $T = U\Psi$.

2.2 Clonal Selection Algorithm (CSA)

In1978, Burnet presented the famous clonal selection theory [6], it is based on such biological feature of antibody clonal selection that make CSA form a new artificial immunity system method [8-10]. In this paper, we design a simple CSA algorithm and the main step of CSA algorithm shows as follows.

Encoding. Let the data prototypes to be an antibody. Concatenate k groups of attributes of the prototype class center *P*, and encode their quantization values into an antibody according to their possible value range.

Antibody-antigen affinity. Antibody-antigen affinity of antibody A is defined by using the clustering cost function.

$$f\left(A\right) = \frac{1}{1 + \sum_{i=1}^{k} \sum_{l=1}^{n} \mu_{il}^{\alpha} \left\| \left(x_l, p_i \right) \right\|^2} \tag{8}$$

The parameter m is a weighting exponent on each fuzzy membership, X is the sample and n is the number of X. $U = \{u_i \mid i = 1, \cdots, k\}$ is the membership matrix [1].

In the CSA algorithm, clonal operators include clonal proliferation, immune genetic operation and clonal selection operation and clonal death.

Clonal proliferation T_c^C. Suppose antibody population at t generation is $A(t) = \{A_1, A_2, \cdots, A_k\}$, Clonal proliferation is defined as follows:

$$T_c^C(A(t)) = [T_c^C(A_1(t)),\cdots,T_c^C(A_i(t)),\cdots,T_c^C(A_k(t))]^T \tag{9}$$

Here $T_c^C(A_i(t)) = I_i \times A_i(t)$, $i = 1,2\cdots,k$, I_i is one vector with a length of q_i, the antibody set $A_{ci}(t)$ has q_i copies of antibody A_i in it.

Immune genetic operations include clonal mutation, clonal crossover and clonal selection operation. Here we define three operations respectively.

Clonal mutation. In order to maintain the information of the original antibody population, the clonal mutation does not act on the initial antibodies $A \in A'$, which can be understood as follows, for example, one clonal pool contains 5 same antibodies, and then the clonal mutation will apply to 4 antibodies in this clonal pool, and leave one antibody as the parent antibody. In this paper, we adopted point mutation.

Clonal crossover. Similar to the clonal mutation, the clonal recombination is unused to the original population, and we adopt the following recombination method. Consider two antibodies: $A_1 = \{x_1, x_2, \cdots, x_L\}$ and $A_2 = \{y_1, y_2, \cdots, y_L\}$, we choose randomly the components between the *i*-th and *j*-th point as the crossover part, namely:

$$T_r^c(A_1, A_2) = \{x_1, \cdots, x_i, y_{i+1}, \cdots, y_j, x_{j+1}, \cdots, x_l\} \tag{10}$$

Clonal selection operator. After the clonal crossover, the best antibody with the maximal affinity in the current population will be kept in the new antibody population.

Clonal death operation. After the immune genetic operation, we obtain the new antibody population $A(t+1) = \left[A_1(t+1), A_2(t+1), \cdots A_k(t+1)\right]$. If there are two antibodies $A_i(t+1)$ and $A_j(t+1)$ in $A(t+1)$, which satisfy $f\left(A_i(t+1)\right) = f\left(A_j(t+1)\right), i \neq j$, then one of $A_i(l+1)$ and $A_j(l+1)$ can be discarded with a probability of p_d.

Apart from above clonal operators, we use a one-step-iteration operator. Each antibody is first decoded into the prototype class center *P*, then the center *P* and the membership matrix *U* is updated by the following iteration.

$$\mu_{ij} = \left(\sum_{l=1}^{k}\left(D\left(x_j, p_l\right)\right)^{-2}\right)^{-1} \Big/ \left(\left\|x_j, p_l\right\|\right)^2$$
$$\forall i, j, p_i = \sum_{j=1}^{n} \mu_{ij}^{\alpha} x_j \Big/ \sum_{j=i}^{n} \mu_{ij}^{\alpha} \tag{11}$$

The prototype class center *P* can be obtained from Eq. (11), and then they are encoded into antibody again for followed clonal operators.

3 The Proposed Algorithm

In our implementation [1], we use the training sample data selected from the images or standard libraries to define the class centers a priori, the data are then classified by calculating the membership values.

The proposed algorithm can be defined in the following steps:

Step1. Choose the training sample data and calculate the linear transformation matrix Ψ and let $T = U\Psi$ according to the UDA.

Step2. For the training sample X and test sample $Xnew$, it becomes $X = X^T T$, $Xnew = Xnew^T T$ through the dimensionality reduction.

Step3. Produce the class centers from the training sample data. In the initial population, all the class centers are randomly generated in a certain range, except one is the mean of the training data.

Step4. Calculate the best center of the training sample data by the CSA algorithm.

Step5. Classify the test sample data according to the best center obtained in Step4. And then calculate the matrix U by the CSA algorithm.

Step6. Classify each test data by hardening the matrix U assigning each point the class of maximum membership.

4 Experimental Results

In this section, six UCI data sets are first used to test the performance of the proposed algorithm, and a comparison is made among the nearest neighbor (NN) based on Principal Component Analysis and linear discrimination analysis (denoted as PCA+LDA for short), nearest neighbor (NN) based on UDA (denoted as UDA+NN for short) and FCM based on UDA (denoted as FCM+UDA). Then, we solve the classification problems of a SAR image by using the four algorithms.

4.1 UCI Data Classification

In this section, six UCI data sets [11], namely Cancer, Diabetes, Wine, Heart-disease, Lung-cancer and Isolet5, are used to test the performance of the proposed algorithm.

The main parameter settings are as follows: For CSA: mutation probability is 0.1, clone size is 20, maximum number of iterations is 100, and the termination conditions is that errand of the affinity of the best individual between two generation is 1 and the value of object function does not improved within 5 generations.

Each test is performed with 10 independent runs by 10-fold cross validation and the right classification rates are averaged. The average classification accuracy and its standard deviation arc given in Table 1.

Table 1. The average classification accuracy and its standard deviations obtained (in %)

Data set	PCA+LDA	UDA+NN	UDA+FCM	UDA+CSA
Cancer	95.56±2.4	95.57±2.5	96.51±2.7	98.68±2.6
Diabetes	69.29±4.6	64.05±4.9	69.01±6.7	70.95±5.7
Wine	76.17±11.8	4.78±10.3	75.36±7.3	80.39±6.5
Heart-disease	42.32±6.7	41.07±10.1	56.07±8.0	57.86±7.4
Lung-cancer	59.00±19.1	62.00±25.3	97.67±1.2	100±0.0
isolet5	96.79±2.4	96.91±4.4	97.63±3.9	98.00±3.3

It can be seen from Table 1, for the classification accuracy, UDA+CSA is better than the three other algorithms on all six data sets. From the view of standard deviation, the classification results of our algorithm are most stable.

4.2 SAR Image Classification

In order to demonstrate the applicability of the proposed algorithm to real SAR image segmentation, experiments on the SAR image are performed. The texture features of SAR images based on the Gray Level Co-occurrence (GLCM) matrix and the energy features from the undecimated wavelet decomposition are applied for SAR image segmentation.

SAR image, as shown in Fig.1 (a), consists of four types of land cover: water and three kinds of corp. We choose randomly sample pixels from four different areas; the numbers of samples of water, white crop, gray crop and tinge crop are respectively 100, 100, 200 and 400; and the test pixels from different areas are 100,100,200 and 200. Table 2 shows average classification accuracy and the Kappa coefficient which are calculated by the confusion matrix for 10 independent runs.

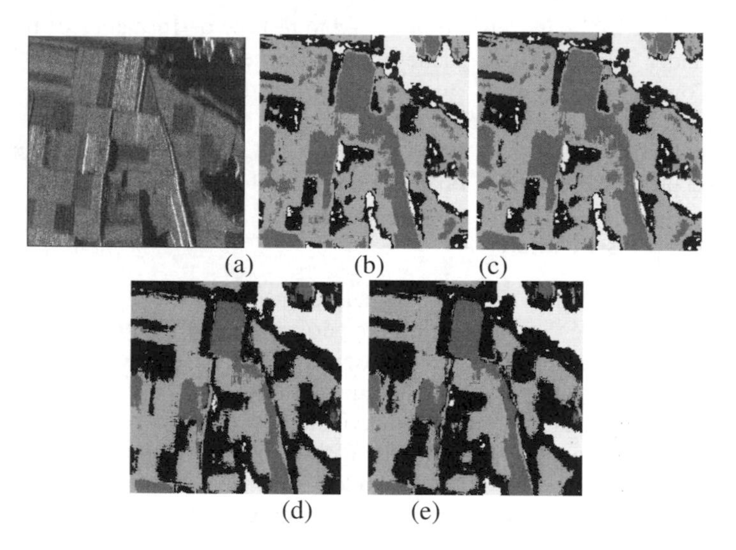

Fig. 1. Segmentation Results of SAR Image. (a) Original image; (b) Segmentation obtained by PCA+LDA classifier; (c) Segmentation obtained by UDA+NN classifier; (d) Segmentation obtained by UDA+FCM classifier; (e) Segmentation obtained by UDA+CSA classifier.

As can be seen from Fig.1, compared with the other three classifiers, UDA+CSA classifier has a better classification result for tinge crop area and gray crop area. The other three classifiers classify wrong more pixels which originally belong to the tinge crop area to the white crop area. From Table 2, we can see that, compared with the the three other classifiers, UDA+CSA classifier gets an increase of 31.83%, 30.16% and 3.33%, on the average classification accuracy respectively; on the Kappa coefficient, UDA+CSA classifier also has a certain degree of improvement, which is 0.3294, 0.2663 and 0.0450.

Table 2. Average classification accuracy (ACA) and Kappa coefficient of SAR image

items	PCA+LDA	UDA+NN	UDA+FCM	UDA+CSA
ACA	74.50	76.17	93.00	**96.33**
Kappa coefficient	0.6600	0.6837	0.9044	**0.9494**

5 Conclusion

In this paper, we present a CSA classification method for high dimensional problems, based on a automatic non-parameter uncorrelated discriminant analysis (UDA). Then we apply the algorithm to the classification of six UCI data sets and a SAR image, and the results show our algorithm is more effective than the three other algorithms. In the future work, we will improve our CSA classification algorithm in the stability and the classification accuracy.

Acknowledgments. This work was supported by the National Natural Science Foundation of China (No. 60703108, No.60703198), Research Fund for the Doctoral Program of Higher Education of China (No. 20070701022); the China Postdoctoral Science Foundation Special funded project (No. 200801426), the Fundamental Research Funds for the Central Universities (No. JY10000902040), and the Fund for Foreign Scholars in University Research and Teaching Programs (the 111 Project) (No. B07048).

References

1. Mostafa, G.H., Timothy, C., Aly, A.: Supervised Fuzzy and Bayesian Classification of High Dimensional Data: a Comparative Study. In: Proceedings of the 2000 International Conference on Image Processing, pp. 772–775. IEEE Press, New York (2000)
2. Jolliffe, I.T.: Principal Component Analysis. Springer, New York (1986)
3. Chen, X., Liang, Y.: A Modified Uncorrelated Linear Discriminant Analysis Model Coupled with Recursive Feature Elimination for the Prediction of bioactivity. SAR and QSAR in Environmental Research 20(1-2), 1–26 (2009)
4. Fukunaga, K.: Introduction to Statistical Pattern Classification. Academic Press, USA (1990)
5. Yang, W.H., Dai, D.Q.: Feature Extraction and Uncorrelated Discriminant Analysis for High-dimensional Data. IEEE Transactions on Knowledge and Data Engineering 20, 601–614 (2008)
6. De, C., Von, L.N., Zuben, F.J.: Learning and Optimization Using the Clonal Selection Principle. IEEE Trans. Evolutionary Computation, Special Issue on Artificial Immune Systems 6, 239–251 (2002)
7. Gao, S.C., Wang, W.: Improved Clonal Selection Combined with Ant Colony Optimization. IEICE Trans. Inf. & Syst. 91(6), 1813–1823 (2008)
8. Li, J., Gao, X.B.: A CSA-based New Fuzzy Clustering Algorithm. Journal of Fudan University (Natural Science) 43(5), 815–818 (2004)
9. Liu, R.C., Sheng, Z.C.: Gene Transposon Based Clonal Selection Algorithm for Clustering. In: The 11th Annual Conference on Genetic and Evolutionary Computation, Montréal, Québec, Canada, pp. 8–12 (2009)
10. Li, F.J., Gao, S.C.: An Adaptive Clonal Selection Algorithm for Edge Linking Problem. IJCSNS International Journal of Computer Science and Network Security 9(7), 57–65 (2009)
11. UCI Machine Learning Repository, `http://archive.ics.uci.edu/ml/`

A General Framework for High-Dimensional Data Reduction Using Unsupervised Bayesian Model

Longcun Jin, Wanggen Wan, Yongliang Wu, Bin Cui, and Xiaoqing Yu

School of Communication and Information Engineering, Shanghai University,
Yanchang Rd 149, Shanghai 200072, China
longcunjin@shu.edu.cn

Abstract. In this paper, we propose a general framework for high-dimensional data reduction using unsupervised Bayesian model. The framework assumes that the pixel reflectance results from linear combinations of pure component spectra contaminated by an additive noise. The constraints are naturally expressed in unsupervised Bayesian literature by using appropriate abundance prior distributions. The posterior distributions of the unknown model parameters are then derived. Experimental results on hyperspectral data demonstrate useful properties of the proposed reduction algorithm.

Keywords: Bayesian modal, inductive cognitive, high-dimensional reduction, unsupervised, general framework.

1 Introduction

The reduction of hyperspectral image has been widely used in remote sensing signal processing for data analysis [1-2]. One is the macrospectral mixture that describes a mixed pixel as a linear mixture endmembers opposed to the other model suggested by Hapke, referred to as intimate mixture that models a mixed pixel as a nonlinear mixture [3-4]. Such statistics are said to be sufficient if they capture all the "relevant information" in the sample about the identity of reduction of hyperspectral image [5]. In this work we proposed a way of quantifying this information using information theoretic notions and show how features which maximize this information can be extracted [6]. It is thus related to a long line of work in statistics [7]. Here, we propose a modification to the kernel functions that can take into account the difference of relative utility of each spectral band by imposing a series of spectral weights [8-9].

The remainder of the paper is organized as follows. An overview of the related work is given in Section 2. The proposed algorithm is discussed in Section 3. Section 4 presents the experimental settings and performance evaluation. Section 5 concludes this paper.

2 Related Work

Classification of hyperspectral data is primarily made on a pixel by pixel basis with classification accuracy figures in the range 79%~84%, and they have not changed

K. Li et al. (Eds.): LSMS/ICSEE 2010, Part II, CCIS 98, pp. 96–101, 2010.

significantly in recent decade [4-5, 11, 30, 46-48]. The natural variability of the material spectra and the noise added by the transmission media and sensor system make necessary the use of statistical methods for information extraction and pattern recognition on hyperspectral data. State of the art, object-based, and object-oriented classification algorithms have been recently used for remotely sensing multispectral image, but little has been done on hyperspectral data due to the large dimensionality of the data. The scale-space framework introduced by the diffusion equation has been also used for image reduction, in conjunction with level sets to detect movement in image sequences, information extraction and image restoration, registration, and classification integrating level sets in a common framework [36, 38-41]. Continuous transformation of the original image into a space of progressively smoother images identified by the scale or level of image smoothing, in terms of pixel resolution.

3 The Proposed Algorithm

The proposed algorithm is based unsupervised Bayesian inductive cognitive model by an approximate inference approach. This can be computed bottom-up as the tree built shown in Figure 1.

initialize each leaf to have $d_i = \alpha, \pi_i = 1$
for each internal node k do

$$d_k = \alpha\Gamma(n_k) + d_{left} d_{right_k}$$

$$\pi_k = \frac{\alpha\Gamma(n_k)}{d_k}$$

end for

Fig. 1. To compute prior on merging, where $right_k (left_k)$ indexes the right (left) subtree of T_k and d_{right_k} (d_{left_k}) is the value of d computed for the right (left) child of internal nodes k

We propose to consider the hyperspectral data reduction problem as in (1), written in vector form as in (2)

$$x_w(r) = \sum_{k=1}^{K} \alpha_{\omega,k} s_k(r) + \varepsilon_\omega(r) \tag{1}$$

$$x(r) = As(r) + \epsilon(r) \tag{2}$$

where $x(r) = \{x_i(r), i = 1,2,\cdots,M\}$ is the set of M observed mixed hyperspectral images, Now, if we note by $x' = \{x(r), r \in R\}, s' = \{s(r),\ r \in R\}$ and $\epsilon' = \{\varepsilon(r), r \in R\}$, then we can write

$$x' = As' + \epsilon \tag{3}$$

In the following, we assume that errors $\epsilon(r)$ are centered, white, Gaussian with the covariance matrix $\sum_\varepsilon = \mathrm{diag}[\sigma_{\epsilon_1}^2, \sigma_{\epsilon_2}^2, \cdots, \sigma_{\epsilon_M}^2]$. This leads to

$$p(x'|s',\Sigma_\epsilon) = \prod_r N(As(r),\Sigma_\epsilon). \tag{4}$$

As we mentioned in the introduction, it is supposed that all the sources of hyperspectral images $s(r)$ are to be piecewise homogeneous and share the same reduction. Then, the homogeneity of the pixels in region k is modeled by

$$p\big(s_j(r)|z'(r) = e_k\big) = N\big(\mu_{jk}, \sigma_{jk}^2\big) \tag{5}$$

and the Bayesian field $z'(r)$ is modeled by

$$p(z'(r), r \in R) \propto \exp\big[\beta \sum_{r\in R} \sum_{r'\in V(r)} z'^t(r) z'(r')\big]. \tag{6}$$

This model for $z'(r)$ is a Bayesian, model which is more commonly represented by

$$p(z(r), r \in R) \propto \exp\big[\beta \sum_{r\in R} \sum_{r'\in V(r)} \delta\big(z(r) - z(r')\big)\big] \tag{7}$$

where $V(r)$ represents the neighbors of r. The parameter β controls the mean size of the regions and has to be $\beta \in [(4/N_v \cdot K), (K/N_v)]$ with N_v number of neighbors of r and K number of classes, to obtain stable solution for z. In this paper, we fix this parameter experimentally.

$$p(s'|z,\theta,x') = \prod_r p\big(s(r)|z'(r),\theta,x(r)\big) = \prod_r N\big(\bar{s}(r), Bb(r)\big)$$

with

$$\begin{cases} B(r) = \big[A^t\Sigma_\epsilon^{-1}A + \Sigma_{z(r)}^{-1}\big]^{-1} \\ \bar{s}(r) = B(r)\big[A^t\Sigma_\epsilon^{-1}x(r) + \Sigma_{z(r)}^{-1} m_{z(r)}\big] \end{cases} \tag{8}$$

Our main goal in our proposed algorithm is to incorporate the idea of multi-scale hyperspectral data reduction into extended Bayesian of inductive cognitive transformations. However, if the reduction patterns do not have regular properties across the hyperspectral data, an adaptative scheme is needed to ensure the wonderful experimental result. In order to extend our algorithm model and closing operations to hyperspectral image, let us consider a hyperspectral image f defined on R^N.

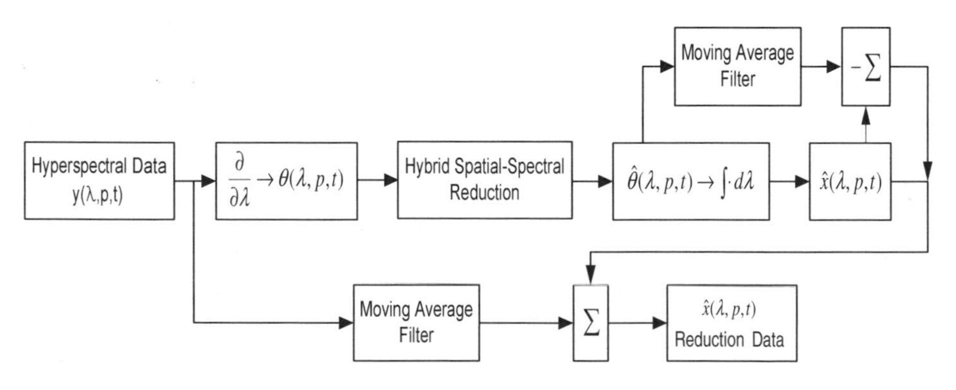

Fig. 2. Architecture of the reduction algorithm

Using Eqs. (9), extended morphological Bayesian models are created as follows. Let the vector $p_k^{\circ}(x,y)$ be the extended opening by reconstruction Bayesian model at the point (x,y) of the hyperspectral image f, defined by

$$p_k^{\circ}(x,y) = \{(f \circ B)^{\lambda}(x,y)\}, \quad \lambda = \{0,1,\cdots,k\} \tag{9}$$

The procedure architecture of the proposed reduction algorithm is shown in Fig. 2.

4 Experimental Results

In this paper, we will present the effectiveness of the dimensional reduction approach based on unsupervised Bayesian model. However, since the data is separated into various discontinuous intervals, the dimensional reduction approach cannot be applied as is and need to be adapted for the hyperspectral data. In order to support hyperspectral data, the dimensional reduction approach based on Bayesian model needs to handle multiple intervals. In one interval, the bands are supposed to be contiguous in the spectral domain.

Table 1. Levels of Bayesian decomposition and associated number of bands for different correlation values (coarse granularity, 3 sub-intervals)

		Correlation Threshold					
		98%	95%	90%	85%	80%	75%
Sub-interval 1 (1-3-2-4) Original: 514 bands	Level of Bayesian decomp.	0	0	0	0	1	1
	# bands after dimension	514	514	514	514	257	257
Sub-interval 2 (6-7-5-8-9) Original: 602 bands	Level of Bayesian decomp.	0	0	0	0	0	0
	# bands after dimension reduction	602	602	602	602	602	602
Sub-interval 3 (10-11-12-13-14-15-16) Original: 1262 bands	Level of Bayesian	0	0	1	1	2	2
	# bands after dimension reduction	1262	1262	631	631	315	315
Total Reduction (in %)		0	0	26.5	26.5	50.6	50.6

Table 1 is considering only three sub-intervals, concatenating close intervals to produce contiguous ranges in frequency domain. On the other hand, Table 1 is considering each interval independently. It can be noticed that by breaking the spectrum into smaller domains, a different level of reduction can be achieved for each interval.

In our work, we conduct some experiments in hyperspectral data dimensionality reduction in order to demonstrate the feasibility of the proposed algorithm. Our datacube was acquired using an Airborne Visible/Infrared Imaging Spectrometer (AVIRIS). The original scene with size of 528×434 pixels, was acquired by the AVIRIS sensor, which is a mixed river/city area in south of china, early in the growing season.

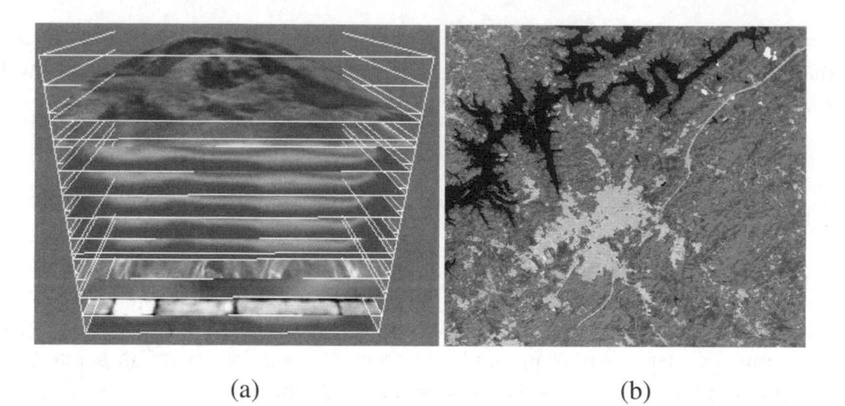

(a) (b)

Fig. 3. (a)High-Dimension Data Expressed by Computer Graphics (b)The Reduction Result of Hyperspectral Image

5 Conclusions

The proposed algorithms have developed depending whether the endmembers belonging to the mixture are known or belong to a known area. Simulation results conducted on real images illustrated the performance of the proposed reduction algorithm based on Bayesian inductive cognitive model. The hierarchical reduction methodologies developed in this paper could be modified to handle more complicated models.

Acknowledgments. This research work is supported by the National Natural Science Foundation of China (60873130, 60872115), the National High Technology Research and Development Program of China (863 program, 2007AA01Z319), the Shanghai's Key Discipline Development Program (J50104), and the Shanghai University's Graduate Student Innovative Fund (08820096).

References

1. Wang, J., Chang, C.-I.: Independent component analysis-based dimensionality reduction with applications in hyperspectral image analysis. IEEE Trans. on Geoscience and Remote Sensing 44(6) (June 2006)
2. Othman, H., Qian, S.-E.: Noise reduction of hyperspectral imagery using hybrid spatial-spectral derivative-domain wavelet shrinkage. IEEE Trans. on Geoscience and Remote Sensing 44(2) (Feburary 2006)
3. Farrell, M.D., Mersereau, R.M.: On the impact of PCA dimension reduction for hyperspectral detection of difficult targets. IEEE Geoscience and Remote Sensing Letters 2(2) (April 2005)
4. Dobigeon, N., Tourneret, J.-Y., Chang, C.-I.: Semi-supervised linear spectral unmixing using a hierarchical Bayesian model for hyperspectral imagery. IEEE Trans. on Signal Processing 56(7) (July 2008)
5. Geng, X., Zhan, D.-C., Zhou, Z.-H.: Supervised nonlinear dimensionality reduction for visualization and classification. IEEE Trans. on Systems, Man, and Cybernetics-Part B: Cybernetics 35(6) (December 2005)

6. Law, M.H.C., Jain, A.K.: Incremental nonlinear dimensionality reduction by manifold learning. IEEE Trans. on Pattern Analysis and Machine Intelligence 28(3) (March 2006)
7. Plaza, A., Martinez, P., Plaza, J., Perez, R.: Dimensionality reduction and classification of hyperspectral image data using sequences of extended morphological transformations. IEEE Trans. on Geoscience and Remote Sensing 43(3) (March 2005)
8. Mohan, A., Sapiro, G., Bosch, E.: Spatially coherent nonlinear dimensionality reduction and segmentation of hyperspectral images. IEEE Geoscience and Remote Sensing Letters 4(2) (April 2007)
9. Keren, O.: Reduction of the average path length in binary decision diagrams by spectral methods. IEEE Trans. on Computers 57(4), 520–531 (2008)

The Model Following Neural Control Applied to Energy-Saving BLDC Air Conditioner System

Ming Huei Chu, Yi Wei Chen, and Zhi Wei Chen

Department of Mechatronic Technology, Tungnan University
No.152, Sec 3, PeiShen Rd, ShenKeng, Taipei, 222, Taiwan, R.O.C.
much48@yahoo.com.tw

Abstract. An AC inverter has been widely used to air conditioner systems for energy saving. But the AC driver will generate high heat dissipation and induce high operating temperature in low speed operation conditions. The modern brushless DC motor (BLDC motor) will improve the high heat generation problem in wide operation speed. This study utilizes the model following neural control applied to modern BLDC driver. A simple approximation of plant Jacobian is proposed, the appropriate speed performance of the BLDC motor for energy saving is defined. The simulation results reveal that the proposed control system is available to control the DC air conditioner system and save energy.

Keywords: Brushless DC motor, Speed control, Neural networks, Energy-saving air conditioner systems.

1 Introduction

Current major trends in air conditioners are saving energy and environmental awareness. The brushless DC motor used in compressors is part of these trends as well. Recent years, the electrical inverter driving a compressor with an AC driver has been widely used to air condition systems for energy saving. But the AC driver will generate high heat dissipation and induce high operating temperature in low speed operation conditions. This will decrease the energy saving efficiency. Attention is being focused on efficiency as air conditioners are consuming less electric power. The air conditioner compressors droved by a DC brush less motor will improve the high heat generation problem in wide operation speed[1]. The air conditioner with DC driver has better energy saving efficiency than that with a conventional AC driver. A conventional PID control is usually applied to the air conditioner with DC driver, but it is difficult to find the appropriate proportion, differential and integer parameters to control the room temperature with specified performance.

The modern precise BLDC systems need to overcome the unknown nonlinear friction, parameters and torque load variations. It is reasonable to apply the adaptive control based on neural networks.

The neural network controls have been used in various fields owing to their capability of on-line learning and adaptability. Tremendous studies for neural network controllers have been conducted to dynamic systems. Psaltis et al. [2] discussed the general learning and specialized learning architectures, populated the input space of

K. Li et al. (Eds.): LSMS/ICSEE 2010, Part II, CCIS 98, pp. 102–109, 2010.

the plant with training samples so that the network can interpolate for intermediate points. The specialized learning architecture doesn't need off-line training connective weights with all data pairs of working region, and which is easily implemented. The error between the actual and desired outputs of the plant is used to update the connective weights. In this sense, the controller learns continuously, and hence it can control plants with time-varying characteristics.

The direct control strategy with specialized learning architectures needs a priori qualitative knowledge or Jacobian of the plant. But it is usually difficult to approximate the Jacobian of an unknown plant. Zhang and Sen [3] presented a direct neural controller for on-line industrial tracking control application, and a simple sign function applied to approximate the Jacobian of a ship track keeping dynamics. The results of a nonlinear ship course-keeping simulation were presented, and the on-line adaptive control was available. But their schematic is not feasible for high performance motion controls. A motion control system needs a neural controller with faster convergent speed. Chu et al. [4,5] proposed a linear combination of error and error's differential to approximate the back propagation error. By this way, the convergent speed will be increased.

This study utilizes the model following direct neural control applied to control the speed of a compressor with BLDC driver. The proposed neural control system needs a well-designed first order reference model, which makes the model following speed control system free of overshoot response. It will have better energy efficiency than a conventional PID control system. In this study, the back propagation error (BPE) is approximated by a linear combination of error and error's differential, so that it is not necessary to build the neural emulator to estimate the Jacobian of plant. By this method, the convergent speed can be improved. The simulation results reveal that the proposed neural control system is available to an air conditioner system with favor performance.

2 Description of the Energy-Saving Air Condition System

The simple BLDC motor system is shown as Fig.1.

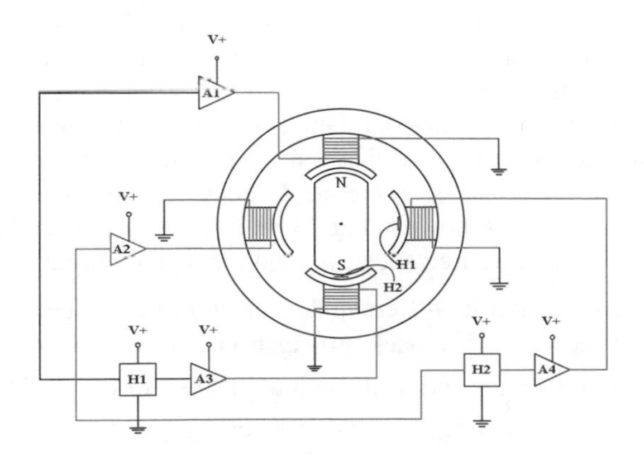

Fig. 1. The BL motor system

The Energy-saving air condition system is shown as Fig.2, where ω_r is the speed command and ω is the actual output speed. T_{comm} is the temperature command and T_S is temperature of room surrounding. A conventional PID controller is used to control the BLDC motor system. The model following direct neural control applied to the BLDC driver speed control is shown in Fig.3, which uses a neural controller to control the BLDC motor system.

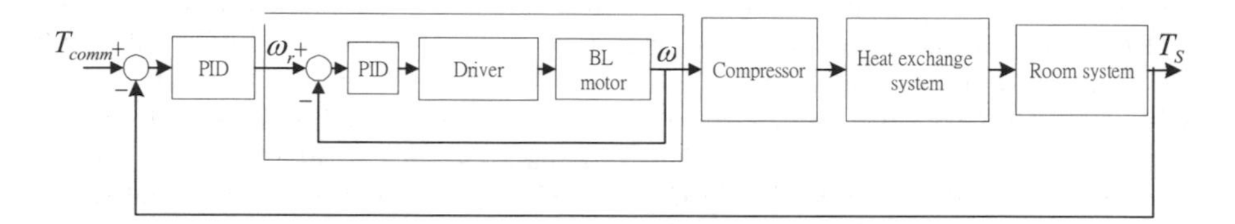

Fig. 2. The Energy-saving air condition system

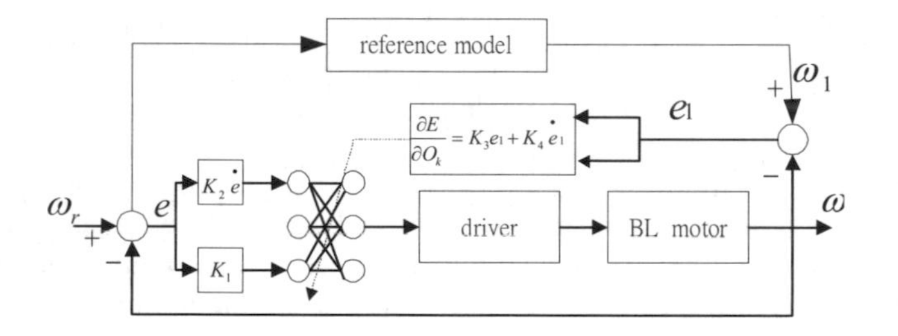

Fig. 3. The block diagram of speed control system

3 Description of the Neural Control

Cybenko [6] has shown that one hidden layer with sigmoid function is sufficient to compute arbitrary decision boundaries for the outputs. Although a network with two hidden layers may give better approximation for some specific problems, but it is more prone to fall into local minima and take more CPU time. In this study, a network with single hidden layer is applied to the speed control system. Another consideration is the right number of units in a hidden layer. Zhang and Sen. [3] have tested different numbers units of the single hidden layer. It was found that a network with three to five hidden units is often enough to give good results. There are 5 hidden neurons in the proposed neural control. The proposed DNC is shown in Fig 2 with a three layers neural network.

The difference between desired speed ω_r and the actual output speed ω is defined as error e. The error e and its differential $\dot{e}$ are normalized between –1 and +1 as the inputs of neural network. The back propagation error term is approximated by the linear combination of error e_1 and its differential shown in Fig. 3. The proposed three layers neural network, including the hidden layer (j), output layer (k) and input layer (i) as illustrated in Fig. 3. The input signals e and $\dot{e}$ are multiplied by the

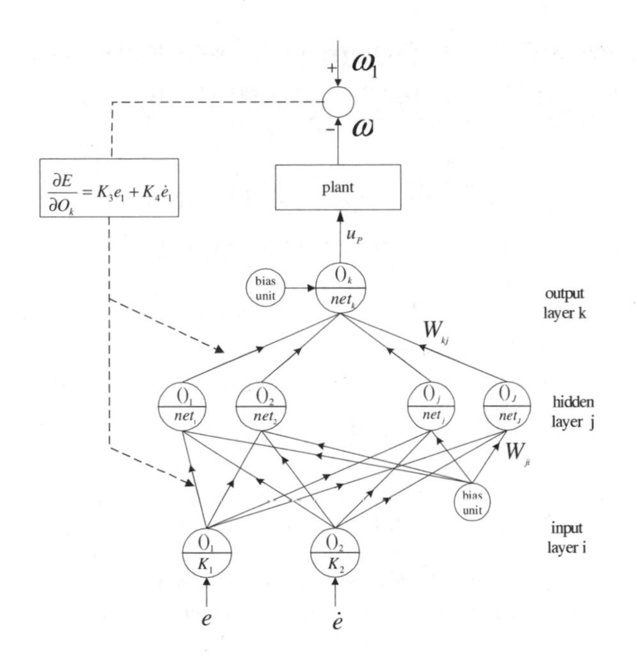

Fig. 4. The structure of proposed neural controller

coefficients K_1 and K_2, respectively, as the normalized signals O_i to hidden neurons. A tangent hyperbolic function is used as the activation function of the nodes in the hidden and output layers. So that the net output in the output layer is bounded between – 1 and +1, and converted into analogous voltage signal between -10V and +10V through a D/A converter. The net input to node j in the hidden layer is

$$net_j = \sum (W_{ji} \cdot O_i) + \theta_j \quad i = 1,2,...I, \ j = 1,2,...J \tag{1}$$

the output of node j is

$$O_j = f(net_j) = \tanh(\beta \cdot net_j) \tag{2}$$

where $\beta > 0$, the net input to node k in the output layer is

$$net_k = \sum (W_{kj} \cdot O_j) + \theta_k \quad j = 1,2,...J, \ k = 1,2,...K \tag{3}$$

the output of node k is

$$O_k = f(net_k) = \tanh(\beta \cdot net_k) \tag{4}$$

The output O_k is treated as control input u_P for the plant. As expressed equations, W_{ji} represent the connective weights between the input and hidden layers and W_{kj}

represent the connective weights between the hidden and output layers, θ_j and θ_k denote the bias of the hidden and output layers, respectively.

The error energy function at the Nth sampling time is defined as

$$E_N = \frac{1}{2}(\omega_{1N} - \omega_N)^2 = \frac{1}{2}e_{1N}^2 \tag{5}$$

The weights matrix is then updated during the time interval from N to N+1.

$$\Delta W_N = W_{N+1} - W_N = -\eta \frac{\partial E_N}{\partial W_N} + \alpha \cdot \Delta W_{N-1} \tag{6}$$

where η is denoted as learning rate and α is the momentum parameter.

$$\frac{\partial E_N}{\partial W_{kj}} = \frac{\partial E_N}{\partial net_k}\frac{\partial net_k}{\partial W_{kj}} = \delta_k O_j \tag{7}$$

And δ_k is defined as

$$\begin{aligned} \delta_k &= \frac{\partial E_N}{\partial net_k} = \sum_n \frac{\partial E_N}{\partial X_P}\frac{\partial X_P}{\partial u_P}\frac{\partial u_P}{\partial O_n}\frac{\partial O_n}{\partial net_k} = \sum_n \frac{\partial E_N}{\partial O_n}\frac{\partial O_n}{\partial net_k} \\ &= \sum_n \frac{\partial E_N}{\partial O_N}\beta(1-O_k^2) \qquad n=1,2,\cdots,K \end{aligned} \tag{8}$$

The sensitivity of E_N with respect to the network output O_k can be approximated by a linear combination of the error e_1 and its differential shown as：

$$\frac{\partial E_N}{\partial O_k} = K_3 e_1 + K_4 \frac{de_1}{dt} \tag{9}$$

where K_3 and K_4 are positive constants. Similarly, the differential of E_N respect to the weights W_{ji} shown as

$$\frac{\partial E_N}{\partial W_{ji}} = \frac{\partial E_N}{\partial net_j}\frac{\partial net_j}{\partial W_{ji}} = \delta_j O_i \tag{10}$$

$$\begin{aligned} \delta_j &= \frac{\partial E_N}{\partial net_j} = \sum_m \frac{\partial E_N}{\partial net_k}\frac{\partial net_k}{\partial O_m}\frac{\partial O_m}{\partial net_j} \\ &= \sum \delta_k W_{km}\beta\left(1-O_j^2\right) \qquad m = 1,2,\cdots,J \end{aligned} \tag{11}$$

The weight-change equations on the output layer and the hidden layer are

$$\begin{aligned} \Delta W_{kj,N} &= -\eta \frac{\partial E_N}{\partial W_{kj,N}} + \alpha \cdot \Delta W_{kj,N-1} \\ &= -\eta \delta_k O_j + \alpha \cdot \Delta W_{kj,N-1} \end{aligned} \tag{12}$$

$$\Delta W_{ji,N} = -\eta \frac{\partial E_N}{\partial W_{ji,N}} + \alpha \cdot \Delta W_{ji,N-1} = -\eta \delta_j O_i + \alpha \cdot \Delta W_{ji,N-1} \quad (13)$$

The weights matrix are updated during the time interval from N to N+1 :

$$W_{kj,N+1} = W_{kj,N} + \Delta W_{kj,N} \quad (14)$$

$$W_{ji,N+1} = W_{ji,N} + \Delta W_{ji,N} \quad (15)$$

4 Dynamic Simulations

The speed of compressor can be changed by changing the correspondent supplied DC voltage. The simulation assumes the supplied DC voltage between 150V and 260V. The block diagram of the BLDC motor speed control system is shown in Fig.3, which consists of a 1.5HP DC compressor, an speed sensor with a gain of 1V/104.67 rad/s(1V/1000rpm), an 12 bits bipolar D/A converter with an output voltage range of -10V to +10V and a driver with voltage gain of 25. The parameters of BLDC motor is shown in Table 1. The open loop simulation assumes the minimum torque load of compressor is 0.33Nm, and DC200V step voltage is supplied to the BLDC motor. The additional torque load of 1Nm is applied at t=1.5s, and an additional torque load of 2Nm is applied at t=3s. The simulation results shown in Fig. 5.(a)

In the design of direct neural controller, a well designed reference model of first order is provided, the number of neurons is defined to be 2, 5 and 1 for the input, hidden and output layers, respectively (see Fig.2). There is only one neuron in the output layer. The output signal of the direct neural controller will be between –1 and +1. In this simulations, the parameters K_1 and K_2 must be adjusted in order to normalize the input signals for the neural controller. In this simulation, the parameters K_3 and K_4 can be determined by $K_3 = K_1$ and $K_4 = K_2$ a step signal of 1V corresponding to 104.67 rad/s is denoted as the speed command, the sampling time is set to be 0.0001s , the learning rate η of the neural network is set to be 0.1 and the coefficient β=0.5 is assigned. Since the maximum error-voltage signal is 1V, the parameters K_1 and K_2 are assigned to be 0.6 and 0.01, respectively, in order to obtain an appropriate normalized input signals to the neural network. The parameters $K_3 = K_1$=0.6 and $K_4 = K_2$ =0.01 are assigned for better convergent speed of the neural network. Assumes a disturbance torque load 1Nm applied to this control system at t=0.5s. The simulation results of the BLDC motor with PI control are shown in Fig.5 (b). Fig.6 (a) represents the speed response of the BLDC motor with proposed neural controller. Fig.6 (b) shows the convergent response of the connective weights. And an extra attention should be taken that the conventional PI controller does not have fast performance of speed regulation as the proposed neural speed control.

Table 1. The parameters of motor

Motor resistance R_a	$4.18\,\Omega$
Motor inductance L_a	$850\,mH$
Inertia of rotor J	$0.005\,kgm^2$
Torque constant K_T	$0.13\,N/A$
Back emf K_B	1.2655V/rpm

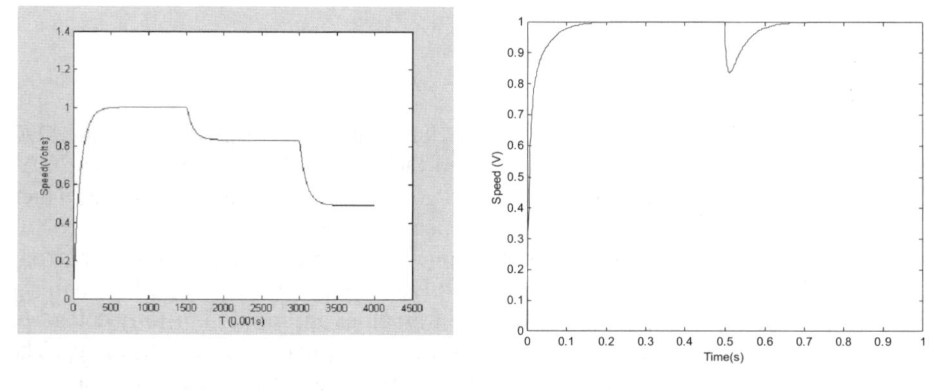

(a)Open loop BLDC motors speed response (b) BLDC motor with PI controller

Fig. 5. Simulation results for speed control of BLDC motor

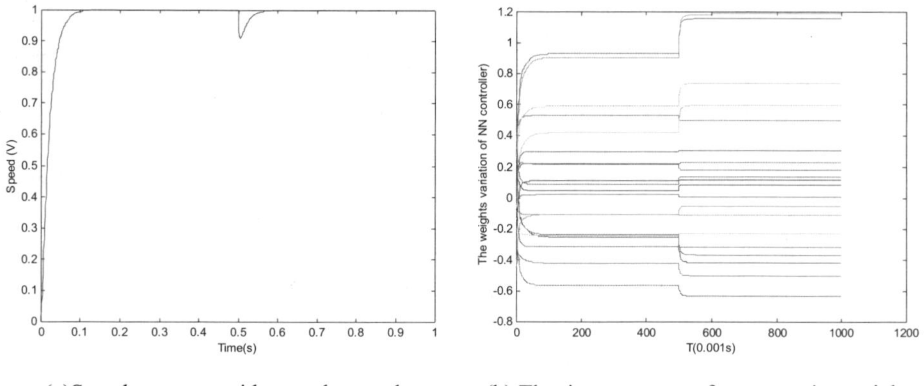

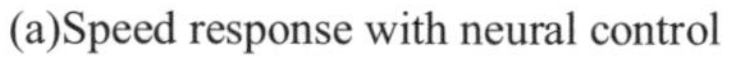

(a)Speed response with neural control (b) The time responses for connective weights

Fig. 6. Simulation results for speed control of BLDC motor with neural network

5 Conclusion

The model following based direct neural speed control is proposed and easily implemented, which has been applied to control the speed of a compressor with BLDC motor. The advantages of this controller are no need of previous knowledge or dynamic model of the plant. The on line learning capability leads the proposed neural controller enhances the adaptability and stability for speed control system. The simulation results show the proposed neural controller has better performance than a conventional PID controller, which reveal the direct neural control is available to control the speed of DC motor with high convergent speed and improve the efficiency of an Energy-saving air condition system.

References

1. Toshiyuki, T., Tetsushi, F.: High Efficiency DC Brushless Motor for Compressor Applications. Matsushita Tech. J. 51(1), 26–29 (2005)
2. Psaltis, D., Sideris, A., Yamamura, A.A.: A Multilayered Neural Network Controller. IEEE Control Systems Magazine 8(2), 17–21 (1988)
3. Zhang, Y., Sen, P., Hearn, G.E.: An On-line Trained Adaptive Neural Controller. IEEE Control Systems Magazine 15(5), 67–75 (1995)
4. Chu, M.H., Kang, Y., Chang, Y.F., Liu, Y.L., Chang, C.W.: Model-Following Controller Based on Neural Network for Variable Displacement Pump. JSME International Journal, Series C 46(1), 176–187 (2003)
5. Kang, Y., Chu, M.H., Liu, Y.L., Chang, C.W., Chien, S.Y.: An Adaptive Control Using Multiple Neural Networks for the Position Control in Hydraulic System. In: Wang, L., Chen, K., S. Ong, Y. (eds.) ICNC 2005. LNCS, vol. 3611, pp. 291–300. Springer, Heidelberg (2005)
6. Cybenko, G.: Approximation by Superpositions of A Sigmoidal Function, Mathematics of Controls. Signals and Systems 2(4), 303–314 (1989)

Develop of Specific Sewage Pretreatment and Network Monitoring System

Rongbao Chen[1], Liyou Qian[2], Yuanxiang Zhou[3], and Xuanyu Li[4]

[1] School of Electrical Engineering and Automation, Hefei University of Technology, Hefei China, 230009
[2] Environmental Protection Office of Huangshan Beauty Spots, Huangshan, China, 245000
[3] School of Resources and Environment, Hefei University of Technology, Hefei, China, 230009
[4] Chizhou University, Chizhou China, 247000

Abstract. After analyzing the current status of sewage treatment system, a construction method of specialization and specificity of sewage treatment is proposed. The huge urban sewage treatment pipe network and processing of mixed water to sewage treatment of specific sources of pollution are simplified, which not only reduces the urban construction scale of urban sewage pipe network but also creates the conditions of reclaimed water using. Therefore, the sewage treatment system can bring many advantages including low cost of investment, incremental regulation, and good treatment effect.

Keywords: Sewage treatment, specific pretreatment, regeneration cycle, network monitoring.

1 Introduction

As we all know, water is indispensable material condition of human survival and social development. At present, a variety of bad water environments and excessive use of water have become a source of damage to water resources. In southwestern, severe drought led to the wells digging depth over 270 m; The construction of Xiaolangdi made the average summer temperature decrease 2℃ in the Luoyang city and the surrounding area. Therefore, the protection of water resources, sewage treatment and water recycling have attracted significant amount of interests from both the industry and academics. Its key issue is the establishment of large-scale real-time monitoring system of hydrological resources, stringent wastewater treatment process and water recycling system.

In the water industry, due to the long time real-time hydrological monitoring of rivers and lakes, we have obtained a more complete real-time data. However, the real-time monitoring construction of eutrophication[1] and sudden changes[2] in water quality developed very slowly. The surface layer pollution in rural areas had become the main "culprit" of causing river pollution[3]. "Sewage treatment" is a very general concept. It involves a number of specially complicated contents such as heavy industry sewage, chemical sewage, food processing wastewater, drugs and hospital sewage. Water recycling is a work that need to be built and carried out universally[4]. Now,

K. Li et al. (Eds.): LSMS/ICSEE 2010, Part II, CCIS 98, pp. 110–116, 2010.

the work is developing and constructing very slowly, and. not all cities and their scale enterprises have waste water treatment[5]in China. It can not be effectively told that whether the wastewater treatment process constructed is scientific and the sewage treatment facilities is operating normally, even the real-time processing data and the emission targets had not been treated seriously. Relevant management of authorities seemed powerless. Therefore, the real-time monitoring network with distinction between industry and multi-level should be established priori.

At present, sources of water pollution have been expanded, in addition to raw sewage and serious industrial waste water, there are agriculture and municipal sewage effluent added. Without any treatment, directly discharged through the pipeline, because of the various of pollutants and the complex components, the sewage is difficult to treat. The living environment is better and better, and the more living communities are increasing the sewage, in which, the components are more complex. In land reclamation drainage, the percentage of pesticides and simultaneous are increasing synchronously. In particular, hospitals and other municipal sewage, are related to disease surveillance and people's livelihood and environmental quality.

China Statistical Yearbook is as follow: (1) in 1980, China's sewage discharge amount is 31 billion cubic meters; (2) in 1997 China's sewage discharge amount was added to 41.6 billion cubic meters, of which the industrial wastewater was 22.7 billion cubic emissions and the living wastewater emissions was 18.9 billion cubic meters; (3) in 2005, China's sewage reached 52.45 billion cubic meters, including industrial effluent emissions 24.31 billion cubic meters and living sewage volume of 28.14 billion cubic[6]. The importance and urgency of sewage treatment are very obvious. Sewage has become a major source of environmental pollution, where the industrial sewage is the most important sources of pollution because the huge volume. Water pollution has caused worldwide attention. Water pollution issue has been put into the work schedule of the World Environmental Organization[7].

Many industries have effluent, such as oil companies, chemical company, dyeing printing, hospital medicine, paper, smelting, tire & rubber company, petrochemical mining Company and drink wine industry, etc., as well as diet, cooking, washing and other municipal sewage. The treatment process are also varied with each other, such as the MBR(Membrane Biologics Reactor) technology, ABR(Anaerobic Baffled Reactor) process, MSBR(Modified Sequencing Batch Reactor) process, SBR (Sequencing Batch Reactor Activated Sludge Process) process, CASS(SBR's improved, Cyclic Activated Sludge System) process. According to different media to be processed, we could use physical method, chemical method, physical chemical method or biological method etc. to do the coagulation, neutralization, oxidation and reduction, electrolysis, extraction or adsorption etc., which could make the processed material back to nature in a physical state or chemical composition(such as COD value in water less than 100) of harmony with nature.

However, there are five important factors that lead to the actual sewage treatment state and the bad operating results as follows: (1) The production plan based on sales is changing; (2) Contaminated media quality is a variation at different times[8]; (3) The influence of environmental temperature's change to the process of sewage treatment is a variation; (4) When new pollution medium generated, the hardware constructed by raw sewage treatment process can not adjust accordingly; (5) Monitoring of toxic and hazardous emissions with other special media. These factors only explain two problems: (1) The sewage treatment process is not a normal amount; (2) Sewage treatment operation must be monitored effectively.

2 Research of Sewage Treatment Specific Pretreatment

Sewage treatment technology can cover the entire community because it covers a wide area. Dealing with sewage from different areas, we must use the knowledge in the specific field and combine with the process requirements to design the sewage treatment process.

According to the mechanism of pollution, the sewage treatment process could be divided into the same type(such as paper mills), essentially the same type (such as power plants), appropriate adjustment type (such as chemical industry, requires a certain capacity scalability and adaptability), adjustment type (such as pharmaceutical companies) and disinfection-type (hospital). Sewage water can also be the extent of contamination and pollution impact on society is divided into livelihood sewage, Scenic sewage, Heavy industry sewage, light industry sewage, hazardous waste water, agricultural effluent and infectious waste water, etc. Livelihood sewage includes sewage from dining and cleaning, the scenic sewage includes dining and bathing water, heavy industry sewage is from steel smelters, etc.; Toxic and hazardous sewage mainly comes from the production of chemical products, printing and dyeing printing, electronic products manufacturing, pharmaceutical production and energy production (such as power plants, oil and gas enterprises); The main component of agricultural waste water is fertilizer; Hospital is a source of infectious waste water.

Classification of wastewater treatment processes explained a fact, which is all sewage treatment must complete the the first stage of pretreatment, then they can enter the second stage of process. Sewage form hematology, radiology, laboratories, infectious diseases rooms of hospitals needs to be disinfected moderately, which could then enter the normal sewage treatment processes. If the classification in accordance with the conventional affiliation, hospital wastewater are also belong to the living sewage. But if the hospital wastewater is treated as the ordinary sewage, pathogens may go directly into the river system. Similarly, scenic sewage treatment must be done before the pretreatment--misplace the rice and vegetable residue, which can not be treated directly.

Wastewater treatment processes need to consider an important parameter that is the cycle of sewage treatment process. The cycle involves capacity and adaptability of hardware. As production increasement and sewage treatment facilities can not accommodate the additional sewage flow in a chemical Plant of Zhejiang, its sewage got a large number of microbial death and emissions pollution. Some enterprises' using water is not much, and the construction of sewage facilities can be designed three-dimensional and just covering very little area. In additional, height increase provides a good water level drop for the use of recycled water. After the processing of targeted sewage treatment of specific contaminated media, as long as the medium excluded, the water quality has been almost the same with running water, then the water is an excellent renewable resources. Within the enterprise, construction of reclaimed water application pipe network just need a small investment, but it can bring a long-term benefit infrastructure as shown in Figure 1.

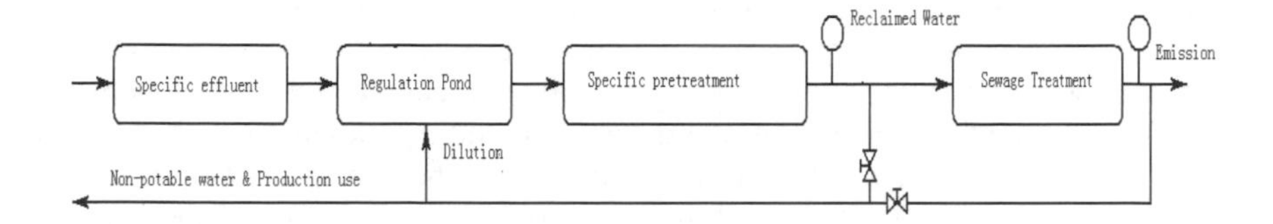

Fig. 1. Schematic diagram of the specific of the sewage treatment process

The original process flow is very simple. Specific sewage in regulation pond flow into "treatment", then let as long as the COD parameter is compliant. This treatment process is feasible in certain areas, such as the sewage treatment of scenic area, where specific pollution is a single medium and more visible; The other are the sewage treatment of enterprise with little change[8]. After the pretreatment of rice, vegetables and slag, the sewage could enter the "treatment" process. Fig. 2 is the flow chart without the pretreatment[7].

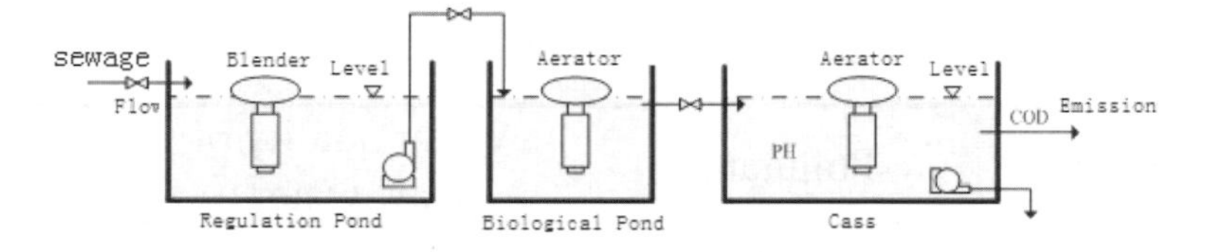

Fig. 2. Schematic diagram of the sewage treatment process

For the hospital sewage, original process also reflects a very obvious problem, for the sewage with chemical composition and special bacteria in hospitals, a variety of laboratory wastewater, etc., the contaminated media content in specific effluent can not show by COD. Therefore, the sewage must be processed by treatment shown in Fig. 1. After pretreatment, wastewater could be reclaimed water after testing and it could be let or recycled after the COD testing. Specific pretreatment of wastewater is actually doing the "reverse processing" according to mechanism, including flotation technology, biotechnology, magnetic technology, carbon technology, biochemical technology, and so on. As for the hospital, the pretreatment means disinfectant. However, in some general hospital, the pass rate of sewage treatment is not high, especially the level of sewage systems and the overall effect of the disinfection treatment is very low, which needs to strengthen supervision and management[9].

3 Develop of Real-Time Monitoring Network of Wastewater Treatment System

The sewage treatment process in many companies is generally automatic monitoring, and urban sewage treatment plant is equipped with advanced, full automation. So far, a lot of sewage treatment processes can be monitored by a computer, b which are however all run independently. The most system's running data are not reported even without data. The establishment of network system real-time monitored by functional departments has become the urgent need for effective sewage treatment. The growing popularity network communication resources provide the facilitate for Sewage network.

Aim at the demand of special sewage disposal for the apartments, COD is not the only monitoring abject but also the disposal situation of the contaminated medium for special sewage, the water include of phosphor, nitrogen, several sorts of metal comes into the environment. We can monitor all of the emission company by long-distance supervisory control center in real time. For example, for the sewage disposal and emission of hospital, on the disinfection situation of the water[10] should be noted. In the departments of hospital, radiological department, stomatology department and clinical department can produce contaminated medium of heavy metal and the content of each one is different, so it should be controlled in pretreatment. In addition, there are some apartments which produce chemistry contamination, include operating room, infection room, etc. The apartments like laboratory, check room, intestinal tract room that have causative agent should put more attention on disinfection, and it is forbidden to come into the part of sewage disposal without pretreatment. Table 1 is a comprehensive hospital sewage discharge data.

Table 1. Hospital sewage discharge data

Number	Pollutants	Water quality of raw water	Unit
1	PH	6-9	
2	SS	150-300	mg/l
3	CODCr	300-450	mg/l
4	BOD5	120-230	mg/l
5	NH3-N	≤50	mg/l
6	TP	≤4	mg/l
7	Total number of bacteria	＞16000	

In the "hospital discharge standard", the disinfection time, volume and residual chlorine disinfectant data are very important. It is sure that the data meets the requirements and the ecological systems could be safe[11].

Sewage treatment process network construction is a key technical points. Construction of the network need to consider the reliability and continuity of run-time. At present, there are a lot of communication network such as LAN, MAN, Internet and other networks. The first three are all required infrastructure, and the popularity of telecommunications has been able to cover any area that can cause pollution.

Real-time network monitoring system of wastewater treatment includes two parts, and we have network facilities of self-control and new network facilities now. Take the sewage treatment construction of Anhui Huangshan scenic[12] as an example, Yupinglou's sewage treatment systems that cover the entire scenic of Wenquan and Tianhai have been upgraded to full automatic control as shown in Fig. 3. Huangshan scenic area sewage treatment monitoring network is shown in Figure 4.

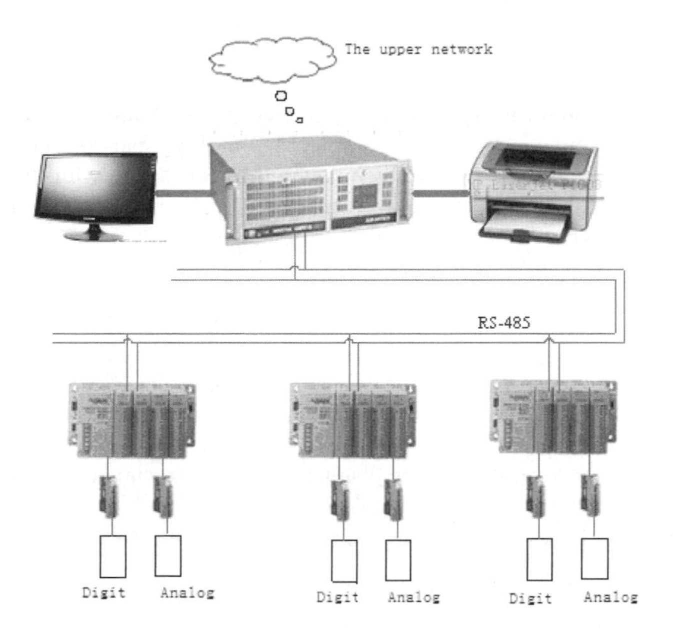

Fig. 3. Full automatic computer control system structure of sewage treatment site

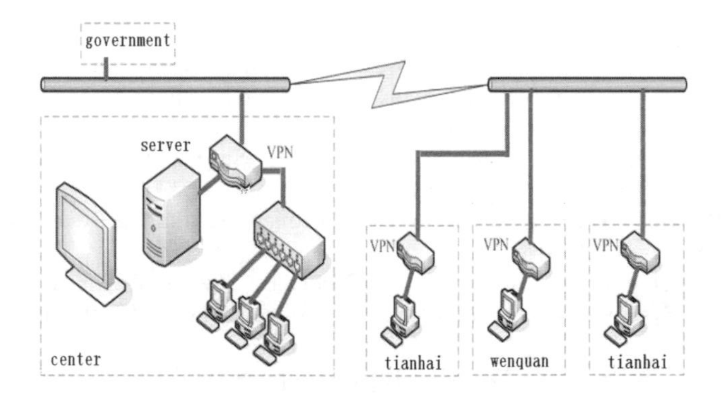

Fig. 4. Huangshan scenic area sewage treatment monitoring system

It is clearly seen from Fig. 4 that as long as the telecommunications ports devices can be linked to the network, any site's(Spa, Tianhai and Yu pinlou area) water changes, equipments states, pH parameters and COD parameter can be monitored by top regulators.

As the improvement of figure transmission environment of the telecommunications network, real-time operating status can also be uploaded by images. For new sewage treatment site, the data to the regulators can be quickly upload by using the telecom's sewage treatment data acquisition devices at the first time.

4 Conclusion

Sewage treatment monitoring system based on telecommunications network is a real-time monitoring system constructed by computer technology, network communication technology and efficient video image techniques. In accordance with the application effects of the scenic, the system can be extended to real-time monitoring of waters such as rivers and lakes, vast areas of rural and reservoir any levels.

References

1. Xiao-ping, Z., Guo-ping, W., Min, T.: Algae bloom control countermea sures for the lakes and reservoirs in the Yangtze Basin. Yangtze River 40(21), 5–9 (2009)
2. Hong-jie, Z., Xiang-zhou, X., Xing-wen, Z.: Treatment Technology for the Organic Waste Water in Suddent Polluted Accident. Journal of Water Resources & Water Engineering 20(5), 67–71 (2009)
3. River Basin Pollution Control Research Center established in Fudan. Wen Wei Po 2010, 4.7 (2010)
4. Wastewater Reuse Potential. China Water Resources News (2001)
5. The situation of urban sewage treatment informed by ministry. China Construction News (2005)
6. National Bureau of Statistics, China Statistical Yearbook (2006)
7. Hui, C.: Research and Development of Distributed Monitoring System of Sewage Treatment. Master, Hefei University of Technology (2009)
8. Yuan-xiang, Z., Kang-ping, C., Feng, W., et al.: Discussion about the problems in the design of the wastewater treatment systems in mounta in beauty spots. Journal of Hefei university of Technology 27(12), 1520–1523 (2004)
9. Dong-jiao, S., Zi-jing, C., Li-na, Z., et al.: Hospital Survey of Sewage Disposal System Discharge. Chinese Journal of Disinfection 26(6), 671–672 (2009)
10. Qi, Z., Rong-bao, C.: Hospital Wastewater Treatment Monitoring System. Electric Engineering (5), 39–41 (2010)
11. Qi, Z.: The Development of the Hospital Secondary Sewage Treatment Monitoring and Controlling System Based on Multi-loop Pre-sterilized. Master, Hefei University of Technology (2010)
12. Rongbao, C., Hui, C., Feng, W., et al.: The Development of Control System for the sewage Treatment Process with Remote Video Monitoring. In: Progress of Computer Technology and Application, Hefei, pp. 589–593 (2008)

Application of Radial Basis Function Neural Network in Modeling Wastewater Sludge Recycle System

Long Luo and Liyou Zhou

Guangzhou Institute of Railway Technology
Guangzhou, 510430, China
gtluolong@163.com

Abstract. Sludge recycle system is an important part of wastewater treatment plants(WWTP), which can ensure the required reactor sludge concentration, maintenance the dynamic balance between secondary sedimentation tanks and sludge reactor sludge concentration. This work proposes development of a Radial Basis Function (RBF) Neural Network model for prediction of the Sludge recycling flowrate, which ultimately affect the Sludge recycling process. Compared with the traditional constant sludge recycle ratio control, new idea is better in response to actual situation. According to analyzing and Evolutionary RBF Neural Network theory, a RBF Neural Network is designed. The data obtained from wastewater treatment were used to train and verify the model. Simulation shows good estimates for the sludge recycling flowrate. So the idea and model is a good way to the sludge recycle flow rate control. It is a meaningful Evolutionary Neural Network application in industry.

Keywords: sludge recycle, Radial Basis Function, wastewater treatment plants.

1 Introduction

In recent years, given the pressure from many areas for intensifying nutrient removal, the wastewater management industry has been facing the challenge of intensifying/upgrading its treatment plants to achieve higher effluent standards with the minimum cost. In order to meet these demands, the use of advanced monitoring and modeling methods is required. Sludge recycling system is an important part of WWTP, which can ensure the required reactor sludge concentration, maintenance the dynamic balance between secondary sedimentation tanks and sludge reactor sludge concentration.Its wastewater treatment plant effluent quality, system stability operations and Operating costs have a major impact. At present, the control of sludge return flow usually two kinds of strategies. [1]The first strategy is to keep sludge recycle flow unchanged; the second strategy is to have recycle flow increasing in the proportion of the system's influent flow. Traditional control method of sludge return are extensive mode of management, resulting in return sludge concentration range, and low concentrations, there is wasted energy.[2] If we could predict sludge recycle flowrate more exactly, we could control sludge recycle process better, which would lead to good saving for energy. This paper mainly discusses how to analyze connected variants and

K. Li et al. (Eds.): LSMS/ICSEE 2010, Part II, CCIS 98, pp. 117–122, 2010.

use RBFNN to get model in sludge recycle process, which ultimately affect the Sludge recycling process. [3]

2 Process Description

2.1 The Pilot Plant

The pilot unit configuration is given in figure 1.When these wastewaters enter the plant they are driven to an aeration tank where they mixed with a sludge made up of bacteria. After they have "eaten" most of the organic matter, water and sludge are driven to a settling tank where they are separated. The sludge flows downwards while the water stays at the top of the tank and flows over an overflow weir to be released to nearby rivers. [4]The sludge is withdrawn from the tank and then split into two streams: one is driven to the aeration tank to keep the sludge concentration at an appropriate level while the other is eliminated. [5]

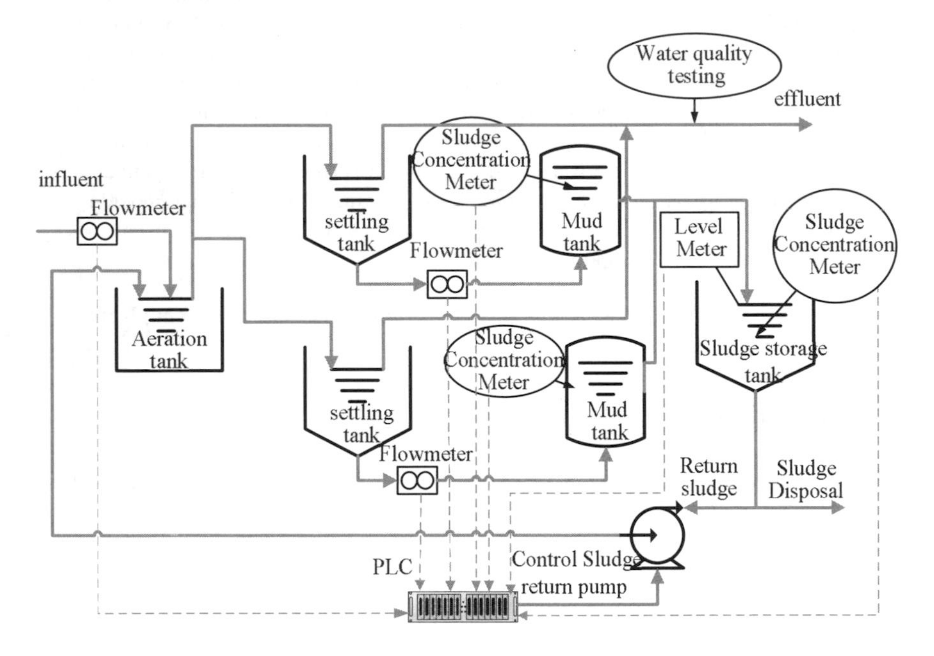

Fig. 1. Diagram of a Wastewater Treatment Plant

2.2 Process Control

Aeration tank of sludge from the secondary settling tank return the sludge return system, biological reactor and secondary settling tank linked to a balanced system . Biological reaction tank MLSS Concentration (MLSS) and the return sludge concentration (RSS) relationship, see equation (1). According to the system dealing with water, return sludge concentration, the concentration of control over biological and chemical mixture pool target calculations obtained the amount of return sludge, see equation (2).

$$MLSS = \frac{R}{1+R} RSS \tag{1}$$

$$Q_R = \frac{Q_{in} \times MLSS}{RSS - MLSS} \tag{2}$$

In this model the input variables are the "Influent Flowrate" Q_{in} , the "Sludge Concentration" in the aeration tank, MLSS and The "return sludge concentration" in the sludge storage tank ,RSS, is assumed to be a process parameter that will be provided at every time step. The output variables is the "return sludge flow" Q_R. It defines the sludge concentration in the settler, a portion of which will be recalculated to the aeration tank to keep the sludge concentration at an appropriate level. [6] [7]This recirculation is defined by a flowrate Q_R, which is adjusted by the technician supervising the plan dynamics. It is represented by a parameter, R, which provides the ratio between this "Recirculation Flowrate" and the "Influent Flowrate". The control of this parameter is the aim of this work. The methods for calculating the set-point for the sludge flow, Q_R, Control variable is the system influent flowrate by the above formula. [8]

3 RBF Neural Network Base on Model

3.1 Architecture of the RBF Neural Network

RBF is a three-layer neural network. The input layer is made up of the signal source node. The second tier is hidden layer, as described in its modules are based on the needs and problems. The third tier is the output layer, which responded to the role of imported models. [9]If the network input is x (x1, x2,…x_n), y= (y1, y2,…y_p) is output vector, RBF implied by the input space to transform the hidden space is non-linear, from the hidden layer and output layer space to transform the output space is linear. The transformation function of hidden units is Radial Basis Functions, and the network can achieve the following mapping between the input and output :

$$y_k = \sum_{l=1}^{m} w_{i,k} R_i(x) \qquad (k=1,2,.....,p) \tag{3}$$

Where network input modules is n, m and p is crackdown module and output modules, w_{ik} is the value from the i hidden units to the k output units. $R_i(x)$ is radial basis functions, Radial Basis Function is a local distribution center of the radial symmetry of the non-negative non-linear attenuation function, Gaussian function is relatively common:

$$R_i(x) = \exp(-\sum \frac{(\overline{x}_i - x_i)^2}{\sigma_i^2}) \qquad (i=1,2, \ldots,m) \tag{4}$$

From above analyzing, RSS and MLSS as input data is the most important for flowrate QR output prediction, except the two important input data, other components containing

Qin also affect QR, and the control strategy is that control of MLSS to a desired value in the aerator, by adjusting sludge recycle flow rate. According to Benchmark Simulation Model no.1 (BSM1), the influence factors for Q_R output prediction could be presented as, so we could design Q_R prediction RBF Neural Network just like Figure 2:

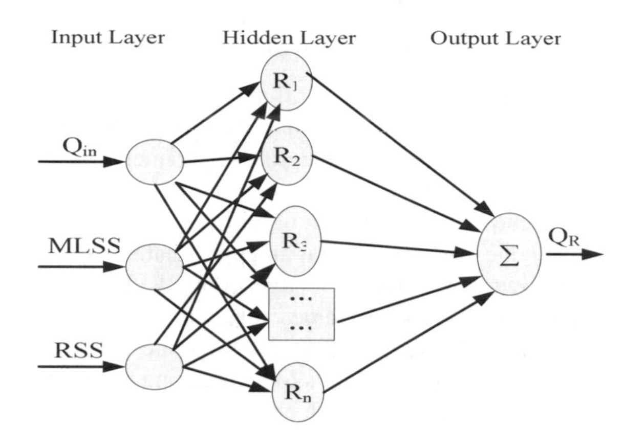

Fig. 2. RBF Neural Network design base on model

Artificial neural network modeling principles sludge recycle system shown in Figure 2, the first sludge recycle system according to the input and output data set data set. In the learning process concentration in the sample data input neural network; according to the input values to calculate the sample Network, the output value; calculated sample the difference between output and network output; basis of the calculated difference from the gradient descent method to adjust the network weight matrix; repeat the process until the whole a sample set of error does not exceed the scope of learning that is the end. [10]

3.2 The Learning Algorithm of the RBF Neural Network

The learning algorithm of RBF network generally involves two processes.

Step1, determine hidden node centers, using Randomly Fixed method .When the number of hidden neurons is smaller than the number of patterns, the centers are randomly chosen from the training data, but in a representative manner. For example, if 13 teaching patterns are loaded and the hidden layer consists of only five neurons, then the patterns with numbers 1, 4, 7, 10, and 13 could be selected.

Step2, Selection of Widths, using Distance Averaging method. A "reasonable" estimate for the global width parameter is the average $\sigma_j = \langle \|\mu_i - \mu_j\| \rangle$,which represents a global average over all Euclidean distances between the center of each unit i and that of its nearest neighbor j. [11]

4 Simulation and Discussion

We sample typical data from china Guangzhou lijiao wastewater plant. The sample of 500 groups Former two weeks data is used to train Evolutionary RBF Neural Network, the other 200 groups data is used to validate RBFNN prediction function. We use MATLAB 7.6 to do the experiment. The maximum archive size was set to 36.The archive is initialized with 18 non-dominated networks of with random number of kernels. The simulation result was shown by Figure 3, Figure 4. Fig. 3 illustrate the forecasting performance of the RBFNN for sludge recycle flowrate respectively, with the application of the test (validation) set. Fig. 4 displays the forecasting performance of RBFNN for sludge recycle flowrate. 'Predicted output' is the NN prediction output and 'Target output' is the expected output. [12]

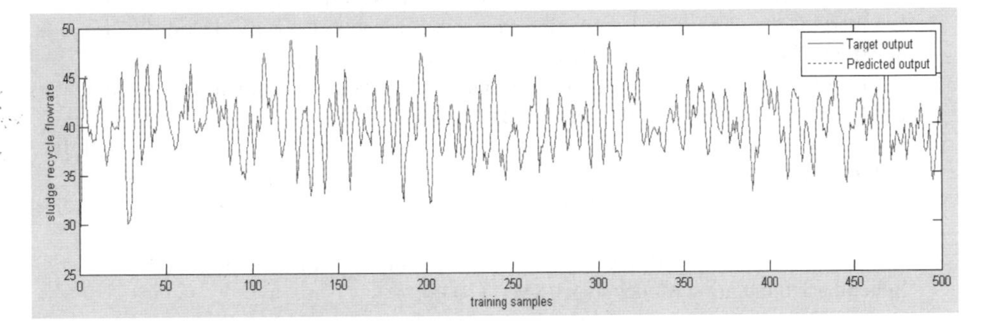

Fig. 3. Simulation results of training samples

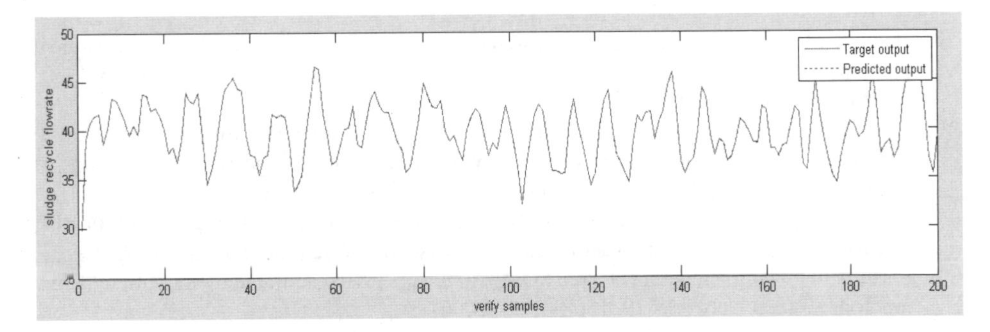

Fig. 4. Simulation results of verify samples

5 Conclusion

This paper presents a process control scheme for a sludge recycling process based on the neural predictive control. A Radial Basis Function Neural Network is developed to model the nonlinear relationships between MLSS and sludge recycling flowrate. The method can adapt the system to a large variety of operating conditions with an enhanced learning ability. [13]The simulation show that the neural network models provided good estimates for the sludge recycling flowrate, which cover a range of data for training and testing purposes. This research has shown that the neural network could with a bit more investigation, be effectively utilized to produce a model for the sludge recycling process to wastewater treatment plants. Also, it is believed that the results obtained can be used in order to improve or develop an actual wastewater treatment plant.

Acknowledgement. This work is get a helping hand from 'Fund of Innovation Creation Academy Group' established by the Guangzhou Education Bureau.

References

1. Ekman, M.: Modeling and Control of Bilinear Systems—Applications to the Activated Sludge Process, Elanders Gotab, Sweden, Stockholm (2005)
2. Lindberg, C.-F.: Control and estimation strategies applied to the actived sludge process. In: Graphics System AB, Sweden Stockholm (1997)
3. Traor, A.: Control of sludge height in a secondary settler using fuzzy algorithms. Computers and Chemical Engineering 30, 1235–1242 (2006)
4. Haykin, S.: Neural Networks: A Comprehensive Foundation, 2nd edn. Prentice-Hall, Inc., Beijing (2001)
5. Garcia, C., Prette, D., Morari, M.: Model Predictive Control: Theory and Practice - a Survey. Automatica 25(3), 335–348 (1989)
6. Kumar, S.: Neural Networks. The McGraw-Hill Companies, Inc., Beijing (2006)
7. Benchmark Simulation Model no.1(BSM1) (2008), `http://www.ensic.u-nancy.fr/COSTWWTP/`
8. Tezel, G., Yel, E., Kaan Sinan, R.: Artificial Neural Network (Ann) Model for Domestic Wastewater Treatment Plant Control. In: BALWOIS 2010 - Ohrid, Republic of Macedonia (2010)
9. Hamed, M.M., Khalafallah, M.G., Hassanien, E.A.: Prediction of wastewater treatment plant performance using artificial neural networks. Environmental Modelling & Software 19, 919–928 (2004)
10. Zeng, G.M., Qin, X.S., Hea, L., Huang, G.H., Liu, H.L., Lin, Y.P.: A neural network predictive control system for paper mill wastewater treatment. Engineering Applications of Artificial Intelligence 16, 121–129 (2003)
11. Liu, J., Lampinen, J.: A Differential Evolution Based Incremental Training Method for RBF Networks. In: GECC0 2005, Washington, DC,USA, June 25-29, pp. 881–888 (2005)
12. Forti, A.: Growing Hierarchical Tree SOM: An unsupervised neural network with dynamic topology. Neural Networks 19, 1568–1580 (2006)
13. Ma, Y., Peng, Y., Wang, S.: New automatic control strategies for sludge recycling and wastage for the optimum operation of predenitrification processes. Journal of Chemical Technology and Biotechnology 81, 41–47 (2006)

Improved Stability Criteria for Delayed Neural Networks

Min Zheng[1,2], Minrui Fei[1,2], Taicheng Yang[3], and Yang Li[1,2]

[1] College of Mechatronic Engineering and Automation, Shanghai University
[2] Shanghai Key Laboratory of Power Station Automation Technology, Shanghai 200072, China
[3] Department of Engineering and Design, University of Sussex, Brighton BN1 9QT, UK

Abstract. This paper is concerned with the stability problem of delayed neural networks. An improved integral inequality Lemma is proposed to handle the cross-product terms occurred in the derivative of Lyapunov functional. By using the new lemma and a novel delay decomposition approach, we propose the new delay-range-dependent stability criteria for time varying delay neural networks. The sufficient conditions obtained in this paper are less conservative than those in the former literature.

1 Introduction

In the past decades, neural networks (NNs) have received considerable attention due to their extensive applications in various areas such as signal processing problems, optimization, pattern recognition, etc. Time delays as a source of instability and bad performance always appear in many NNs, such as Hopfield NNs, cellular NNs and bi-directional associative memory networks. Therefore, increasing interest has been focused on stability analysis of NNs with time delays[1-11]. In [1], the author studies the global robust stability criteria for uncertain NNs with fast time-varying delays, in which the restriction on the derivative of time-varying delay function is removed. An augmented functional is used in [3] for neural networks with mixed time delay system. Paper [5] introduced an augmented Lyapunov-Krasovskii Functional (LKF). A free weighting matrices method based on LMI technique is proposed in [7] to study the delay-dependent stability problems for NNs with time varying delays. However, there still exists high conservatism need further research. In order to reduce the conservatism results from the selection of LKFs, [11-17] presented different kinds of delay decomposing approaches. To best of our knowledge, the two delay intervals $[0, \tau_m]$ and $[\tau_m, \tau_M]$ simultaneous decomposition approach has not been proposed. On the other hand, the magnification of cross-product term $-\int_{t-b_1}^{t-b_2} \dot{x}(s)R\dot{x}(s)ds$ still has some degree of conservatism in many existing references.

In this paper, we consider the stability problem for a class of NNs with time varying delay. An improved lemma with novel decomposition approach is proposed. The sufficient conditions obtained in this paper are less conservative than

K. Li et al. (Eds.): LSMS/ICSEE 2010, Part II, CCIS 98, pp. 123–128, 2010.

those in the former literature. Finally, numerical examples will be given to show the effectiveness of the main results.

2 Problem Formulation

Consider the following NNs with interval time-varying delay:

$$\dot{x}(t) = -Cx(t) + Af(x(t)) + Bf(x(t - d(t))) + J \tag{1}$$

where $x(t) = [x_1(t), x_2(t) \cdots, x_n(t)]^T \in R^n$ is the neuron state vector. $f(x(\cdot)) = [f_1(x_1(\cdot)), f_2(x_2(\cdot)), \cdots, f_n(x_n(\cdot))]^T \in R^n$ denotes the neuron activation function, and $J = [J_1, J_2, \cdots, J_n]^T \in R^n$ are the constant external input vector. $C = diag\{c_1, c_2, \cdots, c_n\}$is a diagonal matrix with $c_j > 0, j = 1, 2, \cdots, n$. A and B are the connection weight matrix and the delayed connection weight matrix, respectively. The time delayed $d(t)$ is a time varying continuous function that satisfies

$$0 \leq d_m \leq d(t) \leq d_M \tag{2}$$

We assume the activation functions $f_j(\cdot)$ are bounded and satisfy the following inequalities

$$0 \leq \frac{f_j(x) - f_j(y)}{x - y} \leq k_j, \quad \forall x, y \in R, x \neq y \tag{3}$$

where k_j are positive constants.The initial conditions are as follows

$$x = \phi(s), \quad s \in [-d, 0]$$

where $\phi(s)$ is a continuous function vector. Assume $x^\star$ is an equilibrium point of system(1), One can derive from (1) that the coordinate transformation $z = x - x^\star$ transforms system (1) into the following system:

$$\dot{z}(t) = -Cz(t) + Ag(z(t)) + Bg(z(t - d(t))) \tag{4}$$

where $z(\cdot) = [z_1(\cdot), z_2(\cdot), \cdots, z_n(\cdot)]^T$ is the state vector of the transformed system. $g(z(\cdot)) = [g_1(z_1(\cdot)), g_2(z_2(\cdot)), \cdots, g_n(z_n(\cdot))]^T$,and $g_j(z_j(\cdot)) = f_j(z_j(\cdot) + x_j^\star) - f_j(x_j^\star)$ with $g_j(0) = 0$. According to (2.2), one can obtain that

$$g_j^2(z_j(t)) \leq k_j z_j g_j(z_j(t)) \leq k_j^2 z_j^2 \tag{5}$$

Therefore, the stability problem of system(1) on equilibrium $x^\star$ is changed into the zero stability problem of system(4).

The purpose of this paper is to establish stability criteria of NNs system (4) based on a novel dual-decomposition LKF approach, which are less conservative than the existing results.For this purpose, the new lemma is introduced.

Lemma 1. *For any constant matrix $R \in R^{n \times n}$, $R = R^T > 0$, scalars $b_1 \leq \tau_t \leq b_2$ and vectors function $\dot{x} : [-b_2, -b_1] \to R^n$ such that the following integration is well defined, then it holds that*

$$-(b_2 - b_1) \int_{t-b_2}^{t-b_1} \dot{x}(s)^T R \dot{x}(s) ds \leq \xi(t)^T \hat{\Omega} \xi(t) \tag{6}$$

where

$$\xi(t) = \begin{bmatrix} x(t-b_1) \\ x(t-\tau(t)) \\ x(t-b_2) \end{bmatrix}, \hat{\Omega} = \Omega + \frac{\tau(t)-b_1}{b_2-b_1}\Omega_1 + \frac{b_2-\tau(t)}{b_2-b_1}\Omega_2 \tag{7}$$

$$\Omega = \begin{bmatrix} -R & R & 0 \\ & -2R & R \\ & * & -R \end{bmatrix}, \Omega_1 = \begin{bmatrix} 0 & 0 & 0 \\ & -R & R \\ & * & -R \end{bmatrix}, \Omega_2 = \begin{bmatrix} -R & R & 0 \\ & -R & 0 \\ & * & 0 \end{bmatrix}$$

3 Main Results

Theorem 1. *The system (4) is asymptotically stable if there exist positive matrices* $P, Q_i, R_i, (i = 1, 2, \cdots, N), W_K, S_K, (K = 1, 2, \cdots, r)$ *and positive diagonal matrices* D_1, D_2, *and* Λ *such that the following LMIs hold*

$$\Pi_K - E_{1K}^T S_K E_{1K} < 0 \tag{8}$$

$$\Pi_K - E_{2K}^T S_K E_{2K} < 0 \tag{9}$$

where

$$\Pi_K = \Pi_0 + \Omega_K$$

$$\Pi_0 = \begin{bmatrix}
\Pi_{11} & R_1 & 0 & \cdots & 0 & 0 & 0 & \cdots & 0 & 0 & \Pi_{111} & \Pi_{112} \\
* & \Pi_{22} & R_2 & \cdots & 0 & 0 & 0 & \cdots & 0 & 0 & 0 & 0 \\
* & * & \Pi_{33} & \cdots & 0 & 0 & 0 & \cdots & 0 & 0 & 0 & 0 \\
* & * & * & \ddots & \vdots & \vdots & \vdots & \vdots & \vdots & \vdots & \vdots & \vdots \\
* & * & * & * & \Pi_{55} & S_1 & 0 & \cdots & 0 & 0 & 0 & 0 \\
* & * & * & * & * & \Pi_{66} & S_2 & \cdots & 0 & 0 & 0 & 0 \\
* & * & * & * & * & * & \Pi_{77} & \cdots & 0 & 0 & 0 & 0 \\
* & * & * & * & * & * & * & \ddots & \vdots & \vdots & \vdots & \vdots \\
* & * & * & * & * & * & * & * & \Pi_{99} & 0 & 0 & 0 \\
* & * & * & * & * & * & * & * & * & 0 & 0 & KD_2 \\
* & * & * & * & * & * & * & * & * & * & \Pi_{aa} & \Pi_{ab} \\
* & * & * & * & * & * & * & * & * & * & * & \Pi_{bb}
\end{bmatrix}$$

$\Pi_{11} = -PC - C^T P + Q_1 - R_1 + C^T Y C; \Pi_{111} = PA + KD_1 - C^T \Lambda - C^T Y A;$
$\Pi_{112} = PB - C^T Y B; \Pi_{22} = -Q_1 + Q_2 - R_1 - R_2$
$\Pi_{33} = -Q_2 + Q_3 - R_2 - R_3; \Pi_{55} = -Q_N - R_N + W_1 - S_1$
$\Pi_{66} = -W_1 + W_2 - S_1 - S_2; \Pi_{77} = -W_2 + W_3 - S_2 - S_3$
$\Pi_{99} = -W_r - S_r; \Pi_{aa} = \Lambda A + A^T \Lambda - 2D_1 + A^T Y A$
$\Pi_{ab} = \Lambda B + A^T Y B; \Pi_{bb} = -2D_2 + B^T Y B$

$$Y = \delta^2 \sum_{i=1}^{N} R_i + h^2 \sum_{j=1}^{r} S_j; \delta = d_m/N, h = (d_M - d_m)/r$$

$$\Omega_K = -E_{1K}^T S_K E_{1K} - E_{2K} S_K E_{2K} + E_{3K} S_K E_{3K}$$
$$E_{1K} = [0_{n\times(N+K-1)n}, I_{n\times n}, 0_{n\times(r-K+1)n}, -I_{n\times n}, 0_{n\times 2n}]$$
$$E_{2K} = [0_{n\times(N+K)n}, -I_{n\times n}, 0_{n\times(r-K)n}, I_{n\times n}, 0_{n\times 2n}]$$
$$E_{3K} = [0_{n\times(N+K-1)n}, I_{n\times n}, -I_{n\times n}, 0_{n\times(r-K+3)n}]$$

Proof. The main idea of this approach is that the delay intervals $[0, d_m]$ and $[d_m, d_M]$ can be divided separately into different subintervals. Constructing LKFs as follows

$$V = V_1 + V_2 + V_3 \tag{10}$$

with

$$V_1 = z(t)^T P z(t) + 2\sum_{i=1}^{n} l_i \int_0^{z_i(t)} g_i(s)ds$$
$$V_2 = \sum_{i=1}^{N} \int_{t-i\delta}^{t-(i-1)\delta} z(s)^T Q_i z(s)ds + \delta \int_{t-i\delta}^{t-(i-1)\delta} \int_v^t \dot{z}(s)^T R_i \dot{z}(s)dsdv$$
$$V_3 = \sum_{j=1}^{r} \int_{t-\tau_m-jh}^{t-\tau_m-(j-1)h} z(s)^T W_j z(s)ds + h \int_{t-\tau_m-jh}^{t-\tau_m-(j-1)h} \int_v^t \dot{z}(s)^T S_j \dot{z}(s)dsdv$$

Considering the derivative of V along the solution of system (4), by using the new lemma1, this theorem can be easily obtained.

4 Numerical Examples

To illustrate the effectiveness of our results, this section will give the following numerical examples.

Example 1. Consider the neural network with time varying delay

$$C = \begin{bmatrix} 0.7 & 0 \\ 0 & 0.7 \end{bmatrix}, A = \begin{bmatrix} -0.3 & 0.3 \\ 0.1 & -0.1 \end{bmatrix}, B = \begin{bmatrix} 0.1 & 0.1 \\ 0.3 & 0.3 \end{bmatrix}, k_1 = 1, k_2 = 1$$

These parameters are from example 2 of [1], the maximal allowable delay $d = 2.03958$. However, the maximal allowable delay $d = 4.19$ according theorem1 of this paper with $N = 3, r = 4$.

Example 2. Consider a neural network with time varying delay

$$C = diag\{1.2769, 06231, 0.9230, 0.4480\}$$

$$A = \begin{bmatrix} -0.0373 & 0.4852 & -0.3351 & 0.2336 \\ -1.6033 & 0.5988 & -0.3224 & 1.2352 \\ 0.3394 & -0.0860 & -0.3824 & -0.5785 \\ -0.1311 & 0.3253 & -0.9534 & -0.5015 \end{bmatrix},$$

Table 1. Allowable upper bounds τ_M for constant delay and interval delay

Method	constant delay	unknown $\dot{\tau}$ $\tau_m = 0$	$\tau_m = 2$
Cho et al.[10]	1.93		
He et al.[9]	3.58	1.54	–
Zhang et al.[11]	3.61	2.07	–
Theorem1(N=3,r=4)	3.63	2.01	2.71
Theorem1(N=3,r=7)	3.63	2.08	2.80

$$B = \begin{bmatrix} 0.8674 & -1.2405 & -0.5325 & 0.0220 \\ 0.0474 & -0.9164 & 0.0360 & 0.9816 \\ 1.8495 & 2.6117 & -0.3788 & 0.8428 \\ -2.0413 & 0.5179 & 1.1734 & -0.2775 \end{bmatrix},$$

$$k_1 = 0.4, k_2 = 0.8$$

Table1 gives the comparison results via the methods in [9-11] and Theorem1 in this paper with different N and r. The results show that our results have less conservatism than the existing results.

5 Conclusions

In this paper, the stability problem for a class of NNs with time varying delay is investigated. Less conservative conditions are derived through an improved integral inequality lemma and a novel decomposition approach. The criteria obtained here are less conservative than those in the former literature.

Acknowledgements

This work is supported by National Nature Science Foundation of China under Grant No.60774059, National 863 key project sub-topics under Grant No. 2006AA04030405, and Shanghai Educational Development Foundation under Grant 06GG10, 08DZ2272400, Shanghai Science Technology Commission No. 09JC1406300, Shanghai University'11th Five-year Plan'211, Innovation Fund of Shanghai University and Research Fund for the Excellent Youth Scholars of Higher Education of Shanghai.

References

1. Qiu, J.Q., Zhang, J.H., Wang, J.F., Xia, Y.Q., Shi, P.: A new global robust stability criteria for uncertain neural networks with fast time-varying delays. Chaos Solitons & Fractals 37, 360–368 (2008)
2. Zhang, H.G., Wang, Z.S., Liu, D.R.: Global asymptotic stability of recurrent neural networks with multiple time-varying delays. IEEE Transactions on Neural Networks 19(5), 855–873 (2008)

3. Liu, Z.W., Zhang, H.G.: Robust stability of interval neural networks with mixed time-delays via augmented functional. Control Theory & Applications 26(12), 1325–1330 (2009)
4. Hua, C.C., Long, C.N., Guan, X.P.: New results on stability analysis of neural networks with time-varying delays. Physics Letters A 352, 335–340 (2006)
5. Singh, V.: A generalized LMI-based approach to the global asymptotic stability of delayed cellular neural networks. IEEE Transactions on Neural Networks 15(1), 223–225 (2004)
6. Xu, S., Lam, J., Ho, D.W.C.: Novel global robust stability criteria for interval neural networks with multiple time-varying delays. Physics Letters A 342(4), 322–330 (2005)
7. Li, T., Guo, L., Sun, C.Y., Lin, C.: Further results on delay-dependent stability criteria of neural networks with time-varying delays. IEEE Transactions on Neural Networks 19(4), 726–730 (2008)
8. Chen, W.H., Zheng, W.X.: Improved Delay-Dependent Asymptotic Stability Criteria for Delayed Neural Networks. IEEE Transactions on Neural Networks 19(12), 2154–2161 (2008)
9. He, Y., Liu, G., Rees, D.: New delay-dependent stability criteria for neural networks with time-varying delay. IEEE Transactions on Neural Networks 18(1), 310–314 (2007)
10. Cho, H., Park, J.: Novel delay-dependent robust stability criterion of delayed cellular neural networks. Chaos Solitons Fractals 32, 1194–1200 (2007)
11. Zhang, H.G., Liu, Z.W., Huang, G.B., Wang, Z.S.: Novel weighting-delay-based stability criteria for recurrent neural networks with time-varying delay. IEEE Transactions on Neural Networks 31(1), 91–106 (2010)
12. Gu, K., Han, Q.L.: Discretized Lyapunov functional for linear uncertain systems with time-varying delay. In: 2000 American Control Conference, Chicago, IL (2000)
13. Gu, K., Kharitonov, V.L., Chen, J.: Stability of time-delay systems. Birkhäuser, Basel (2003)
14. Zheng, M., Fei, S.M.: Stability of linear systems with time delay:a new delay fractioning based Lyapunov-Krasovskii approach. In: Sixth IEEE International Conference on Control and Automation, Guangzhou, CN, pp. 937–941 (2007)
15. Zheng, M., Fei, S.M.: H_∞ State Feedback Control with Memory for Uncertain Linear Systems with Interval Time-varying Delay. Acta Automatica Sinica 33(11), 1211–1215 (2007)
16. Du, B., Lam, J., Shu, Z., Wang, Z.: A delay-partitioning projection approach to stability analysis of continuous systems with multiple delay components. IET Control Theory Applications 3(4), 383–390 (2009)
17. Yue, D., Tian, E.G., Zhang, Y.J.: A piecewise analysis method to stability analysis of linear continuous/discrete systems with time-varying delay. International Journal of Robust and Nonlinear Control 19, 1493–1518 (2009)

Aplication of the Single Neuron PID Controller on the Simulated Chassis Dynamometer

Weichun Zhang, Bingbing Ma, Peng Yu, and Baohao Pei

Shandong University of Technology
Zibo, Shandong Province 255049, China
Zhangwc@sdut.edu.cn

Abstract. The single neuron self-adaptive PID controller was introduced after analyzing MCG-200 simulated chassis dynamometer control system. The simulation process of running resistance in the laboratory was improved. The suitable single neuron self-adaptive PID controller was designed, the new control system using single neuron self-adaptive PID algorithm was simulated and laboratory dates obtained on the improved chassis dynamometer was compared with datas conducted on the real road. Results show that: the single neuron self-adaptive PID controller has simpler structure, stronger self-adaptive ability and can replace the traditional PID controller.

Keywords: Vehicle chassis dynamometer, the measurement and control system, the single neuron self-adaptive PID.

1 Analysis of MCG-200-Type Vehicle Simulated Chassis Dynamometer Control System

MCG-200-type Vehicle Simulated chassis dynamometer can be used to achieve these tests include the fuel economy test, the performance test, the emission behavior and analysis, the reliability test and some special experiments related with the vehicle driveline, the dynamometer can test a lot of different tests and the simulated operating conditions is very complex, the corresponding software platform conveniently operate, have very high stability and reliability, is easy to upgrade and so on.

With Windows XP as the software platform and Advantech IPC-610 for the hardware platform, an developed open platform for measurement and control which is suitable for MCG-200-type Simulated chassis dynamometer require some main functions and technical characteristics, which are as follows:

It should ensure operational safety and that the entire system is always stable and safe ; According to different test requirements, it should accurately simulated different operating conditions of the vehicle to complete vehicle performance and fuel economy testing requirements;

It should be easy to interact between man and machine, the test pilot also can adjust some parameters of the system to meet the different needs of the project;

It should be have operating condition tips;

Acquiring speed, torque, fuel consumption and other signal;

Outputting excitation voltage control signal;

K. Li et al. (Eds.): LSMS/ICSEE 2010, Part II, CCIS 98, pp. 129–138, 2010.

Displaying data including speed, torque, fuel consumption, output voltage, etc.;
It should complete some function including database operations, data storage and querying;
Data analysis;
Help information, auxiliary functions.

2 The Analysis of Controller Structure

2.1 The Control of the Running Resistance in the Dynamometer Test

According to vehicle speed, the control system of the running resistance can control the loading system to simulate the actual road resistances of motion to correctly carry out a variety of road test and can ensure that the measurement and control is reliable, repeatable and reproducible.

In the course of the road simulation test, on the dynamometer resistance should be imposed as follows:

$$F_{djx} = F_{lsx} - F_{gsx} + \left(M - I_{xt} / R^2\right)\frac{dv}{dt} \qquad (1\}$$

Where:

F_{djx} is the resistance that should be loaded;
F_{lsx} is the running resistance on the road;
F_{gsx} is the running resistance on the Drum;
M is the weight of the vehicle;
I_{xt} is the moment of inertia of the Drum system;
R is the radius of the Drum.

The loaded control of the simulated dynamomete is achieved by changing the exciting current of the dynamometer. In the test, the exciting current is changed by tracking the measured instantaneous speed, instantaneous acceleration and comprehensively calculating the static parameters of the measured structural parameters of vehicles, weight parameters and the resistance parameters on the drum test rig , on the road or on the drum, then we get the difference ΔF ($\Delta F = F_{djl} - F_{djc}$) between the loaded theoretical resistance of the dynamometer and the current measured resistance convert an simulated signal to fed back to the SCR controller, contact the SCR conduction angle, debug the excitation voltage of test-bed electric dynamometer to change the dynamic load of the electric dynamometer to make them consistent, so we can simulate the resistance of a working condition. In practice, we need to convert ΔF to the theoretical value of torque that should be loaded of the direct-current dynamometer ($\Delta M_{djl} = \frac{\Delta F \times R}{\eta_d I}$, where: η_d is the transmission efficiency of the various Speed-up stalls, I is the Speed-up ratio of transmission, R is the radius of the drum). When loaded, we firstly calculated the corresponding voltage value of the loaded torque which is outputted by IPC under the test speed to load the direct-current dynamometer, then we use PID to regulate output voltage increment Δu ,

$\Delta u = K_P(\Delta M_i - \Delta M_{i-1}) + K_I \Delta M_i + K_D(\Delta M_i - 2\Delta M_{i-1} + \Delta M_{i-2})$ where: Δu is the feedback value based on the measured value of torque, ΔM_i is the deviation between the theoretical torque value and the actual at the given time i, ΔM_{i-1} is the deviation between the theoretical torque value and the actual at the given time i-1, ΔM_{i-2} is the deviation between the theoretical torque value and the actual at the given time i-2. When loaded, the software block diagram is as follows:

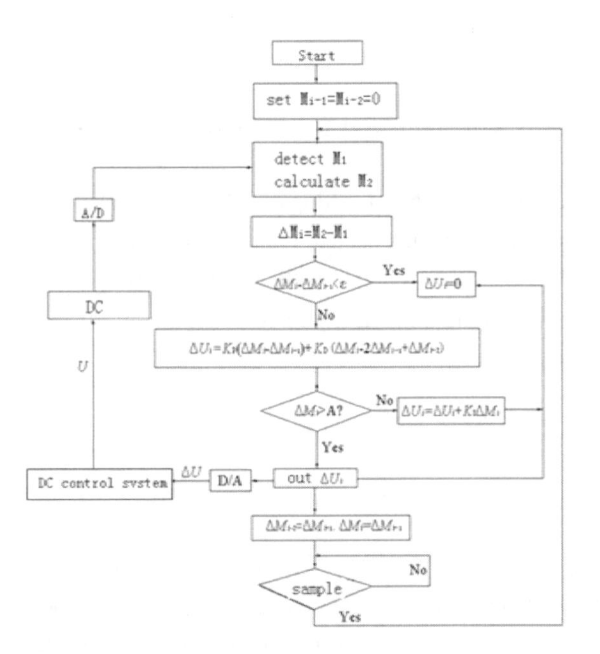

Fig. 1. Thc loadcd flowchart of the traditional PID

2.2 The SingleNeuron Control Model

The conventional PID control algorithm is simple, robust, reliable and easy to realize, ae long an wo corrcctly sct thc parameters K_P, K_I and K_D according to the mathematical model of the control object, it will be working. Nevertheless, the actual chassis dynamometer analog system is a complex non-linear, uncertain systems, it is difficult to establish an accurate mathematical model and the parameters of the model change with the operating conditions of the dynamometer, so it requires that the tuning of PID control parameters does not depend on the mathematical models and can adjust PID parameters online according to the dynamic information of the Dynamometer system to meet the requirements of timely control. Neural networks have the self-adaptative, self-learning, parallel processing and stronger fault-tolerant capability [1, 2]. If using neural networks into the dynamometer PID control process, it can give intelligence to the control and adjust PID parameters on-line, which will let dynamometer simulation system get strong reproduction ability and robustness [3]. In front, according to the analysis of the chassis dynamometer, we can establish the single neuron PID model, the control block diagram is shown in Fig.2.

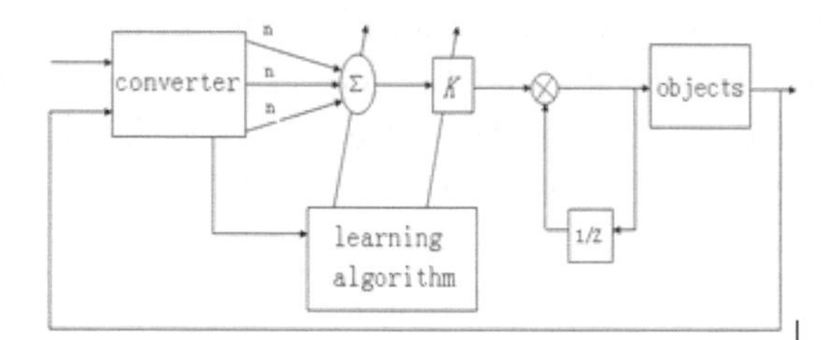

Fig. 2. The block diagram of single neuron PID

In the Fig.2, the input and output settings of the converter is $r(k)$ and $y(k)$, $r(k)$ is the loaded theoretical torque which need to loaded by dynamometer, $y(k)$ is the current measured torque. The converter's output is the state needed by the neuron which learns control (getting from the state transition method of the Chapter IV) , x_1 , x_2 and x_3 is the input of the neurons. Here:

$$\left.\begin{aligned} x_1(k) &= r(k) - y(k) = e(k) \\ x_2(k) &= \Delta e(k) \\ x_3(k) &= e(k) - 2e(k-1) + e(k-2) \end{aligned}\right\} \tag{2}$$

Proportional, integral and differential coefficient is K_P, K_I and K_D,which is equivalent to three weights of neurons. $Z(k) = x_1(k) = r(k) - y(k) = e(k)$ is the performance indicators or the progressive signal. $\Delta u(k)$ is the output value of the controller at the given time k, that is the incremental of the control variables .In the actual control process, the voltage variation which output to excitation cubicle is $\pm\Delta u(k)$, but in the computer simulation process, to get the resistance which need be loaded by dynamometer, we should use the voltage value but not the voltage variation, so the actual voltage value is $u(k) = u(k-1) + \Delta u(k)$. In the figure, K is scale coefficient of the neurons which generate control signals through linked search, that is:

$$u(k) = u(k-1) + K \sum_{j=1}^{3} w_j(k) x_j(k) \tag{3}$$

Where: w_i is the network weights which correspond with the PID controller parameters K_P , K_I and K_D; Δu is the incremental of the control variables voltage at the given time k, $u(k)$ is the voltage value which is outputted by excitation cubicle to dynamometer at the given time k.

It can be seen that the neural network controller have a PID controller structure and the neural network weighting coefficient w_1 , w_2 and w_3 which correspond with the

PID controller parameters K_P, K_I and K_D, the single neuron self-adaptive PID controller exactly achieve the self-adaptive, self-learning function by adjusting weighting coefficient or the PID controller parameters K_P, K_I and K_D, The adjustment of the weighting coefficient can adopt different learning rules to constitute different control algorithms. Here we introduce in the quadratic performance index of the output error from the single neuron controller(that is the discussed back-propagation learning algorithm in the previous), so we can minimize the performance by modifying the network weights to achieve the optimal control of self-adaptive PID [4, 5].

The controller has the functions of the online learning and self-adaptive adjustment of PID parameters, but the gain K does not has the adjusting function of the online learning, so we adopt the improved self-adaptive PSD algorithm to increase the automatic adjustment algorithm of the gain.

The options of K is very important, it has been found that the greater K is, the better the rapidity is, but the overshoot is too large so that it can make the system unstable. When the controlled object latency increases, k must be reduced in order to ensure the system stability, but if the selection of K is small, the rapidity will become bad, if too small, the system response will exist steady-state error. Therefore, the value of K will have a significant impact on the performance of the neuron control system, especially the uncertain object of of the loop gain, the value of K should automatically adjust as the change of the object's open-loop gain, it requires that controller should have the ability to automatically adjust the gain and make PSD algorithm use the gain of the neuron self-adaptive PID controlle by means of the control thinking of PSD (Proportional, Summation, Derivative) to constitute the neuron PID control of self-adjusting gain.

After combining PSD algorithm, the fixed value of K is replaced by the variable $K(k)$ by using equation (3). The PSD self-adaptive on-line adjustment algorithm of $K(k)$ is as follows:

On the point of $\mathrm{sign}\, e(k) = \mathrm{sign}\, e(k-1)$;

$$K(k)=K(k-1)+c*K(k-1)/T_v(k-1) \tag{4}$$

On the point of $\mathrm{sign}\, e(k) \neq \mathrm{sign}\, e(k-1)$;

$$K(k)=0.75K(k-1) \tag{5}$$

Where: $T_v(k)=\overline{|\Delta e(k)|}/\overline{|\Delta e^2(k)|}$

$$T_v(k)=T_v(k-1)+L*\mathrm{sign}\left[|\Delta e(k)|-T_v(k-1)|\Delta e^2(k)|\right] \tag{6}$$

$$0.025 \le c \le 0.05; 0.05 \le L \le 0.1$$

$$\Delta e(k)=x_1(k)=e(k)-e(k-1)$$

$$\Delta e^2(k)=x_3(k)=e(k)-2e(k-1)+e(k-2)$$

In the actual control of the constant value, due to the various effects of noise and interference, the error can not be zero but repeatly fluctuate in the vicinity of zero. When the fluctuation is very frequent, it is possible to make become a very small value because of the frequent attenuation which attenuate 0.75 times of the former moment,that will result the controller not to guarantee a certain degree of dynamic performance; at the same time, the value of k may be adjusted to a too large value when the time lag increases. Therefore, we must set a certain conditions for the joining of PSD adjustment algorithm to limit the adjustment range of K values. In this paper, the solution is setting a lower limit K_{min} ($K_{min} > 0$) and a upper limit K_{max} for K, on the point of $K < K_{min}$ and $\mathrm{signe}(k) \neq \mathrm{signe}(k-1)$,it should skip (5)which is for the attenuation calculation and (6) which is the calculation for adjusting $T_v(k)$;on the point of $K \geq K_{max}$ and $\mathrm{signe}(k) = \mathrm{signe}(k-1)$, it should skip (4)which is the calculation for increasing the value of *K* and (6). In this way, we not only effectively limit the adjustment range of *K*, but we also ensure the integrity of the PSD algorithm not to be destroyed and enhance the stability of the control algorithm in practical application.

The introduction of self-adaptive algorithm is for adjusting the value of *K*, after that, *K* can be self-adaptive to change the size to improve the control performance. When the error *e (k)* is larger, the value of *K* can gradually increase to make errors quickly reduce;

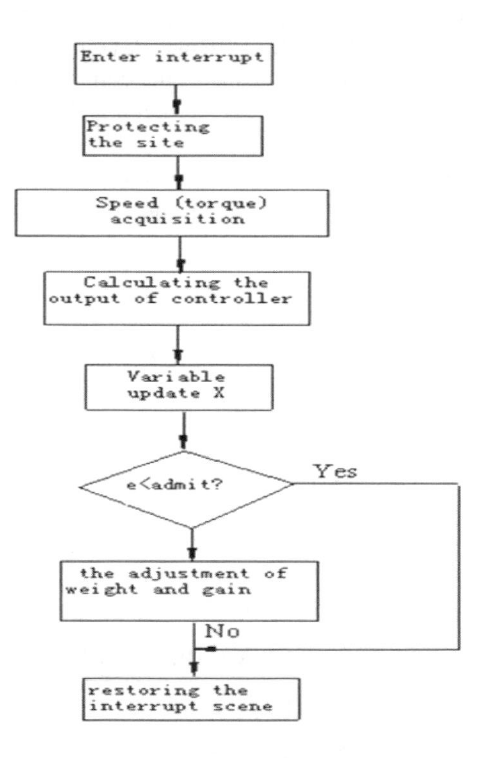

Fig. 3. The flowchart of the gain self-tuning single neuron PID algorithm

when the error value is sign-changing, that is the system just emerge overshoot, the value of K will attenuate 0.75 times of the former moment to curb the increase of the overshoot, so the control effect, self-learning, self-organizing capacity and robustness of the algorithm are higher than the average single neuron control algorithms.

In the speed interrupting program(that is the implementation process of the single neuron PID), the first initialization include interrupt mask bit, interrupt standard bit, parameters and mark assignments, etc.; then we need collect the dynamometer speed (torque) and calculate the two input variables and output of the single neuron; then we check whether the speed (torque) error is in the allowed limits or not, if there is, we do not amend the gain and weights; if not, we need amend the gain and weights; finally, we should restore the site and then return. Fig.3 shows the flowchart of the gain self-tuning single neuron PID algorithm.

3 The Simulation of the Single Neuron PID Control

PIDNN control system simulation package compiled by using VB control and simulate the established controller. In the simulation process, the input and output data are "normalized".

3.1 The Control of the First-Order Objects with Time-Lag Come from PIDNN

Considered the sample holder, the first-order transfer function with time-lag is:

$$G(s)=\frac{Y(s)}{V(s)}=\frac{1-e^{Ts}}{s}\frac{K_m}{1+T_m s}e^{-\tau s} \tag{7}$$

Where: $Y(s)$ is the outputting object; $V(s)$ is the inputting object; K_m is the magnifying coefficient of the object; T_m is the time constant of the object; T is the system sampling period; τ is the lag time.

If $\tau = NT$, N is a constant, the equation (7) were transformed by the z to be:

$$G(z)=\frac{Y(z)}{V(z)}=\frac{K_m z^{-1-N}\left(1-e^{-T/T_m}\right)}{1-e^{-T/T_m}z^{-1}} \tag{8}$$

If $K_m = 2/\min$, $T = T_m = 0.25\min$ and $\tau = 0.25\min$, so $N = 10$, that is to say the lag time is 10 times of the time constant of the object, it is the typical first-order objects with time-lag. According to the above parameters, we can get the discretization formulaL:

$$y(k+1)=0.368y(k)+1.264v(k-10) \tag{9}$$

The sampling square wave function is given by the system as follows:

$$r(k)=1(k)-0.8(k-100) \tag{10}$$

The learning step of SPIDNN is $\eta = 0.02$, each step have $m = 200$ sampling points, according to the initial selection of the network connection weights and the principles which is stated in the previous chapter, we choice the initial connection weights of the proportionality factor and the differential element which are from the network input layer to the hidden layer, the initial connection weights is:

$$w_{11}(0) = w_{12}(0) = 1, \quad w_{21}(0) = w_{22}(0) = -1$$

The initial connection weights from the input layer to the hidden layer is:

$$w_{13}(0) = 0.1, w_{23}(0) = -0.1$$

The initial connection weights of the network input layer is: $w_{j1}(0) = 0.1, \ j = 1, 2, 3$.

Then we carry out the real-time simulation according to the above parameters. In the Fig.4, (a) is the system response curve after 10-step of Networked Learning; (b) is the convergence curve of the system objective function in the former 10-step learning process; (c) is the system response curve after 100-step of Networked Learning; (d) is the convergence curve of the system objective function E in the former 100-step learning process.

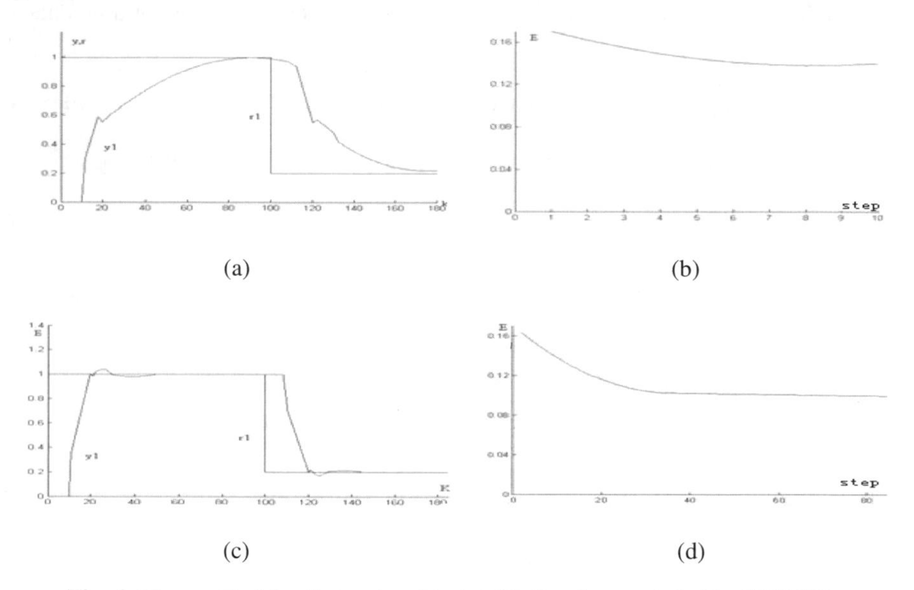

Fig. 4. The result of the first-order objects with time-lag controled by SPIDNN

From the simulation results we can see that PIDNN neuron control successfully conducted the control of the objects with time-lag through the self-learning and the parameter adjustment, there is no steady-state error and the responding speed is very high.

3.2 The Control of the Non-linear Time-Varying Single-Variable Systems

Non-linear object is:

$$y(k+1) = \frac{y(k)}{1 + y^2(k)} + v^2(k) \tag{11}$$

The structure and initial connection weights of the controller is same to the controller which is used to control the objects in the front, the system's output is the square wave signals (11), the sample points of each group is 200, it is for learning step. The square-wave responding curve after that the network learn 50 times is Fig.5.

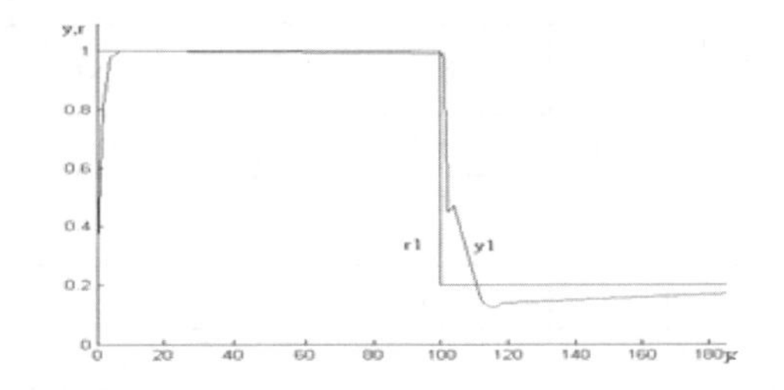

Fig. 5. The square-wave simulation curve of SPIDNN control

From the figure we can see that both of the dynamic and static performance of the system are very good under the control of the single neuron controller.

4 The Test Analysis of Simulated Chassis Dynamometer

According to the test data, the absolute error and relative error of the vehicle road test and the test on the traditional PID-controlled, the test on the single neuron self-adaptive PID-controlled dynamometer is in the Table 1, Table 2.

Table 1. The Error Comporison List of the Test on the Tiditional PID-controled Dynamometer

		Tests on road	Tests on the dynamometer	The absolute error	The relative error
The accelerated test	Acceleration times	38.03(s)	38.93.(s)	0.90	2.37%
	Acceleration distance	645.86(m)	650.04(m)	4.18	0.65%
Acceleration fuel consumption test		22.56/100(L/ km)	22.68/100(L/k m)	0.12	0.53%
Isokinetic fuel consumption test	30(km/h)	8.2	8.4	0.2	2.44%
	40(km/h)	8.4	8.5	0.1	1.2%
	50(km/h)	9.2	9.4	0.2	2.17%
	60(km/h)	10.8	10.5	-0.3	2.78%
	70(km/h)	12.3	12.4	0.1	0.81%

Table 2. The Error Comporison List of the Test on the Singleneuron Self-adaptive PID-controled Dynamometer

		Tests on road	Tests on the dynamometer	The absolute error	The relative error
The accelerated test	Acceleration times	38.03(s)	38.65.(s)	0.63	1.66%
	Acceleration distance	645.86(m)	647.29(m)	1.43	0.22%
Acceleration fuel consumption test		22.56/100(L/km)	22.68/100(L/km)	0.04	0.18%
Isokinetic fuel consumption test	30(km/h)	8.2	8.4	0.2	2.44%
	40(km/h)	8.4	8.5	0.1	1.2%
	50(km/h)	9.2	9.1	-0.1	1.1%
	60(km/h)	10.8	10.6	-0.2	1.85%
	70(km/h)	12.3	12.4	0.1	0.81%

5 Conclusion

When we conduct the accelerated test and acceleration fuel consumption test on the road or on the dynamometer, the starting speed is 30km / h, the terminating speed is 90km/h.From Table 1 and Table 2 we can see that conducting the vehicle test of the dynamometer on a single neuron self-adaptive PID control dynamomete have the smaller deviation compared with the same test which conduct under the simulated chassis dynamometer controled by the traditional PID, its relative error of the acceleration times is 1.66%, the relative error of the accelerating distance is 0.22%, the maximum relative error of Isokinetic fuel consumption is 2.44%, all of these are in the limitation, and the test data have the high repetitivity, it fully validated the feasibility of control methods and further improve the test accuracy of the performance.

References

1. Zhang, M.H., Xia, C.L., Tian, Y., Liu, D., Li, Z.Q.: Speed control of brushless DC motor based on single neuron PID and wavelet neural network. In: 2007 IEEE International Conference on Control and Automation, ICCA, pp. 617–620 (2008)
2. An, S.J., Shao, L.M.: Diesel engine common rail pressure control based on neuron adaptive PID. In: 2008 IEEE International Conference on Cybernetics and Intelligent Systems, CIS (2008)
3. Zhang, M.G., Li, W.H.: Single neuron PID model reference adaptive control based on RBF neural network. In: Proceedings of the 2006 International Conference on Machine Learning and Cybernetics, pp. 3021–3025 (2006)
4. Du, Y.P., Wang, N.: A PID controller with neuron tuning parameters for multi-model plants. In: Proceedings of 2004 International Conference on Machine Learning and Cybernetics, vol. 6, pp. 3408–3411 (2004)
5. Wang, N., Ji, P.: Neuron PID variable structure control. In: Proceedings of IEEE Asia-Pacific Conference on Circuits and Systems, pp. 319–322 (2000)

Research on the Neural Network Information Fusion Technology for Distinguishing Chemical Agents

Minghu Zhang[1,*], Dehu Wang[1], Lv Shijun[1], Jian Song[2], and Yi Huang[1]

[1] Dept. of Shipboard Weaponry, Dalian Naval Academy, 116018 Dalian, China
[2] Arms Tactics Research Center, Naval Arms Command Academy, 510430 Guangzhou, China
zmhly20070618@163.com, zmhly20040318@sina.com

Abstract. For implementing effectively detection and rapid exact identification for the chemical agents in sea-battlefield, firstly, the conception, treatment model and system structure of the information fusion, are introduced; secondly, the neural network(NN) information fusion system model are built by the multi-sensors information fusion (MSIF) technology; At the same time, connecting the wavelet analysis with the NN organically, and based on the wavelet transfer and the NN, the system of the speedy features extraction and identification for chemical agents -the NN Distinguishing Chemical Agents (NNDCA) system- is founded. The model of the NNDCA and the method of the feature extraction for the chemical agents based on the wavelet analysis are established, and the hardware accomplishment and the software structure of the NNDCA system are put forward; lastly, the experimental and simulated results show: it is feasible that the analyses for the chemical agents with the NNDCA system based on the MSIF technology and the wavelet analysis. The method can remarkably heighten the accuracy and credibility of the measurement results, and the results are of repeatability.

Keywords: Information fusion, hiberarchy and architecture, NN information fusion system, system establishment and realization.

1 Introduction

So far, the reports and references on the Neural Networks Distinguishing Chemical Agents (NNDCA) at home and abroad are very fewer but the attitude of making use of the NNDCA in the western developed countries is positive. After analyzing and discussing, the results show that this is a kind of valid method to realize the automation and intelligence of distinguishing chemical agents using the neural network (NN) combined with chemical sensors, and it can, rapidly and accurately, identify thc kinds of the chemical agents.

The multi-sensors information fusion (MSIF) is a new technology which develops at 1980s, and it belongs to a new type of data processing technologies. It's also named

* Supported by the National Defense Basic Research Foundation of China (435B956), the National Defense Pre-Research Found of China (41101050403), and the Scientific Research Development Foundation of Dalian Naval Academy (KF201003018).

K. Li et al. (Eds.): LSMS/ICSEE 2010, Part II, CCIS 98, pp. 139–147, 2010.

the multi-sensors data fusion technology or distributed sensors. The MSIF technology based on the multidisciplinary theory, and it processes information by applying computer technology, sensor technology, mathematical tools and signal processing technology etc.

Takes example for the Naval Ships Chemical Detection (NSCD), for implementing effectively detection and rapid exact identification for the chemical agents in sea-battlefield, the automatic detection technique -NN information fusion technology- for the chemical agents, is studied. The model of the NN information fusion system is built. At the same time, connecting the wavelet analysis with the NN organically, and based on the wavelet transfer and NN, the system of the speedy features extraction and identification for chemical agents, the NNDCA system, is founded. In order to establish the NSCD system of high speed decision-making, advanced principle, higher precision and sensitivity, automation, and intelligence, and heightening the NSCD technology.

2 The Conception and Treatment Model of Information Fusion

2.1 Conception

The conception of information fusion began in 1970s' initial stages, rooted in the requirement of the C^3I (Command, Control, Communication and Intelligence) system in military domain, then intituled the multi-sources correlation or multi-sensors mix data fusion, and the technologies were established in 1980s. The JDL(Joint Directors of Laboratories) of the DOD(department of defense) of the US defined information fusion as such process from military application: those data and information from multi-sensors and multi-info-sources add association, correlation and combination, to obtain the accurate position estimation, identity estimation, and the proper integrity estimation for the battlefield instance and menace and its importance degree. Waltz and Llinas put up suppiement and modification the above definition: the information fusion is a sort of multilayer and/or multiside treatment process, the process is detection, association, correlation, estimation and combination to obtain the accurate state estimation, identity recognition, the whole situation estimation and menace estimation.

2.2 Treatment Model

The information fusion treatment model is for a set of the treatment processes' description, which is the constitution of the system function units, not refered the physics frame and software realization of each of the units, and the treatment processes also allow feedback, viz regulate the fusion frames based on the decision-making. Fig.1 is the information fusion treatment model that was put forward by the data fusion work group in the US, which it plays important influence to understand the basic conception of the information fusion.

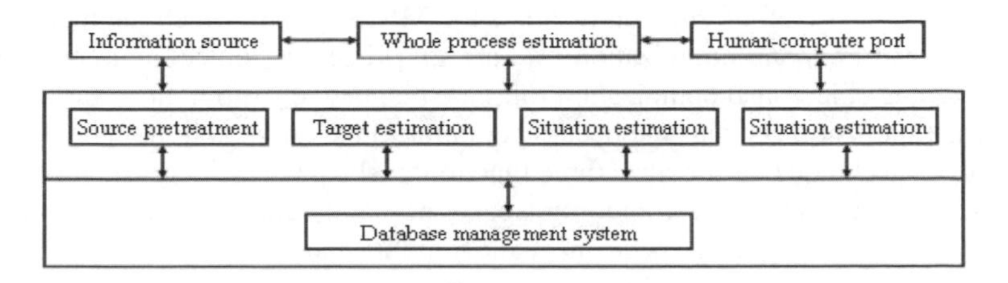

Fig. 1. Information fusion treatment model

3 The System Frame of the Information Fusion

The research of the system frame of the information fusion includes two parts: hiberarchy and architecture. The hiberarchy mostly based on the information to analyze the fusion system; and the architecture mostly based on the hardware.

3.1 Hiberarchy

The information fusion system may be compartmentalized according to administrative levels, but there are more opinions. At present, the universal acceptant opinion is three layers fusion structure, namely data layer, feature layer and decision-making layer.

The hiberarchy of the information fusion is compartmentalized according to the information abstract degree. In the practice engineering application of the MSIF system, the hiberarchy model, which was established and adopted, should synthetically consider the sensors performance, system calculation ability, communications bandwidth, expectation nicety rate, and existence fund power. Moreover, the establishment of the hiberarchy based on the information can better confirm the architecture based on the system hardware.

3.2 Architecture

The system hardware architecture are approximately divided to three species: concentration style, distribution style and admixture style.

The architecture adopted is entirely the various practice requirement. When the architecture is designed, it's confirmed by the established system hiberarchy, at the same time, the many supportion technologies must also be considered, such as data traffic, database management, human-computer port, and sensor management.

4 Constructing the NN Information Fusion System

4.1 Fusion Thought

The NNDCA is the intelligent system of distinguishing chemical agent using the sensor technology as a method, taking the signal processing and modeling processing as foundation.

The hardware base of information fusion is the MSIF system, which the multi-sources information and harmony optimization rule are its processing object and core. Though their processing objects and optimization rules are respectively different in various fields, the information fusion contains the deep-layer commonness link and thought. The information fusion technique must realize three functions, which are as follows:

(1) Sensors management: Including space management, working management, time management, frequency spectrum management, and so on, which are driven by those multitudinous information requirement factors, the purpose is to make the best of the resources of the limited sensors, to cover all possible hunt airspaces using the multi-sensors system, and to offer the more exact, reliable and precise decision-making information with the lesser cost and lower error. Its core is how to select the sensors, sensors working modes, and sensors optimizing strategies, in order to optimize the whole performance of the system. The information fusion, which asks for implementing the multi-sources information harmony and optimization, must enhance the sensors management. Therefore, the sensors management becomes the indispensable link of the information fusion.

(2) Multi-sources information harmony management: Because the expression forms of the space-time characteristic, which are expressed the multi-sources information, are respectively different, we must unify their space-time reference frames and adopt the uniform expression format, in order to use and dispose the multi-sources information under the common frame, which are sustained by the coordinates transform and database.

(3) Multi-sources information harmony processing: The information fusion process includes check-test, relevancy, track, estimate, and synthesis. The multi-sources information harmony processing must be achieved under certain structure.

4.2 Model Construction

NN is a new technology for pattern recognition, which can distinguish nonlinear complex objects using self-adapting mode. It has advantages of high legitimacy and strong anti-interference capacity. At presents, there are many applied studies on the BP NN and RBF NN in the chemical agents and environment monitoring, the BP is the most comprehensive application in the aspect of spectrum discrimination and is the most successful technology, and appears many improved BP algorithm. And the RBF NN has become the research hotspot at present, which has higher discrimination capability and faster learning speed. Kohonen network and adaptive resonance NN are belonging to self-organizing and self-adapting NN, which can cultivate independent learning ability, and be applied in the adaptability training of the equipment to new surroundings. NN have high recognition capability and mature theory, and the algorithm is complex, only combined with the method of the efficient feature extraction or selection, can we get ahead in the practical application.

Selecting the sample data of the chemical agents to carry on the feature extraction, and this is the premise of the selecting recognition parameter and the structuring distinguishing NN. Carries on the wavelet energy spectrum analysis for the chemical agents sample parameter, to obtain the corresponding wavelet energy spectrum characteristic vector of the parameter residual error data, which are available in structuring the study sample of the recognition network.

Firstly, founded on certain quantity training samples collection (usually is the symptom, chemical agent of data sct), to carry on the training for the NN to obtain the expectation recognition network, then to comply with the current recognition input, the chemical agent is distinguished for the system. Before study and recognition, it usually does justice to the primary distinguishing data and the training sample data. The goal is to provide the appropriate recognition input for the recognition network and the training sample, and the usable feature vector for the NN Distinguishing. When the Chemical Agent enters the chemical sensors, the complex film in the sensors adsorb the chemical agent molecule. According to the differences of the chemical agent compound, it may cause the resonance frequencies change through the mini-sensors. The mini-sensors have the different frequency feature response signal for the different chemical agent. This kind of response is recorded and transmitted to the NN, which has trained and joins the chemical sensors, to examine the hairlike distinction in these signals. The chemical agent feature is recorded in the database with comparison, distinction and confirmation the chemical agent type.

The NNDCA for the NSCD system based on the MSIF is to extract or select the feature of the chemical agent in sea battlefield, according as the feature of the every chemical agent which is measured by the multi-sensors in the system, and make use of the NN' ability of the non-linear mapping and pattern distinguishing, eventually distinguish the chemical agent to be measured by synthesis and analysis. Before using the NN to distinguish the chemical agents, the NN is trained to anticipant recognition network by some quantitative training samples sets.

The NNDCA was built up by the numerical value calculating, neural network's learning and training. The input network's node-number is the dimension of the input characteristic vectors; the output's is amount of the chemical agents; the middle-tier's should be selected by the request of the training sample-set size and training error. The precondition of the distinguishing preferences and constructing recognition neural network is the characteristic extracting for these samples data, the eigenvector of the normalized wavelet energy distribution can be obtained by the analyzing wavelet energy distribution for the data of the chemical agent sample parameters, and which may be regarded as the calculation basis of the parameter recognition confidence-degree or samples comprehensive discrete degree. The corresponding wavelet energy distribution eigenvector of the parameter residual data can be obtained by treating the chemical agent sample residual data, which can construct the learning sample of the recognition neural network. If the training effect is imperfect, it need readjust the parameters of the neural network, repartition the training samples, and retrain the neural network, till attain to the satisfying effect.

The sketch map of the NNDCA model based on the MSIF is showed in Fig.2.

4.3 Chemical Agents Feature Extraction

In the process of distinguishing the chemical agents, we can get the practical data of many parameters by chemical sensors. These data contain rich information of the chemical agents. Feature extraction or selection is the powerful tool to reduce the mode dimension. Reducing dimension of the primitive features can help minimize the wrong recognition rate of the classified device. Via wavelet package analysis technology, we can record the signals energy data of various frequencies, thus we get the

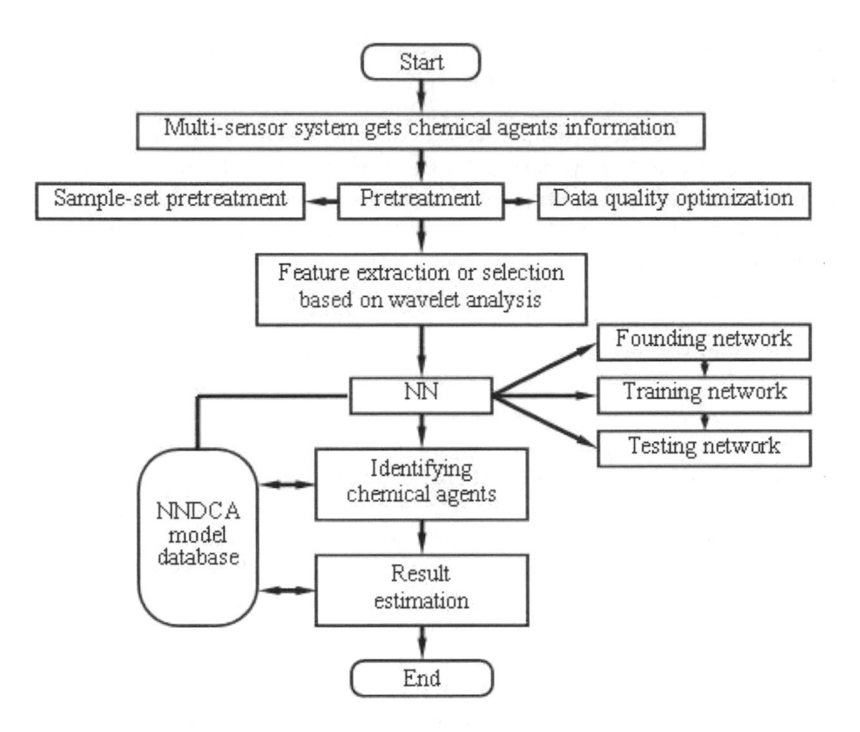

Fig. 2. The sketch map of the NNDCA model

frequency energy feature, which distinguishes one from another. Choosing feature parameter is based on choosing the sample data's features. With parameter features' comprehensive discrete degree or recognition credit degree as a criterion, we can get a group of better feature parameters to recognize. We have known that the different chemical agents obtain the different frequency energy feature vectors. They reflect the different chemical agents, and which provides the practical feature vectors for the recognition of chemical agents based on the neural network. According to the feature vector to work out the sort decision-making, i.e., to confirm the chemical agent that it is which kind of chemical agents.

4.4 System Realization

(1)The realization of hardware

In order to realize the NNDCA, to heighten the stability and reliability of the NSCD system, and to adapt to the requirement of miniaturization, the calculating method of the NNDCA should be transplanted to the chip of the hardware, and get the practical application. For the interface-control part of the NNDCA and NSCD system, the embedded designing thought may be used, the various single-chip or embedded computers can be adopted as the running platform. For the system of the biggish data quantity, it can adopt the DSP device to realize the beginning pretreatment function, and the embedded computer to realize the data traffic and interface management of the subassemblies in the NNDCA classifiers and NSCD system. In order to make the system run speedily and obtain the finer real-time ability, the neural network classifiers can also adopt the realization project of the digital circuit based on FPGA.

(2)The component of software
The software exploitation of the NNDCA is the quite complex software engineering. The software is mostly composed of the many modules, such as the chemical agent recognition process evaluation module, NNDCA module, data processing module, data management module, maintenance module, and database. The module of the NNDCA and the database are the cores of the wholly formed software.

5 Experiment and Simulation

To verify the validity of the NNDCA system, using the multi-sensors fusion system, the response data of six different chemical agents, like GB, GD, VX, HD, L, AC, are obtained by experiment under normal temperature condition with the multi-sensor array. In experiment, the concentration of neural agent ranges from 0.01mg/m^3 to 100mg/m^3, vesicant agent ranges from 1.0mg/m^3 to 100mg/m^3, and AC from 10mg/m^3 to 1000mg/m^3; based on the simulation natural environment of the sea battlefield, the experiment is divided into the following four cases: the single chemical agent without interference, single chemical agent with interference, mixed chemical agents without interference, and mixed chemical agents with inference.

Firstly, the wavelet package analysis technology is executed for the observed data of the single chemical agent without interference, the energy of the chemical agent's signals at various frequency zones is counted, and a kind of the frequency zone energy feature using recognition is obtained, in which a group of better feature parameters is selected to recognize the chemical agent, using the comprehensive discrete degree or recognition confidence-degree of features parameter samples as a criterion. Secondly, the wavelet package analysis technology is executed for the observed data of other three cases, and the feature parameters of the chemical agents are extracted to distinguish the chemical agents efficiently. Lastly, on the basis of obtaining experiment data to a certain extent, the simulation experiment is set up; in the processes of experimentation and simulation, the mass correlative data of the chemical agent recognition system can be obtained, and these data reflected directly or indirectly the rich information of the chemical agent form different aspects; based on the extraction or selection of the features information and data processing relatively for the chemical agent, the chemical agent can be distinguished by the NNDCA system. An example of the experiment is as follows: the twenty samples of the single chemical agent were measured, contrasting the methods of the single sensor and neutral networks based on multi-sensor, the measured data and the results show that measurement accuracy is significantly enhanced by the method of the neural networks based on multi-sensor, as shows in Fig. 3 (Take example for the HD). The testing results prove that it is of the better recognition effect using the NN to analyze the chemical agents.

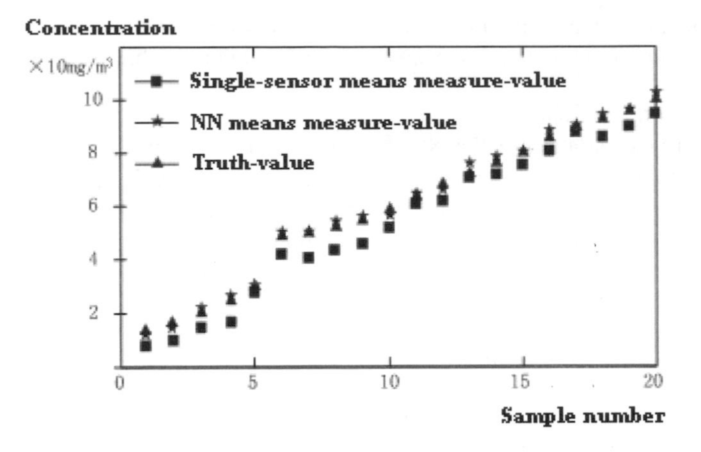

Fig. 3. The measurement results contrast of the different samples for the single HD

6 Conclusions

Based on the experimental research, through the experiment and simulation, using the wavelet signal for the chemical agent to extract or select the features, and the experimental conclusions are as follows:

(1) The multi-dimensional information of the chemical agents can be obtained by using the measurement system based on the multi-sensor fusion; the data from the multi-sensor can be fused by the NN, the experimental results show that the method can remarkably fall the effects of the factors, interference, concentration, and condition, etc, for the measurement results, and heighten the accuracy and credibility of the measurement results.

(2) The anticipant recognition NN can be gained by training, testing, and verifying, on the basis of the relatively processing for the learning sample data and its feature extraction or selection, it is completely feasible that the analyses for the chemical agents with the NNDCA system based on the MSIF technology.

References

1. Zhu, Y.M., Li, X.R.: Unified fusion rules in the multi-sensor multi-hypothesis network decision systems. IEEE Transactions on Systems, Man and Cybernetics, Part A 33(4), 502–513 (2003)
2. Zhu, Y.M., Song, E.B., Zhou, J., et al.: Optimal dimensionality reduction of sensor data in multi-sensor estimation fusion. IEEE Transactions on Signal Processing 53(5), 1631–1639 (2005)
3. Song, S.P., Que, P.W., Zhang, W.Y.: Design of Line-focusing Ultrasonic Transducer Array Used for Flaw Detection in Seabed Pipelines. Chinese Journal of Scientific Instrument 27(5), 547–550 (2006) (in Chinese)
4. Sun, S.L.: Distributed optimal component fusion weighted by scalars for fixed-lag Kalman smoother. Automatica 41(12), 2153–2159 (2005) (in Chinese)

5. Mirjalily, G.: Blind Adaptive Decision Fusion for Distributed Detection. IEEE Transactions on Aerospace and Electronic Systems 39(1), 34–51 (2003)
6. Han, C.Z., Zh, H.Y.: Multi-sensor information fusion and automation. Acta Automaticsa Sinica (28), 117–124 (2002) (in Chinese)
7. Liu, Y., Gao, X.G., Lu, G.S., et al.: Weighted Attribute Information Fusion Based on OWA Aggregation Operator. Chinese Journal of Scientific Instrument 27(3), 322–325 (2006) (in Chinese)
8. Zhang, M.H., Ding, Z.H., Bai, Y.: Study on multi-sensor fusion technology in naval ships chemical detection system. Chinese Journal of Scientific Instrument 27(Suppl. 12), 119–123 (2006) (in Chinese)
9. Luo, Z.Q.: Universal decentralized estimation in a bandwidth constrained sensor network. IEEE Transactions on Information Theory 51(6), 2210–2219 (2005)
10. Feng, Z.Q., Yang, Y.H.: Fusion recognition of dot target multi-spectrum data based on ANN. Optics and Precision Engineering 11(4), 412–415 (2003) (in Chinese)
11. Wang, L.G., Zhang, L.H., Kang, J.L.: Study on the neural networks for distinguishing chemical agents based on the wavelet analysis. Transactions of Beijing Institute of Technology 26(2), 120–124 (2006) (in Chinese)
12. Zhang, M.H., Pang, G.C., Liu, H., et al.: Chemical agent identification technology for naval ships chemical detection. Journal of Guangxi Normal University: Natural Science Edition 26(1), 220–2223 (2008) (in chinese)

Simulating Energy Requirements for an MDF Production Plant

Cristina Maria Luminea and David Tormey

Centre For Design Innovation, Institute of Technology Sligo,
Ballinode, Sligo, Ireland
cristina@designinnovation.ie

Abstract. The main focus of this paper is to look at production management in a manufacturing facility and correlate it with the energy consumption. The end result of this process will be a better understanding of the production system and the energy loses. This will be closely followed by the creation of different scenarios that ideally will lead to a lowering in the energy consumption. So far, simulation has been used in manufacturing facilities for modelling supply chain management, production management and business processes. This research brings a novel approach to investigating the adaptability of industrial simulation processes and tools for modelling the energy consumption with respect to a variable production output.

Keywords: Simulation, Production Management, Energy Management.

1 Introduction

The case study company considered for this research project is a Medium Density Fibreboard (MDF) manufacturer. This paper will outline the development of a simulation project that will model the energy usage along with the production process of the case study company. Modelling the energy consumption with respect to the production output is a relatively new and innovative research domain. The majority of automated manufacturing plants consume high amounts of energy and they usually rely on sophisticated data management systems which monitor, record and control production resources. This results into high amounts of data being stored. Unfortunately the majority of these plants have neither the time, resources nor skills necessary to analyze and utilise the gathered data to its full extent. The novelty of this research lies in the application of traditional industrial simulation tools for modelling and optimizing raw material processing to include energy consumption for sustainable and cost efficient production.

The author considered the study of Professor Jingge Ling and Professor Shusheng Pang [1] on the modelling of energy demand in an MDF plant, along with the paper published by Carvalho et. al. [2] which looked at the hot pressing of MDF. These studies demonstrate the effective use of simulation in predicting the energy consumption in Medium Density Fiberboard production facilities, by looking at production management. This encouraged the author to take the research to the next level and consider industrial simulation software tools for modelling the correlation between energy consumption and production process in the case study company.

K. Li et al. (Eds.): LSMS/ICSEE 2010, Part II, CCIS 98, pp. 148–156, 2010.

The conceptual model, data collection and analysis and the model development are presented in this research paper, along with the validation of the model and the results. These show that the model is able to predict the energy consumption and the production in satisfied accuracy with discrepancy of -5.61% to 1.62% for energy consumption and 0.86% to 3.06% for the level of production.

1.1 Literature Review

In 2006 Professor Jingge Ling and Professor Shusheng Pang from the University of Canterbury, New Zealand published the paper 'Modelling of Energy Demand in an MDF Plant'. This paper looks at taking inputs like MDF production, log moisture content and fibre drying method and predicting the energy demand for both heat and electricity.

The authors developed a computer model in order to quantify the energy demand in the medium density fibreboard (MDF) plant '*based on the production process from chip preparation, refining, fibre drying, mat forming, hot pressing to product finishing*'.

The model was than validated using energy audit results from three commercial plants. This showed that the model was able to 'simulate the energy demand with a discrepancy of -5 to +7% for thermal energy and ±4% for electricity'.

According to the authors' research, this is the first effort to predict both heat energy and electricity consumption in an MDF plant. A few years earlier though, in 2003, Carvalho et. al. published the paper 'A global model for the hot-pressing of MDF' in the Wood Science and Technology Journal.

Their model was used to 'predict the evolution of the variables relating to heat and mass transfer (temperature, moisture content, gas pressure and relative humidity), as well as the variables relating to mechanical behaviour (pressing pressure, strain, modulus of elasticity and density).' Also, they reached the conclusion that the model they developed is able to facilitate the scheduling of the press cycle 'to fulfil objectives of minimization of energy consumption, better quality of the board and increased process flexibility.'

In September 2007 Professors Ling and Pang published another paper: 'Modelling of thermal energy demand in MDF production' [3]. Through their research they created a model which can be used to '*examine the effects of various production capacities, product grades, operation times and fiber drying methods on thermal energy demand and distribution.*' They also proved that the model is able to predict the energy demand with an accuracy of -17% to +6%.

The above cited research demonstrates the effective use of simulation in predicting the energy consumption in Medium Density Fibreboard production facilities, by looking at production management. The work of Ling and Pang and Carvalho et. al. proves that energy consumption can be modelled with an accuracy ranging between -17% to +6%, using spreadsheets for simulation. This encouraged the author to take the research to the next step and consider industrial simulation software tools for modelling the correlation between energy consumption and production process in the case study company.

1.2 The Process of MDF Manufacturing

The Process of MDF manufacturing is presented in figure 1 below.

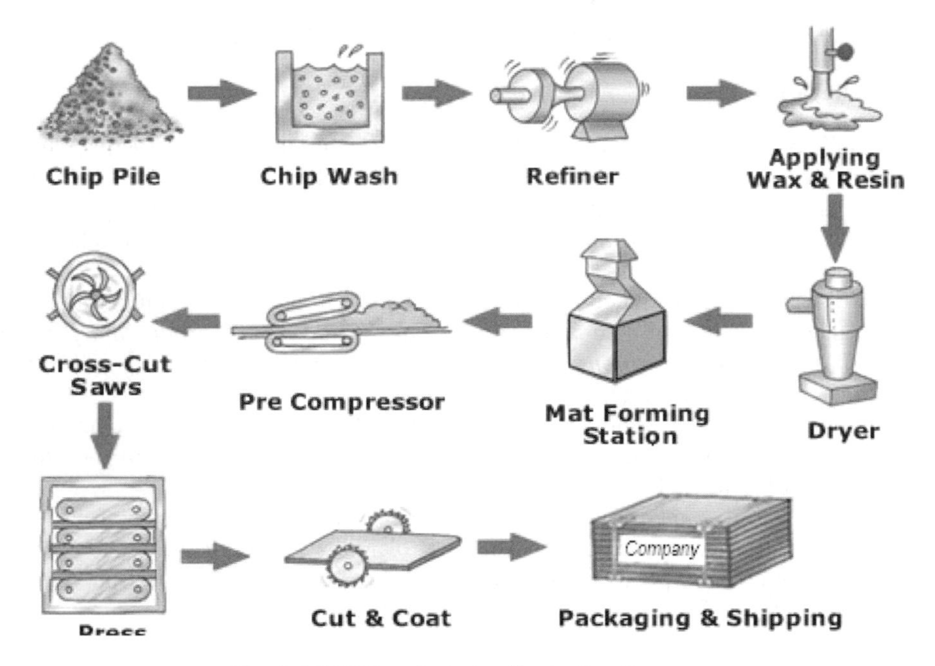

Fig. 1. MDF Manufacturing Production Process

The raw material is usually arriving to an MDF manufacturing company under the form of wood chips. These are washed and then processed into a Refiner which transforms them into wood fibre. From the refiner they go to the blownline where wax and resin solution is injected. They will than arrive into a tube drier where their moisture will be reduced to a target of 10 – 12%. From here the fibres are sent to storage bins before being transferred to the mat forming station. The next step in the production process is reducing the mat thickness through a continuous cold press, followed by cutting the mat and then compressing it in a batch press.

After the hot pressing the panels are directed into the work in progress (WIP) storage waiting for the cut and coat process. Through this process the panels are sanded, trimmed and cut to the standard sizes for packaging.

The manufacturing process of the case study company includes two main production lines and a cut and coat line, which combined, requires 85% of the company's energy needs.

2 Model Development

For the purpose of this study the simulation model was developed in ProModel [4], a discrete event simulator designed to model systems where events occur at defined points in time. Its typical applications include: assembly lines, transfer lines and flexible manufacturing systems.

ProModel is an intuitive software which requires little to no programming skills. The models are built only by defining the way a system operates, mostly through part flow and operation logic. During the simulation ProModel displays an animated representation of the system on the screen. When the simulation is over, performance measures are being presented to the user in both tabular and graphed forms for evaluation.

2.1 Defining the Conceptual Model

As stated above, the production process of the case study company consists in two main production lines and a Cut and Coat line which account for almost 80% of their total energy consumption. Their main problem consists in their energy consumption remaining constant despite a drop in production therefore, the main objective of this project is to lower the energy consumption and more specifically, lower the amount of kilowatts consumed per every unit produced. The figure below represents a diagram illustrating the conceptual mode.

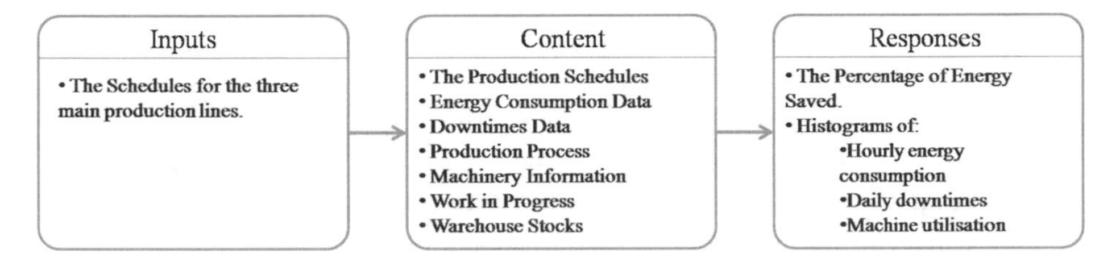

Fig. 2. Conceptual Model

In the case of this project, the inputs or the experimental factors consist in the schedules for the three main production lines. The responses, in terms of defining the achievement of the objective will be the percentage of saved energy which translates into the amount of money being saved.

Having identified the model's inputs and outputs, the author proceeded in identifying the content of the model itself. It is easy to recognize that the model has to be able to accept the experimental factors and to provide the required responses. Therefore the model must be able to represent the schedules for the three main production lines and to provide the relevant reports: the hourly energy consumption, the daily downtimes and the machine utilization. The model also needs to include the production process along with information on the conveyors, the downtimes, the energy consumption data, the work in progress and the warehouse stock information. All of these are being presented in more detail in the next section of this chapter.

2.2 Performing Data Collection and Analysis

The type of data needed to be collected is usually dictated by the objectives of a project. In the case of this research project the input data includes, but is not limited to production schedule, conveyor speeds and length and historical energy data. This data was collected from the two monitoring systems available on the site.

PI, developed by OSIsoft [5], is one of the data collection systems operating in the case study company. It is an operational, event and real – time data management infrastructure, which brings together different types of data from a variety of sources like: systems, equipment, solutions, applications, locations and networks. PI gathers and archives large volumes of data on servers, it converts the real time data into actionable information offering access to real – time or historical data for the entire enterprise at any time, it provides notifications and it allows for anyone to view the data graphically, identify problems and take corrective actions.

The second monitoring system is an energy monitoring system called eSight [6]. This is an intranet solution which provides energy analysis and reporting. It enables the user to monitor standard utilities such as electricity, gas and water. It also facilitates the presentation of data in a wide range of graphs, tables, reports and exports. In addition, reports may be run on an ad hoc basis, saved as templates for later use or scheduled for automatic production and distribution by email.

Since much of the raw data collected from the systems described above cannot be inputted into a simulation model directly, some input analysis such as probability distribution analysis were performed. All the data that was collected and analysed was recorded in an Excel document called Interface.xls.

This data includes:

- The necessary time for changing the dies in the press.
- The data for all the machinery in the model (e.g. conveyor length and speed).
- The work in progress (WIP) and the warehouse stock levels and demand information.
- The production schedules for all the production lines.
- Raw data of downtimes and energy consumption which is being used to produce user distributions for the model.

The energy data collected for this model divides the lines into regions as it is collected by different energy meters.

Most of the analysis on the collected data has been performed already and the data has been introduced in the model. The author has decided to use an Excel document as an Interface for the purpose of familiarity. Most of the staff in the case study company that will be working with this model is familiar with the Excel format. The Interface also replicates the look and feel of most of the documents currently generated in the company. From this point the data is being imported in the simulation model and used to predict stocks and energy consumption as well as performing various scenarios with the purpose of lowering the energy consumption.

The data generated by the model is being exported in the Results excel document. This will contain general results on production and energy consumption.

2.3 Developing the Simulation Model

The author started the development of the simulation model by mapping the three main production lines. You can find a screenshot of the ProModel simulation model in Fig. 3 below. After defining the locations the author proceeded in defining the entities of the simulation model representative of the wood fibre, the panels, and the skins. For the purpose of modelling the energy consumption, the production schedules and the production downtimes a control entity has also been defined. As an example, in the case of energy consumption, the control entity cycles inside the location every hour, determining the times at which energy consumption needs to be recorded.

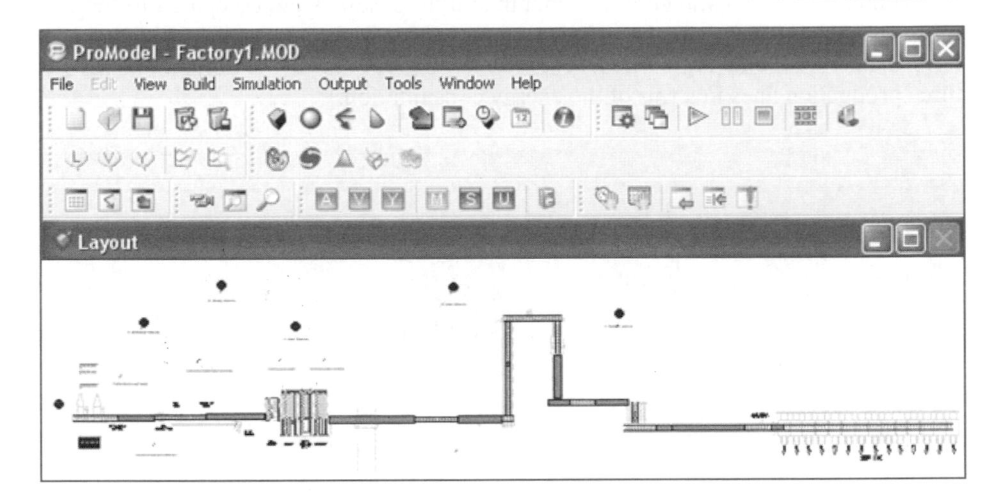

Fig. 3. The ProModel Model of the Case Study Company

The next step in the simulation model development consisted in the definition of the processes that the entities need to follow within the system. This resulted in a basic representation of the production process of the case study company, which was followed by the implementation of the production schedules. The data representing the three production schedules for the three main production lines was imported from the excel spreadsheet. The control entity created above is used to determine each day's production schedule by cycling inside each location at an interval of 24 hours.

After the implementation of the production schedule for the three production lines, the author proceeded in the implementation of the production downtimes and the energy consumption. This part of the model uses the distributions created from the raw data gathered in the case study company. In terms of the energy consumption, the model records the data every hour. For the data to reflect the reality, the model samples three different user distributions depending on the schedule of the production line.

The final part in the model coding was the implementation of the work in progress and the warehouse stock.

During the simulation run, the model is populating seven different arrays with results of the daily production, hourly energy consumption and work in progress as well as warehouse stock and demand.

All of these results are exported in the Results.xlsx file where, for a better understanding of them reports, charts and diagrams are being created.

3 Model Validation and Results

For the purpose of model validation the author populated the Interface file with data gathered during the course of one month. The most important data consisted in the planned production schedule, the planned number of press loads and the energy data. The simulation model was run for a month and the results were compared to the real life situation. This has been done three times for three different months and the results have been consistent. Figure 4 below presents the results of the simulation for the three production lines in terms of the energy consumption.

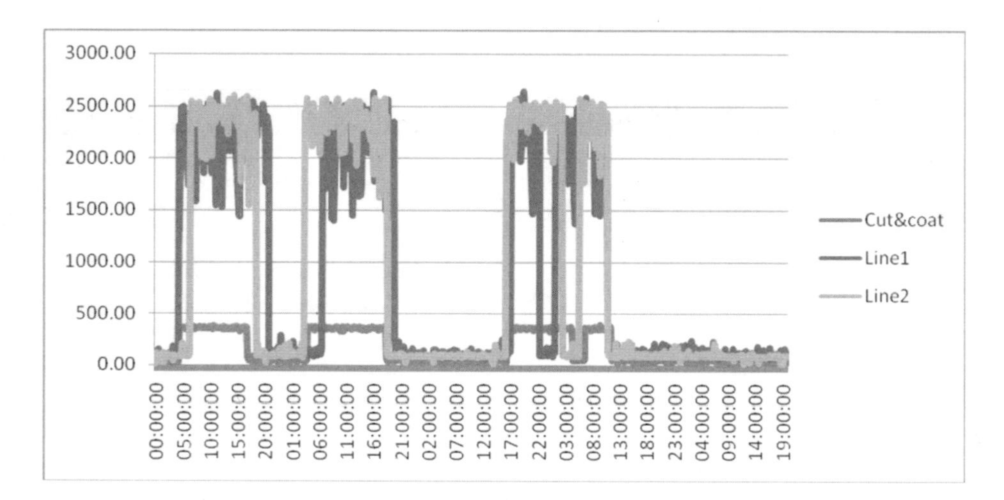

Fig. 4. The total energy consumption for the three production lines resulted from the simulation model

The model validation concluded that the model is able to predict energy consumption and production in satisfied accuracy with discrepancy of -5.61% to 1.62% for energy consumption and 0.86% to 3.06% for production.

After the model validation the author proceeded in defining six simulation scenarios with the intention of lowering the energy consumption. These scenarios have been determined in collaboration with the case study company which decided on the

Table 1. The results of the six simulated scenarios

ID	Scenario	kW/Unit saved	Energy (kW) saved
1	10% increase in energy efficiency of the refiners	3.30%	3.58%
2	10% increase in energy efficiency of the driers	1.74%	1.89%
3	15% increase in energy efficiency of the driers	2.62%	2.89%
4	58 second cycle time on the press	4.10%	0.46%
5	55 second cycle time on the press	9.12%	1.12%
6	New production schedule scenario	0.05%	5.86%
	Total	**20.93%**	**15.80%**

changes in the production process that can be easily achieved. The main areas of interest for the company were changing the production schedule, lowering the press cycle and increasing the energy efficiency of the refiners and driers. All of these scenarios have been separately implemented in the model, resulting in energy savings of 0.46% to 5.86% per month.

After the implementation of the six scenarios in the simulation model the collected data showed that the fifth scenario of decreasing the cycle time on the presses of the two production lines to 55 seconds would represent the most effective way for the case study company to decrease the kw/Unit consumed by 9.12%. Even though the overall energy consumption would be decreased by only 1.12% the simulation model showed an increase in productivity by 11.13%. Another scenario that should not be neglected is the 10% energy savings on the refiners which lowered the overall energy consumption as well as the kW/Unit by over 3%

As a total the implementation of the 6 scenarios would decrease the case study company's energy consumption by over 15%.

4 Conclusion

The study presented above modelled the energy demand in an MDF manufacturing company. For this purpose the author reviewed a case study which demonstrates the benefits of using energy modelling in an MDF plant. This case study has been conducted using spreadsheet simulation and, to the author's knowledge it represents the first effort to predict both heat energy and electricity consumption in an MDF plant. Following this study the author has decided to consider a novel approach of using an industrial simulation tool to model energy consumption coupled with production management in an MDF manufacturing facility.

The conceptual model presented in this paper helped the author in defining and understanding the objectives and the main requirements of the project. It also identified the inputs and outputs, along with the actual content of the model itself.

The model coding included details of the production schedule implementation, the monitoring of the energy consumption and the development of the work in progress and customer demand which influences the warehouse stock. The developed simulation model represents an accurate representation of the production facility in the case study company, allowing the users to analyse the model in detail.

The final point discussed in this paper is the validation and the results of the simu lation project. This showed that the model is able to simulate the energy demand with a discrepancy of -5.61% to 1.62% and the production with a discrepancy of 0.86% to 3.06%. Further simulation scenarios resulted in the lowering of energy consumption of 0.46% to 5.86% per month.

The developed model has a great potential in predicting the energy consumption in an MDF manufacturing facility with the scope of improving the kilowatt per unit ratio.

References

1. Li, J., Pang, S.: Modelling of Energy Demand in an MDF Plant. In: Conference Proceedings of CHEMECA 2006: Knowledge and Innovation, Auckland, New Zealand, September 17-20, 6 p. (2006)

2. Carvalho, L.M.H., Costa, M.R.N., Costa, C.A.V.: A global model for the hot pressing of MDF. Wood Science and Technology 37, 241–258 (2003)
3. Li, J., Pang, S.: Modelling of thermal energy demand in MDF production. Forst Products Journal 57(9), 97–104 (2007)
4. ProModel, `http://www.promodel.com`
5. OSIsoft, `http://www.osisoft.com`
6. eSightenergy, `http://www.esightenergy.com`

Three-Dimensional Mesh Generation for Human Heart Model

Dongdong Deng, Junjie Zhang, and Ling Xia

Department of Biomedical Engineering, Zhejiang University, Hangzhou 310027, China
xialing@zju.edu.cn

Abstract. Mesh generation is the precondition of finite element analysis. The quality of the mesh determines the precision of the computational results, and low-quality meshes might lead to incorrect results. Therefore, it is necessary to produce high-quality meshes for finite element analysis. Most commercial software generates meshes on the basis of the entity of an object, while seldom uses the discrete point data to produce meshes directly. Furthermore, the compatibility problem among different software always slows down the progress of research. This paper aims at producing Constrained Delaunay Tetrahedral meshes for human heart anatomy model with TetGen.

Keywords: Mesh generation, TetGen, Human Heart model, Constrained Delaunay Tetrahedralization.

1 Introduction

Heart plays an important role in the human body, lots of people died of cardiac asynchrony and the disorder of the mechanical properties each year. Therefore, understanding the mechanism of various cardiac disease becomes extremely important for the prevention and treatment of heart disease [1].

Heart modeling and simulation is an effective means of studying the cardiac problems. Current heart modeling can be divided into three levels: sub-cellular and cellular level models [2], tissue and organ level models [3] and the complex models level [4]. At the level of tissue and organ modeling, the finite element method (FEM) is usually used to study the cardiac mechanical properties [5], which requires mesh generation in the first place. The quality of mesh determines the accuracy of calculation results: poor grid may lead to erroneous results. Thus, it is necessary to design a good grid. There are many software for mesh generation, but there are compatibility problems among different software, such as unmatched data format. Therefore, we are more concerned about designing an automatic mesh generation program that meets the computational accuracy for cardiac mechanical modeling.

Automatic meshing generation has been the hotspot in the research of FEM, and a lot of different algorithms was put forth, such as mapping methods [6-8], grid-based method [9] etc. This paper mainly aims at the constrained Delaunay tetrahedron method.

K. Li et al. (Eds.): LSMS/ICSEE 2010, Part II, CCIS 98, pp. 157–162, 2010.

2 Materials and Method

2.1 Heart Anatomic Model

The heart anatomic model used in this article was taken from a healthy adult male in Zhujiang Hospital, Southern Medical University, P.R. China. It was scanned using a spiral computer tomography (Philips / Brilliance 64). The size of raw computer tomography data were 512×512 with the spatial resolution 0.3574mm×0.3574mm× 0.33mm. After data acquisition, the anatomic model is reconstructed using the software ScanIP (Simpleware Inc.) .

2.2 Constrained Delaunay Tetrahedron Meshing

A constrained tetrahedralization is a decomposition of a three-dimensional domain Ω into a tetrahedral mesh, such that the boundary $\partial\Omega$ is presented in the faces of the mesh. Generally, Ω may be arbitrarily complicated in shape, contain internal boundaries (separating different regions) and holes. Suppose that X is a Piecewise Linear Complex (PLC), t is a tetrahedron formed by the four vertices in X. Given a constraint plane, if two points are located on both sides of the constraint plane, then these two points are not visible, otherwise they are visible. If the circumsphere of t does not contain any point which is visible from the interior of t, then t satisfies the constrained Delaunay condition. It is a constrained Delaunay tetrahedron [10].

Constrained Delaunay tetrahedralization is an ideal structure. It can be well expressed by the space which is surrounded by piecewise linear constraints of surface and internal constraint surface. It not only follows the character of boundaries of piecewise linear complex, but also has good mathematical properties. It satisfies the Delaunay conditions, and suitable for solving many partial differential equations, such as finite element analysis.

Suppose X_0 is the initial piecewise linear complex, then the process step of constrained Delaunay tetrahedral meshing can be summarized as follows:

1. Establish the initial Delaunay tetrahedral meshes D_0 for the vertexes of X_0
2. Restoration the lines belonging to X_0 in D_0 through Inserting points on the missing segment , then revise X_0 to X_1, D_0 to D_1
3. Elimination the degradation factor by fine-tuning or insert new vertexes, then revise X_1 to X_2, D_1 to D_2
4. Using a cavity tetrahedral meshing method to recover the surfaces belong to X_2 in D_2

In step 1, a lot of standard algorithm is available to compute D_0, but it may not meet some of the lines or surfaces in X_0, after step 2, D_1 contains all the segments in

X_1, however, some may not meet the surface in X_1. Then the step 3 was processed to ensure the existence of constrained Delaunay tetrahedral mesh. When all partial degradation factors were excluded, D_2 is the only Delaunay tetrahedral mesh in X_2. Although the existence of constrained Delaunay tetrahedral mesh is guaranteed, still D_2 might not satisfy part of the surfaces in X_2. Step 4 can restore part of the faces in X_2, then the required constraint Delaunay tetrahedral mesh was obtained.

The tetrahedral Meshing is based on grid point data, so some relatively flat grid maybe exist when the point data were not optimized before meshing, and the grid density cannot be controlled, thus affecting the accuracy of finite element analysis. In order to improve the accuracy, it is necessary to optimize the grid.

The ratio of radius - edge length is commonly used to measure the quality of the grid, it was proposed by Miller et al's [11]. R is the circumradius of tetrahedron, L is the shortest side length of tetrahedron, and Q is the ratio of radius - edge length of the tetrahedron, Q = R / L. The smaller of the value of Q, the more suitable the mesh is for the finite element analysis.

In order to save time in the modeling of cardiac mechanics, we used the functions provided by TetGen (http://tetgen.berlios.de/) to carry out the mesh generation.

3 Results

The human atria and ventricles were meshed using the constrained Delaunay tetrahedron method. And the results were shown in Fig. 1 and Fig. 2. The points of input surface mesh of atria were 25428, the surface polygon was 8476, and the segments were 12714. After tetrahedron meshing, the points of atria were 7962, the tetrahedron were 26679, triangulated surfaces were 61332, subsurfaces were 15948, and the segments were 16450. The points of input surface mesh of ventricle were 46830, the surface polygon was 15610, and the segments were 23415, after tetrahedron meshing, the points of ventricle were 12583, the tetrahedron was 42375, triangulated surfaces were 97306, subsurfaces were 25158, and the segments were 28189.

From Fig. 1 (a) it is clear that the size and shape of the tetrahedron were not uniform: the tetrahedron in pulmonary veins, superior vena cava and inferior vena cava were slender, the tetrahedron size in frontal and lateral wall of atria was larger and shape was more likely equilateral triangle. The inhomogeneities in the internal atrium were shown in Fig. 1 (c). As for the atrial appendage, it is thicker than atrial wall, the tetrahedron was larger and the shape was more slender.

From Fig.1 and Fig. 2, we can see that the tetrahedron in ventricle surface was more homogenous than that of the atria, because the shape of ventricle is more regular than the atria, but the shape of tetrahedron in the apex of ventricle was very slender and the size was not uniform. In the left ventricular internal, the shape of tetrahedron was too slender and inhomogeneous, so that they are not suitable for simulation.

In order to get high quality tetrahedron, we optimized the meshes using ratio of radius - edge length. In our simulation, we set the Q as 1.4 and the max tetrahedron volume as 1.2. Fig. 1 and Fig. 2 show the optimized meshing results. After optimization, the points of atria were 13452, the tetrahedron was 44212, triangulated surfaces were 101431, subsurfaces were 26014, and the segments were 21294. The points of ventricle were 40333, the tetrahedron were 191220, triangulated surfaces were 402474, subsurfaces were 40122, and the segments were 35310.

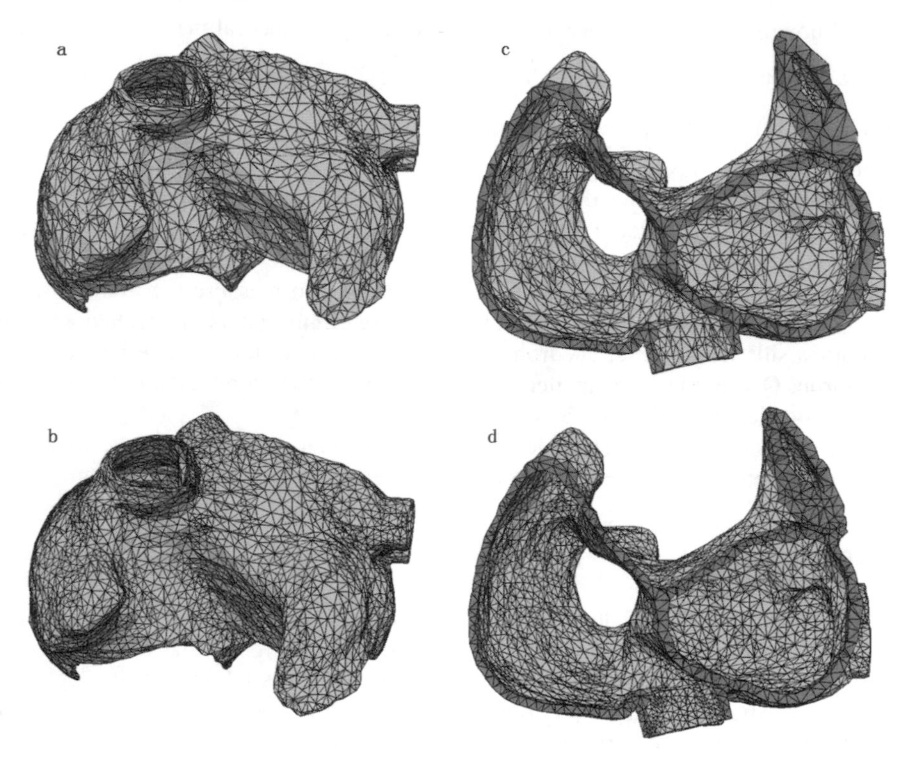

Fig. 1. The tetrahedron meshes of atria, (a) the tetrahedron mesh before optimizing; (b) the mesh after optimizing; (c) the sectional view of atria before optimizing; (d) the sectional view after optimizing

After the optimization, the shape of tetrahedron was more uniform in both atria and ventricles, the slender shape tetrahedron was reduced a lot, and the apex of the ventricle was well meshed, the internal wall of ventricle was uniformly meshed and the slender tetrahedrons disappear.

4 Discussion

Tetrahedral mesh has the character of simplicity, flexibility and strong adaptation for complex boundary. It has become a commonly used unit in finite element analysis, but there are some problems in tetrahedral meshing. The amount of generated triangles is too large. For a general object, there are approximately tens of thousands of triangles, or even one hundred thousand triangles, it is difficult to achieve rotation, translation, scaling, etc, and in the subsequent calculations, it will consume more CPU time, so the auto-adaptive meshing method is our future research priorities..

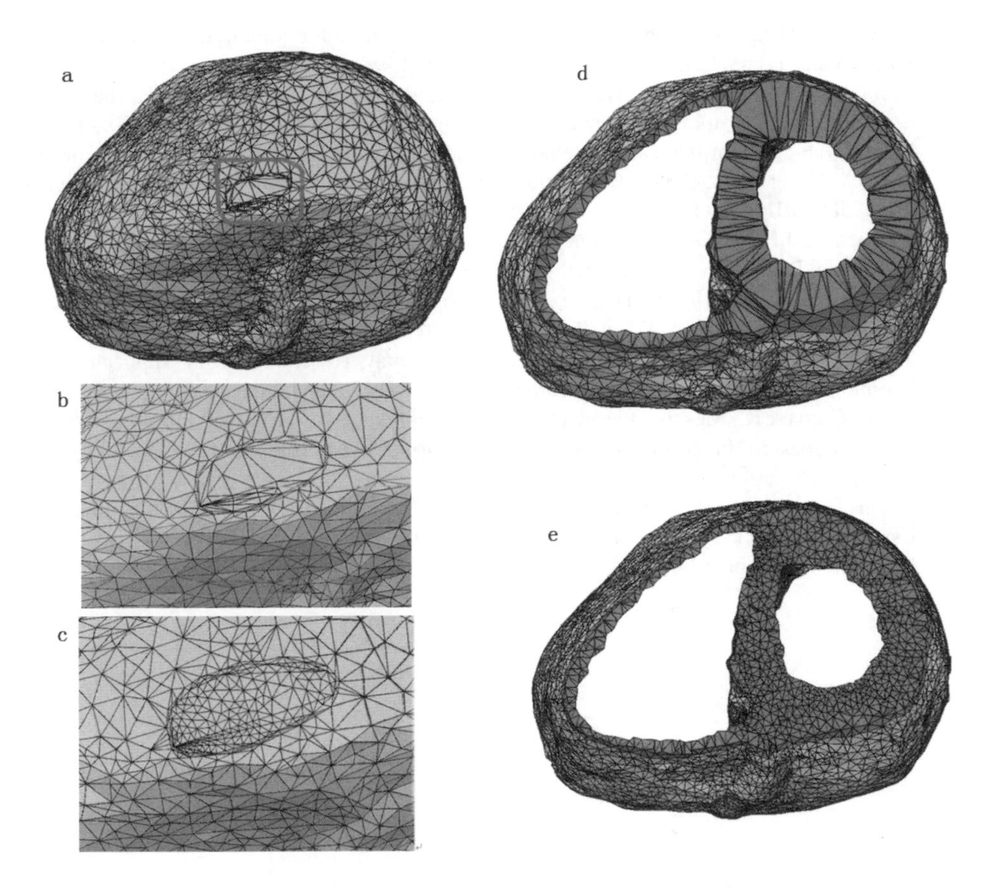

Fig. 2. The tetrahedron meshes of ventricle, (a) the tetrahedron mesh before optimizing, (b) and (c) the amplification of the red rectangle surrounded part in (a), (b) is before optimizing and (c) is after optimizing; (d) the sectional view of ventricle before optimizing; (e) the sectional view after optimizing

Acknowledgments. This work is supported in part by the 973 National Key Basic Research & Development Program (2007CB512100) of China

References

1. Xia, L., Huo, M., Wei, Q., Liu, F., Crozier, S.: Analysis of Cardiac Ventricular Wall Motion Based on a Three-Dimensional Electromechanical Biventricular Model. Phys. Med. Biol. 50(8), 1901–1917 (2005)
2. Rudy, Y., Silva, J.R.: Computational Biology in the Study of Cardiac Ion Channels and Cell Electrophysiology. Q. Rev. Biophys. 39(1), 57–116 (2006)
3. Winslow, R.L., Scollan, D.F., Holmes, A., Yung, C.K., Zhang, J., Jafri, M.S.: Electrophysiological Modeling of Cardiac Ventricular Function: From Cell to Organ. Annu. Rev. Biomed. Eng. 2, 119–155 (2000)

4. Ackerman, M.J.: The Visible Human Project: A Resource for Anatomical Visualization. Stud. Health Technol. Inform. 52 Pt. 2, 1030–1032 (1998)
5. Crampin, E.J., Halstead, M., Hunter, P., Nielsen, P., Noble, D., Smith, N., Tawhai, M.: Computational Physiology and the Physiome Project. Exp. Physiol. 89(1), 1–26 (2004)
6. Brown, P.R.: A Non-Interactive Method for the Automatic Generation of Finite Element Meshes Using the Schwarz-Christoffel Transformation. Comput. Meth. Appl. Mech. Eng. 25(1), 101–126 (1981)
7. Thompson, J.F.: A General Three-Dimensional Elliptic Grid Generation System on a Composite Block Structure. Comput. Meth. Appl. Mech. Eng. 64(1-3), 377–411 (1987)
8. Kadivar, M.H., Sharifi, H.: Double Mapping of Isoparametric Mesh Generation. Computers & Structures 59(3), 471–477 (1996)
9. Watson, D.F.: Computing the N-Dimensional Delaunay Tessellation with Application to Voronoi Polytopes. The Computer Journal 24(2), 167–172 (1981)
10. Si, H., Gärtner, K.: Meshing Piecewise Linear Complexes by Constrained Delaunay Tetrahedralizations. In: Proceeding of the 14th International Meshing Roundtable, pp. 147–163 (2005)
11. Miller, G.L., Talmor, D., Teng, S.-H., Walkington, N., Wang, H.: Control Volume Meshes Using Sphere Packing: Generation, Refinement and Coarsening. In: Proceeding of the 5th International Meshing Roundtable, pp. 47–62 (1996)

Interactive Identification Method for Box-Jenkins Models*

Li Xie, Huizhong Yang, and Feng Ding

Control Science and Engineering Research Center, Jiangnan University,
Wuxi, P.R. China 214122
xieli2412@126.com, yhz@jiangnan.edu.cn, dingf@mail.tsinghua.edu.cn

Abstract. This paper converts a Box-Jenkins model into two identification submodels with the system model parameters and the noise model parameters, respectively. However, the information vectors in the submodels contain unmeasurable variables, which leads the conventional recursive least squares algorithm impossible to generate the parameter estimates. In order to overcome this difficulty, the interactive least squares algorithm is derived by using the auxiliary model identification idea and the hierarchical identification principle. The simulation results indicate that the proposed algorithm has less computational burden and more accurate parameter estimation compared with the auxiliary model based recursive generalized extended least squares algorithm.

Keywords: Parameter estimation, interactive, Box-Jenkins models, auxiliary model, hierarchical identification.

1 Introduction

System identification in the stochastic framework has received much attention and many new approaches have emerged in the past few decades [1,2]. For example, Wang and Ding presented a filtering based recursive least squares algorithm for controlled autoregressive autoregressive moving average (CARARMA) systems [3]; Han and Ding proposed a multi-innovation stochastic gradient algorithm for multi input multi-output systems [4]; Wang et al. considers identification for Hammerstein output error moving average systems using the auxiliary model identification idea [5]; Han et al. developed a hierarchical least squares based iterative algorithm for multivariable systems with moving average noises [6].

The Box-Jenkins models are widely used for prediction and control [7], which can be viewed as a class of general stochastic models, including the controlled autoregressive models, output error models, output error moving average models and so on as their special cases. For the Box-Jenkins models, the bias compensation least squares algorithm can provide unbiased system model parameter

* This work was supported by the National Natural Science Foundation of China (No. 60674092).

K. Li et al. (Eds.): LSMS/ICSEE 2010, Part II, CCIS 98, pp. 163–169, 2010.

estimates, but it fails to estimate the noise model parameters [8]. Furthermore, Liu et al. developed a least squares based iterative algorithm, which can produce more accurate estimates compared with the recursive approaches, but the limitation was that it required large computational load [9].

Recently, Xiao et al. derived a residual based interactive least squares identification algorithm for controlled autoregressive moving average (CARMA) models [10]; Wang et al. proposed a residual based interactive stochastic gradient algorithm for controlled moving average (CMA) models [11]. This paper extends the interactive estimation idea in [10] and [11] to study identification problems of the Box-Jenkins model, the basic idea is to convert such a system into two identification submodels; and then use the auxiliary model based identification idea [5,12] and the hierarchical identification principle [6] to interactively estimate the system model parameters and the noise model parameters.

The rest of the paper is organized as follows. In Section 2, we present the interactive least squares algorithm for Box-Jenkins models, and compare the computational efficiency of which with the auxiliary model based recursive generalized extended least squares algorithm. In Section 3, we present an simulation example to illustrate the results in this paper, and give some concluding remarks in Section 4.

2 The Interactive LS Algorithm

Consider parameter identification problems for the Box-Jenkins model,

$$y(t) = \frac{B(z)}{A(z)} u(t) + \frac{D(z)}{C(z)} v(t). \tag{1}$$

where $\{u(t)\}$ and $\{y(t)\}$ are the system input and output, respectively, $\{v(t)\}$ is an uncorrelated white noise with zero mean and variance σ^2, $A(z)$, $B(z)$, $C(z)$ and $D(z)$ are polynomials in the unit delay operator z^{-1} $[z^{-1}y(t) = y(t-1)]$ with known orders n_a, n_b, n_c and n_d, represented as

$$\begin{aligned}
A(z) &= 1 + a_1 z^{-1} + a_2 z^{-2} + \cdots + a_{n_a} z^{-n_a}, \\
B(z) &= b_1 z^{-1} + b_2 z^{-2} + \cdots + b_{n_b} z^{-n_b}, \\
C(z) &= 1 + c_1 z^{-1} + c_2 z^{-2} + \cdots + c_{n_c} z^{-n_c}, \\
D(z) &= 1 + d_1 z^{-1} + d_2 z^{-2} + \cdots + d_{n_d} z^{-n_d}.
\end{aligned}$$

The objective of this paper is to present an interactive least squares algorithm to estimate the parameters a_i, b_i, c_i and d_i in system (1) by using available input-output data $\{u(t), y(t) : t = 1, 2, 3, \cdots\}$, and test the effectiveness of the proposed algorithm by simulation.

Define two inner variables,

$$x(t) := \frac{B(z)}{A(z)} u(t), \tag{2}$$

$$w(t) := \frac{D(z)}{C(z)} v(t), \tag{3}$$

the Box-Jenkins model in (1) can be written as

$$y(t) = x(t) + w(t). \tag{4}$$

Define the parameter vectors $\boldsymbol{\theta}_{\rm s}$ and $\boldsymbol{\theta}_{\rm n}$ of the system model and the noise model, respectively, and the information vectors $\boldsymbol{\varphi}_{\rm s}(t)$ and $\boldsymbol{\varphi}_{\rm n}(t)$ as

$$\begin{aligned}
\boldsymbol{\theta}_{\rm s} &:= [a_1, a_2, \cdots, a_{n_a}, b_1, b_2, \cdots, b_{n_b}]^{\rm T} \in \mathbb{R}^{n_a+n_b}, \\
\boldsymbol{\theta}_{\rm n} &:= [c_1, c_2, \cdots, c_{n_c}, d_1, d_2, \cdots, d_{n_d}]^{\rm T} \in \mathbb{R}^{n_c+n_d}, \\
\boldsymbol{\varphi}_{\rm s}(t) &:= [-x(t-1), -x(t-2), \cdots, -x(t-n_a), \\
&\quad u(t-1), u(t-2), \cdots, u(t-n_b)]^{\rm T} \in \mathbb{R}^{n_a+n_b}, \\
\boldsymbol{\varphi}_{\rm n}(t) &:= [-w(t-1), -w(t-2), \cdots, -w(t-n_c), \\
&\quad v(t-1), v(t-2), \cdots, v(t-n_d)]^{\rm T} \in \mathbb{R}^{n_c+n_d}.
\end{aligned}$$

Using the above definitions, Equations (2)-(4) can be written as,

$$x(t) = \boldsymbol{\varphi}_{\rm s}^{\rm T}(t)\boldsymbol{\theta}_{\rm s}, \tag{5}$$

$$w(t) = \boldsymbol{\varphi}_{\rm n}^{\rm T}(t)\boldsymbol{\theta}_{\rm n} + v(t), \tag{6}$$

$$y(t) = \boldsymbol{\varphi}_{\rm s}^{\rm T}(t)\boldsymbol{\theta}_{\rm s} + \boldsymbol{\varphi}_{\rm n}^{\rm T}(t)\boldsymbol{\theta}_{\rm n} + v(t). \tag{7}$$

Define

$$\overline{y}(t) := y(t) - \boldsymbol{\varphi}_{\rm n}^{\rm T}(t)\boldsymbol{\theta}_{\rm n}. \tag{8}$$

Equation (7) is equivalently written as

$$\overline{y}(t) = \boldsymbol{\varphi}_{\rm s}^{\rm T}(t)\boldsymbol{\theta}_{\rm s} + v(t). \tag{9}$$

Thus we obtain two identification submodels (9) and (6), one including the system model parameters $\boldsymbol{\theta}_{\rm s}$ and the other including the noise model parameters $\boldsymbol{\theta}_{\rm n}$. However, a difficulty is arising here, because the information vectors $\boldsymbol{\varphi}_{\rm n}(t)$ and $\boldsymbol{\varphi}_{\rm s}(t)$ contain unmeasurable variables $x(t)$, $w(t)$, $\overline{y}(t)$ and $v(t)$, which leads the conventional recursive least squares algorithm impossible to estimate $\boldsymbol{\theta}_{\rm s}$ and $\boldsymbol{\theta}_{\rm n}$. In order to overcome this difficulty, this paper is to study an interactive least squares algorithm by applying the auxiliary model identification ideal and the hierarchical identification principle.

First of all, using the auxiliary model identification ideal to construct the following auxiliary model,

$$x_a(t) = \boldsymbol{\varphi}_a^{\rm T}(t)\boldsymbol{\theta}_a, \tag{10}$$

where $\boldsymbol{\varphi}_a(t)$ and $\boldsymbol{\theta}_a$ are the information vector and parameter vector of the auxiliary model, respectively. Replace the unmeasurable variables $x(t-i)$ in $\boldsymbol{\varphi}_{\rm s}(t)$ with $x_a(t)$ and define

$$\begin{aligned}
\hat{\boldsymbol{\varphi}}_{\rm s}(t) &:= [-x_a(t-1), -x_a(t-2), \cdots, -x_a(t-n_a), \\
&\quad u(t-1), u(t-2), \cdots, u(t-n_b)]^{\rm T} \in \mathbb{R}^{n_a+n_b}.
\end{aligned} \tag{11}$$

Use $\hat{\boldsymbol{\varphi}}_{\rm s}(t)$ and $\hat{\boldsymbol{\theta}}_{\rm s}(t)$ to replace $\boldsymbol{\varphi}_a(t)$ and $\boldsymbol{\theta}_a$ in (10), $x_a(t)$ can be computed by

$$x_a(t) = \hat{\boldsymbol{\varphi}}_{\rm s}^{\rm T}(t)\hat{\boldsymbol{\theta}}_{\rm s}(t). \tag{12}$$

Replacing the unmeasurable inner variable $x(t)$ in (4) with $x_a(t)$ and using the above formula, the estimate of $w(t)$ can be computed by

$$\hat{w}(t)=y(t)-\hat{\boldsymbol{\varphi}}_{\rm s}^{\rm T}(t)\hat{\boldsymbol{\theta}}_{\rm s}(t). \tag{13}$$

Let $\hat{v}(t-i)$ be the estimate of $v(t-i)$, define

$$\begin{aligned}\hat{\boldsymbol{\varphi}}_{\rm n}(t):=[&-\hat{w}(t-1),-\hat{w}(t-2),\cdots,-\hat{w}(t-n_c),\\ &\hat{v}(t-1),\hat{v}(t-2),\cdots,\hat{v}(t-n_d)]^{\rm T}\in\mathbb{R}^{n_c+n_d}.\end{aligned} \tag{14}$$

Replacing $\boldsymbol{\varphi}_{\rm n}(t)$ and $\boldsymbol{\theta}_{\rm n}$ in (6) with their estimates $\hat{\boldsymbol{\varphi}}_{\rm n}(t)$ and $\hat{\boldsymbol{\theta}}_{\rm n}(t)$, the estimate of $v(t)$ can be computed by

$$\hat{v}(t)=\hat{w}(t)-\hat{\boldsymbol{\varphi}}_{\rm n}^{\rm T}(t)\hat{\boldsymbol{\theta}}_{\rm n}(t). \tag{15}$$

Then, using the hierarchical identification principle to compute the estimate of $\overline{y}(t)$ in (8), i.e., replacing $\boldsymbol{\varphi}_{\rm n}(t)$ with $\hat{\boldsymbol{\varphi}}_{\rm n}(t)$ and $\hat{\boldsymbol{\theta}}_{\rm n}$ with its estimate $\hat{\boldsymbol{\theta}}_{\rm n}(t-1)$, we have

$$\hat{\overline{y}}(t)=y(t)-\hat{\boldsymbol{\varphi}}_{\rm n}^{\rm T}(t)\hat{\boldsymbol{\theta}}_{\rm n}(t-1). \tag{16}$$

Defining and minimizing two quadratic criterion functions,

$$J_1(\boldsymbol{\theta}_{\rm s}):=\sum_{i=1}^{t}[\hat{\overline{y}}(t)-\hat{\boldsymbol{\varphi}}_{\rm s}^{\rm T}(t)\boldsymbol{\theta}_{\rm s}]^2, \tag{17}$$

$$J_2(\boldsymbol{\theta}_{\rm n}):=\sum_{i=1}^{t}[\hat{w}(t)-\hat{\boldsymbol{\varphi}}_{\rm n}^{\rm T}(t)\boldsymbol{\theta}_{\rm n}]^2, \tag{18}$$

gives the following interactive least squares (ILS) algorithm for the Box-Jenkins model:

$$\hat{\boldsymbol{\theta}}_{\rm s}(t)=\hat{\boldsymbol{\theta}}_{\rm s}(t-1)+\boldsymbol{L}_{\rm s}(t)[\hat{\overline{y}}(t)-\hat{\boldsymbol{\varphi}}_{\rm s}^{\rm T}(t)\hat{\boldsymbol{\theta}}_{\rm s}(t-1)], \tag{19}$$

$$\boldsymbol{L}_{\rm s}(t)=\frac{\boldsymbol{P}_{\rm s}(t-1)\hat{\boldsymbol{\varphi}}_{\rm s}(t)}{1+\hat{\boldsymbol{\varphi}}_{\rm s}^{\rm T}(t)\boldsymbol{P}_{\rm s}(t-1)\hat{\boldsymbol{\varphi}}_{\rm s}(t)}, \tag{20}$$

$$\boldsymbol{P}_{\rm s}(t)=[\boldsymbol{I}-\boldsymbol{L}_{\rm s}(t)\hat{\boldsymbol{\varphi}}_{\rm s}(t)]\boldsymbol{P}_{\rm s}(t-1), \tag{21}$$

$$\begin{aligned}\hat{\boldsymbol{\varphi}}_{\rm s}(t)=[&-x_a(t-1),-x_a(t-2),\cdots,-x_a(t-n_a),\\ &u(t-1),u(t-2),\cdots,u(t-n_b)]^{\rm T},\end{aligned} \tag{22}$$

$$x_a(t)=\hat{\boldsymbol{\varphi}}_{\rm s}^{\rm T}(t)\hat{\boldsymbol{\theta}}_{\rm s}(t), \tag{23}$$

$$\hat{\overline{y}}(t)=y(t)-\hat{\boldsymbol{\varphi}}_{\rm n}^{\rm T}(t)\hat{\boldsymbol{\theta}}_{\rm n}(t-1), \tag{24}$$

$$\hat{\boldsymbol{\theta}}_{\rm n}(t)=\hat{\boldsymbol{\theta}}_{\rm n}(t-1)+\boldsymbol{L}_{\rm n}(t)[\hat{w}(t)-\hat{\boldsymbol{\varphi}}_{\rm n}^{\rm T}(t)\hat{\boldsymbol{\theta}}_{\rm n}(t-1)], \tag{25}$$

$$\boldsymbol{L}_{\rm n}(t)=\frac{\boldsymbol{P}_{\rm n}(t-1)\hat{\boldsymbol{\varphi}}_{\rm n}(t)}{1+\hat{\boldsymbol{\varphi}}_{\rm n}^{\rm T}(t)\boldsymbol{P}_{\rm n}(t-1)\hat{\boldsymbol{\varphi}}_{\rm n}(t)}, \tag{26}$$

$$\boldsymbol{P}_{\rm n}(t)=[\boldsymbol{I}-\boldsymbol{L}_{\rm n}(t)\hat{\boldsymbol{\varphi}}_{\rm n}(t)]\boldsymbol{P}_{\rm n}(t-1), \tag{27}$$

$$\begin{aligned}\hat{\boldsymbol{\varphi}}_{\rm n}(t)=[&-\hat{w}(t-1),-\hat{w}(t-2),\cdots,-\hat{w}(t-n_c),\\ &\hat{v}(t-1),\hat{v}(t-2),\cdots,\hat{v}(t-n_d)]^{\rm T},\end{aligned} \tag{28}$$

$$\hat{w}(t)=y(t)-\hat{\boldsymbol{\varphi}}_{\rm s}^{\rm T}(t)\hat{\boldsymbol{\theta}}_{\rm s}(t), \tag{29}$$

$$\hat{v}(t)=\hat{w}(t)-\hat{\boldsymbol{\varphi}}_{\rm n}^{\rm T}(t)\hat{\boldsymbol{\theta}}_{\rm n}(t). \tag{30}$$

The ILS algorithm performs an interactive estimation process. On the one hand, the estimate of the system model parameter vector $\hat{\boldsymbol{\theta}}_{\rm s}(t)$ relies on the estimate of the noise model parameter vector $\hat{\boldsymbol{\theta}}_{\rm n}(t-1)$, see (19) and (24); on the other hand, the estimate of the noise model parameter vector $\hat{\boldsymbol{\theta}}_{\rm n}(t)$ depends on $\hat{\boldsymbol{\theta}}_{\rm s}(t)$, see (25)-(30). The steps of computing the parameter estimates using the ILS algorithm (19)-(30) are listed below:

1. Set $u(t)=1/p_0$, $y(t)=1/p_0$, $\hat{w}(t)=1/p_0$, $\hat{v}(t)=1/p_0$ and $\hat{x}(t)=1/p_0$ for $t\leqslant 0$ and $p_0=10^6$.
2. Let $t=1$, set the initial values $\hat{\boldsymbol{\theta}}_{\rm n}(0)=\mathbf{1}/p_0$, $\hat{\boldsymbol{\theta}}_{\rm s}(0)=\mathbf{1}/p_0$, $\boldsymbol{P}_{\rm n}(0)=p_0\boldsymbol{I}$, $\boldsymbol{P}_{\rm s}(0)=p_0\boldsymbol{I}$, where $\mathbf{1}$ is a column vector with appropriate dimensions whose elements are all 1, $\boldsymbol{I}$ is an identity matrix with appropriate dimensions.
3. Collect the input-output data $u(t)$ and $y(t)$, form $\hat{\boldsymbol{\varphi}}_{\rm n}(t)$ by (28) and compute $\bar{\bar{y}}(t)$ by (24).
4. Form $\hat{\boldsymbol{\varphi}}_{\rm s}(t)$ by (22), compute the gain vector $\boldsymbol{L}_{\rm s}(t)$ by (20), update the parameter estimation $\hat{\boldsymbol{\theta}}_{\rm s}(t)$ by (19).
5. Compute the covariance matrix $\boldsymbol{P}_{\rm s}(t)$ and the auxiliary model output $x_a(t)$ by (21) and (23), respectively.
6. Compute the noise model output $\hat{w}(t)$ by (29) and the gain vector $\boldsymbol{L}_{\rm n}(t)$ by (26), update the parameter estimation $\hat{\boldsymbol{\theta}}_{\rm n}(t)$ by (25).
7. Compute the covariance matrix $\boldsymbol{P}_{\rm n}(t)$ by(27) and the white noise $\hat{v}(t)$ by (30).
8. Increase t by 1 and go to step 3.

By estimating the parameters of the system model and the noise model separately, the ILS algorithm can reduce the dimensions of the involved covariance matrices and require small computational load and enhance the parameter estimation accuracy. To illustrate the advantages of the proposed algorithm, the numbers of multiplications and additions for each computation step of the ILS algorithm and the auxiliary model based recursive generalized extended least squares (AM-RGELS) algorithm [9] are compared in Table 1, where the numbers in the brackets denote the computation loads for a system with $n_a=n_b=n_c=n_d=5$ at each step. From Table 1, it is clear that the ILS algorithm has smaller computational load than the AM-RGELS algorithm, where $n:=n_a+n_b+n_c+n_d$.

Table 1. Comparing computational efficiency

Algorithms	Number of multiplications	Number of additions
AM-RGELS	$2n^2+5n+n_a+n_b$ [910]	$2n^2+3n+n_a+n_b$ [870]
ILS	$2(n_a+n_b)^2+2(n_c+n_d)^2+5n$ [500]	$2(n_a+n_b)^2+2(n_c+n_d)^2+3n+1$ [461]

3 Example

Consider the following Box-Jenkins model,

$$y(t) = \frac{B(z)}{A(z)}u(t) + \frac{D(z)}{C(z)}v(t),$$
$$A(z) = 1 + 1.60z^{-1} + 0.80z^{-2}, \quad B(z) = 0.30z^{-1} - 0.70z^{-2},$$
$$C(z) = 1 - 0.40z^{-1}, \quad D(z) = 1 + 0.50z^{-1},$$
$$\boldsymbol{\vartheta} = [a_1, a_2, b_1, b_2, c_1, d_1]^{\mathrm{T}} = [-1.60, 0.80, 0.30, -0.70, -0.40, 0.50]^{\mathrm{T}}.$$

In simulation, the input $\{u(t)\}$ is taken as a persistent excitation signal sequence with zero mean and unit variance, and $\{v(t)\}$ as a white noise sequence with zero mean and variance σ^2. Applying the AM-RGELS algorithm and the ILS algorithm to estimate the parameters of this system, the parameter estimates and their errors $\delta := \|\hat{\boldsymbol{\vartheta}}(t) - \boldsymbol{\vartheta}\|/\|\boldsymbol{\vartheta}\|$ with different variances are shown in Tables 2 to 3. When $\sigma^2 = 0.30^2$ and $\sigma^2 = 0.50^2$, the corresponding noise-to-signal ratios are 25.42% and 42.37%, respectively.

From Tables 2-3, we can see that the ILS algorithm can provide more accurate parameter estimates than the AM-RGELS algorithm for the same data length; and the lower the noise level is, the higher the estimation accuracies are.

Table 2. The parameter estimates and errors ($\sigma^2 = 0.30^2$)

Algorithm	t	a_1	a_2	b_1	b_2	c_1	d_1	δ (%)
AM-RGELS	100	-1.60808	0.83152	0.29933	-0.68904	-0.61668	0.41159	11.55519
	200	-1.61266	0.81669	0.29055	-0.67041	-0.54426	0.46614	7.46686
	2000	-1.59977	0.80240	0.30722	-0.70139	-0.46144	0.47924	3.19085
	3000	-1.60295	0.80444	0.30655	-0.70119	-0.43588	0.48994	1.86759
ILS	100	-1.58683	0.81153	0.31303	-0.70762	-0.50012	0.42678	6.16376
	200	-1.60204	0.80316	0.29353	-0.68508	-0.44408	0.52978	2.72375
	2000	-1.59869	0.80092	0.30738	-0.70329	-0.43158	0.50101	1.59521
	3000	-1.60220	0.80332	0.30673	-0.70242	-0.41129	0.50905	0.81200
True values		-1.60000	0.80000	0.30000	-0.70000	-0.40000	0.50000	

Table 3. The parameter estimates and errors ($\sigma^2 = 0.50^2$)

Algorithm	t	a_1	a_2	b_1	b_2	c_1	d_1	δ (%)
AM-RGELS	100	-1.59088	0.81877	0.30947	-0.64868	-0.68020	0.37538	15.23095
	200	-1.62840	0.83680	0.28827	-0.64112	-0.59611	0.42584	10.89378
	2000	-1.59992	0.80424	0.31203	-0.70180	-0.47915	0.46553	4.26427
	3000	-1.60522	0.80777	0.31090	-0.70179	-0.45055	0.47775	2.78958
ILS	100	-1.58472	0.81331	0.32302	-0.65533	-0.63030	0.41217	12.32897
	200	-1.62498	0.83447	0.29391	-0.64756	-0.56614	0.44658	9.14661
	2000	-1.59897	0.80319	0.31310	-0.70346	-0.46901	0.47245	3.69340
	3000	-1.60463	0.80701	0.31203	-0.70296	-0.44221	0.48387	2.32547
True values		-1.60000	0.80000	0.30000	-0.70000	-0.40000	0.50000	

4 Conclusions

This paper derives an interactive least squares algorithm for the Box-Jenkins model by using the auxiliary model identification idea and the hierarchical identification principle. Compared with the AM-RGELS algorithm, the proposed algorithm has small computational load and can provide high parameter estimation accuracy.

References

1. Eng, F., Gustafsson, F.: Identification with stochastic sampling time jitter. Automatical 44, 637–646 (2008)
2. Liu, X.G., Lu, J.: Least squares based iterative identification for a class of multirate systems. Automatica 46, 549–554 (2010)
3. Wang, D.Q., Ding, F.: Input-output data filtering based recursive least squares identification for CARARMA systems. Digital Signal Processing 20, 991–999 (2010)
4. Han, L.L., Ding, F.: Multi-innovation stochastic gradient algorithms for multi-input multi-output systems. Digital Signal Processing 19, 545–554 (2009)
5. Wang, D.Q., Chu, Y.Y., Ding, F.: Auxiliary model-based RELS and MI-ELS algorithm for Hammerstein OEMA systems. Computers & Mathematics with Applications 59, 3092–3098 (2010)
6. Han, H.Q., Xie, L., Liu, X.G., Ding, F.: Hierarchical least squares based iterative identification for multivariable systems with moving average noises. Mathematical and Computer Modelling 51, 1213–1220 (2010)
7. Tufa, L.D., Ramasamy, M., Patwardhan, S.C., Shuhaimi, M.: Development of Box-Jenkins type time series models by combining conventional and orthonormal basis filter approaches. Journal of Process Control 20, 108–120 (2010)
8. Zheng, W.X.: Least squares identification of a class of multivariable systems with correlated disturbances. Journal of Franklin Institute 336, 1309–1324 (1999)
9. Liu, Y.J., Wang, D.Q., Ding, F.: Least squares based iterative identification for identifying Box-Jenkins models with finite measurement data. Digital Signal Processing 20, 1458–1467 (2010)
10. Xiao, Y.S., Zhang, Y., Ding, J., Dai, J.Y.: The residual based interactive least squares algorithms and simulation studies. Computers and Mathematics with Applications 58, 1190–1197 (2009)
11. Wang, L.Y., Xie, L., Wang, X.F.: The residual based interactive stochastic gradient algorithms for controlled moving average models. Applied Mathematics and Computation 211, 442–449 (2009)
12. Ding, F., Liu, P.X., Liu, G.J.: Auxiliary model based multi-innovation extended stochastic gradient parameter estimation with colored measurement noises. Signal Processing 89, 1883–1890 (2009)

Research on Nano-repositioning of Atomic Force Microscopy Based on Nano-manipulation

Sun Xin* and Jin Xiaoping

School of Mechatronics Engineering and Automation, Shanghai University, yanchang road, 149 200072 Shanghai, P.R. China

Abstract. Nano-manipulation technology is an emerging field in the development of modern science and technology. Thus, the improvement of its positioning and repositioning precision has become each nano worker's dream and ultimate goal. However, due to the hysteresis, creep, and other nonlinearity of piezoelectric ceramics tube (PZT) as well as the probe's tip deviations caused by cantilever deformation, it leads larger error of relative displacement between probe and sample, which adds enormous inconvenience to the nano-manipulation and repositioning. The subject is to research and design a 3-D repositioning control technology to improve repositioning accuracy.

Keywords: Atomic Force Microscopy, Nano-manipulation, Nano-repositioning, neural network, genetic algorithm.

1 Introduction

Since the Nano-technology has been developing so far, there have occurred several reposition technologies of scanning probe microscope, some of them still in common use. These technologies more or less has certain shortcomings due to varied application fields as well as differential instrument utility. These reposition technologies are approximately classified into three: reposition based on marked point、reposition based on composition of scanning probe microscope、reposition based on template matching.

For the sake of improving precision in reposition of scanning probe microscope, many experts or scientists aim at platform mechanical structure designing as a result, nano-positioning precision has been improved greatly. Given these improvements, influence there still be many disadvantages existing in nano-repositioning due to uncertain what scanning probe microscope's manipulation and driving affected by slow-moving、nonlinearity、excursion、external environment and so forth bring about, so nano-repositioning is more vulnerable to affection than nano-positioning. Regard this fact as the prerequisite, this thesis study how to model the nano-repositioning system as well as apply control arithmetic and compensation strategy into the system aim to improve XY dimensional nano-repositioning precision of scanning probe microscope efficiently.

---

* This research is supported by program of Shanghai Science and Technology Commission (No.0852nm06800).

K. Li et al. (Eds.): LSMS/ICSEE 2010, Part II, CCIS 98, pp. 170–176, 2010.

2 Modeling of AFM's Control System

Due to be characteristics of piezoelectric ceramic materials, whether it is open-loop control or closed loop control with sophisticated sensors, the AFM table are indispensable from studying the precise mathematical model. However, due to the piezoelectric ceramic has a hysteresis, creep and other non-linear natural features, to obtain a more accurate model of it is quite difficult.

Static BP is applied in this section to identify and adjust on-line structure and parameters of our AFM table.

AFM used for this micro-displacement working platform is a single-input single-output system. Table-driven voltage as input, displacement of table as output, using the following formula for the model that

$$y(k+1) = f[y(k),\cdots,y(k-n+1);u(k),\cdots,u(k-m+1)] \quad (1)$$

As Hecht[2] and Hornik[3] had proved, feed-forward BP neural network with a hidden layer of three layers can approximate nonlinear function with any precision. The three parallel BP network are used here.

In order to taking into account the fit accuracy and shorter training time of the network requirements, select the network with a layer of hidden layer only one, but by adjusting the number of hidden layer neurons to determine and select the best group which of identification error and the optimal number of steps.

Because the AFM table of nonlinear hysteresis and creep are more serious, the selection of the initial value of the network is particularly important after a direct bearing on whether the convergence of learning and training time are.

We select L-M BP algorithm for training network.

LM algorithm has the global characteristics of the gradient descent method with Gauss-Newton method has the local convergence. Hidden layer transfer function is taken as tansig function, the output layer purelin function, target error is taken as 0.000015, damping factor productivity is taken as 0.005, set hidden nodes were 6 to 12. Order to train the network, the network output square error as shown in Table 1.

Table 1. The error of BP training with different number of neurons

neurons	6	7	8	9	10	11	12
error	0.1835	0.1750	0.1729	0.1562	0.1509	0.1528	0.1519

Considering the speed and training error, the number of neurons selected 12 for network training. As can be seen from the table, when the number of hidden layer neurons changes from 6 to 12, the training error is close to. Therefore, comprehensive consideration of training error and the training step, the number of neurons selected 9 for network training.

After 23 steps, the training error on the target value is reduced to the target error, the network output mean square error of 0.1519. The average displacement of the value of 112.9nm, the maximum error in 85V of 390.5nm.

In order to reduce post-line identification of time, after repeated training we select a set of optimal weights as initial weights for training and validation for use. The static obtained AFM platform LM BP neural network model parameters are:

Parameters of the hidden layer are,

$$W^H = [16.8000\ 16.8000\ 16.8000\ -16.8000\ -16.8000\ 16.8000\ 16.8000\ -16.8000\ 16.8000\ 16.8000\ 16.8000\ -16.8000]^T$$

$$B_1 = [-16.8000\ -13.7455\ -10.6909\ 7.6364\ 4.5818\ -1.5273\ 1.5273\ -4.5818\ 7.6364\ -10.6909\ 13.7455\ -16.8000]$$

Output layer parameters are,

$$W^O = [-0.4960\ 0.1372\ -0.6378\ -0.6012\ 0.5157\ -0.8605\ 0.5661\ 0.4099\ -0.6229\ -0.9224\ 0.8575\ 0.9165]$$

$$B_2 = -0.0843$$

3 Control Theory and Methods of AFM Table

Traditional control theory and methods, such as classical control theory, modern control theory and adaptive control theory, both require precise control object model, but obtain an accurate model of actual control system is difficult, even impossible, object uncertainties and external disturbances are often not satisfied with the special assumptions. Therefore, the control system must be considered in the case of uncertainty, it still can control the feedback control system stability and to meet the desired requirements. This leads to a special analysis and processing uncertain system control theory - the emergence of modern control theory.

Neural network is a network with topology, can fully approximate any nonlinear function, while the combined neural network and adaptive control are combined to form a neural network adaptive control. Neural network identification and the Mathematical model, adaptive control according to the results of recognition by a law directly regulating the parameters of the controller, the system to meet the established performance indicators. In many practical process control case prove that the control precision, stability has been greatly improved.

3.1 Adaptive PID Control Theory Based on DRNN

Adaptive PID based DRNN control system consists of neural network identifier (DRNI), traditional PID controller and neural network controller(DRNC), as shown in Figure.1.

3.2 Results Analysis

DRNN-based Self-Tuning PID control system has been discussed in previous chapters . DRNI inputs are $\{u(k), y(k-1)\}$, output is the expected output of the current displacement of the AFM. DRNC inputs are $\{r(k), y(k-1), u(k-1)\}$, PID controller output is a 3 factor of K_p、K_i、K_d. DRNI and DRNC weights are modified using the dynamic BP algorithm.

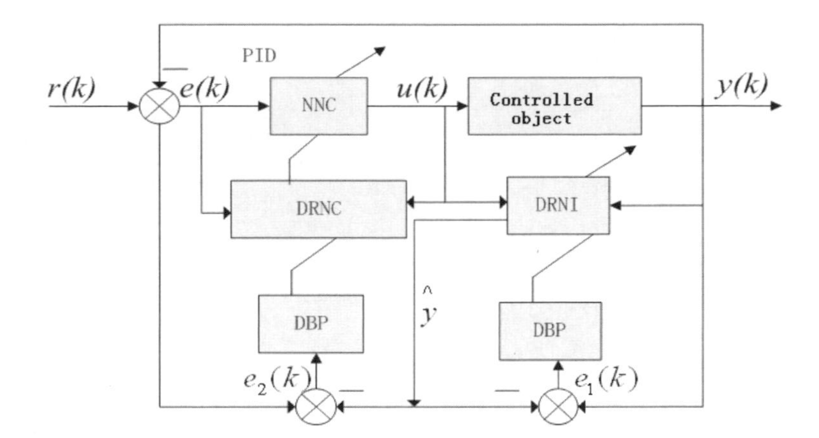

Fig. 1. Adaptive PID control based on DRNN diagram

Beginning of the experiment assigned to the initial weights coefficients of the network in the range $[-1,1]$, select the initial learning rate 0.2, momentum coefficient of 0.8, sampling time set to 1s. At this time of the PID controller output control signal curve shown in Figure.3

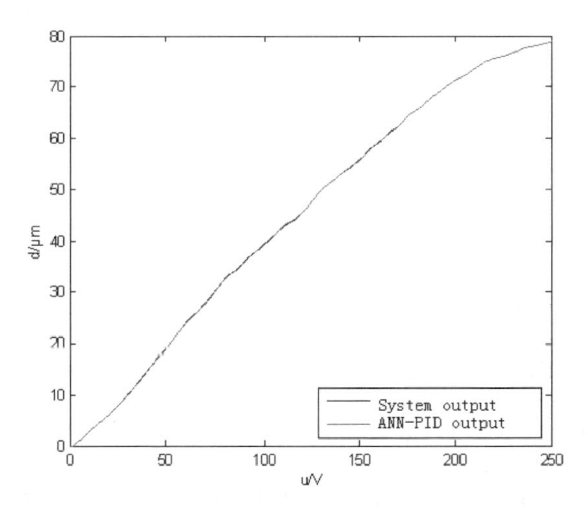

Fig. 2. ANN adaptive PID controller displacement - voltage characteristic curve test

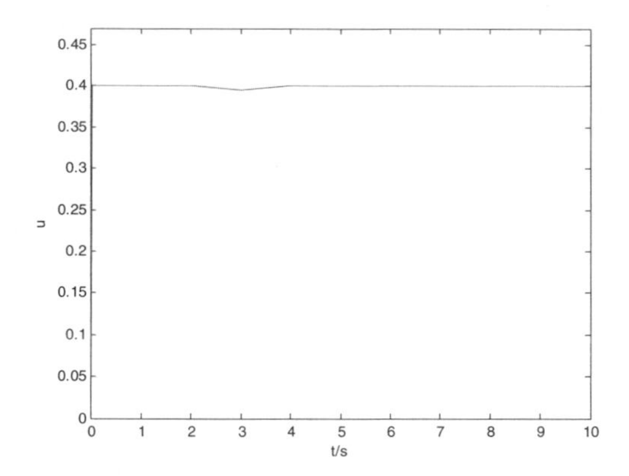

Fig. 3. Output of controller based on dynamic neural network PID

Dynamic neural network based PID tuning shown in Figure as Figure .4,

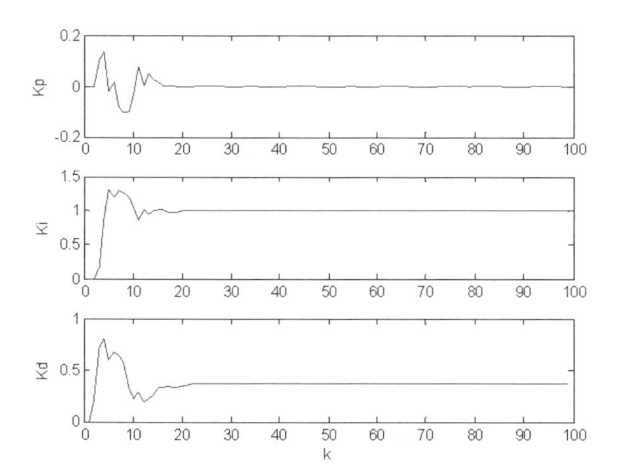

Fig. 4. Tuning coefficient map of controller based on dynamic neural network PID

BP self-tuning PID control of dynamic displacement under the maximum error of 350.6nm and the average positioning accuracy of 87nm.

To compare the experimental results, it is also doing traditional PID and the self-tuning PID control performance compared experiment. In the displacement - voltage characteristics test, displacement maximum error of conventional PID is 838.8nm ,while self-tuning PID test is 347.7nm; average positioning accuracy of the former is 341.4nm, the latter is 114.6nm.

Traditional PID and the self-tuning PID control performance comparison results are shown in below Fig. 5and Fig. 6.

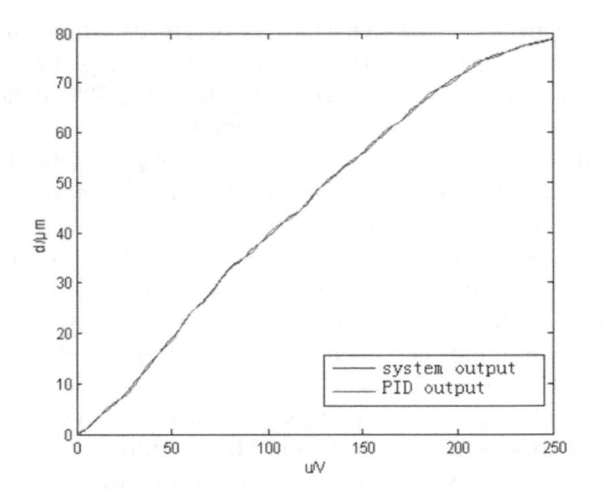

Fig. 5. Output of controller based on conventional PID

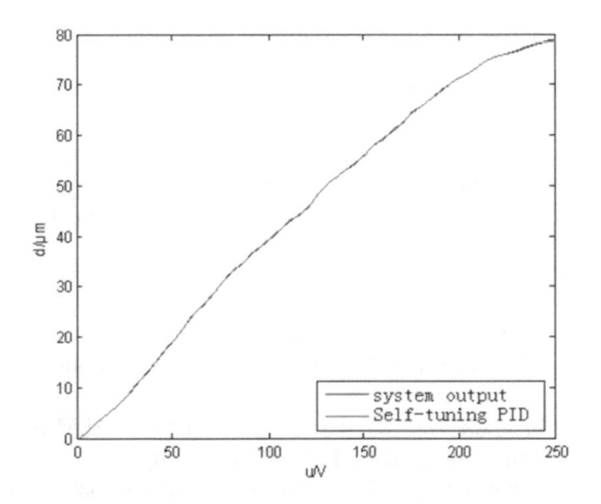

Fig. 6. Output of controller based on Self-tuning PID

4 Conclusion

We adopted self-tuning PID control and neural network based on dynamic self-tuning PID control and carried out experiments on three-dimensional positioning AFM table.

Traditional PID control algorithm is simple and easy to implement, the value of our experience in reference to the basis of trial in the experiments of its parameter setting. When using this method in the displacement of the drive within the tracking control, the maximum error is 830.8nm in the range of 170V-190V, the average positioning accuracy of 341.4nm. This shows that the traditional PID control algorithm for PZT table difficult to achieve satisfactory control performance requirements.

Self-tuning PID controller model on the table of online identification, PID controller is obtained by calculating the device parameters to achieve a model with changes in the drive line adjustment. The results showed that self-tuning PID control to the displacement in the context of the whole tracking control, the maximum displacement error is 347.7nm in the range of 165V-175V, the average positioning accuracy of 114.6nm. So from the experimental data, control results improved and errors are reduced, positioning accuracy is improved.

Dynamic recurrent neural network based self-tuning PID control method in neural networks and self-tuning PID control combines the learning capabilities of neural networks, while PID controller parameters are modified on-line. Similarly, the displacement in the context of the entire table tracking control, the maximum error in the 85V-95V range is 350.6nm, the average positioning accuracy of 87nm. Dynamic recurrent neural network based self-tuning PID controller has also been greatly improved positioning accuracy compared with self-tuning PID controller.

Comprehensive experiments show that neural network based self-tuning PID algorithm able to achieve good performance.

References

1. Shen, F., et al.: Design and test of the nanostructure reaccessible UHVSTM system. Journal of Chinese Electron Microscopy Society 1, 151–156 (1999)
2. Hecht-Nielsen, R.: Theory of the back-propagation neural networks. In: Proceeding of the International Joint Conference on Neural Networks, vol. 1, pp. 593–611 (1989)
3. Hornik, K.M., Stinchcombe, White, H.: Multiplayer feed forward networks are universal approximators. Neural Networks 2, 359–366 (1989)
4. Yong-yi, W., Jian, T.: Neural network control, pp. 1–21. Mechanical Industry Press, Beijing (1998)
5. Xin, S., Da-qin, Y.: The Research of the Scanner X-Y Direction Modeling Method of Piezoelectricity Only and Analysis of Demarcated Precision. Instrumentation Technology 7, 51–53 (2007)
6. Xin, S., Xiao-Ping, J., et al.: 2-D Nano-positioning System for AFM Based on H∞ Control. Journal of Shanghai University (English Edition), Received
7. Maeda, Y.: Study of the nanoscopic deformation of an annealed nafion film by using atomic force microscopy and a patterned substrate. Ultramicroscopy 108, 529–535 (2008)

Research on Expression Method of a Unified Constraint Multi-domain Model for Complex Products

Chen Guojin, Su Shaohui, Gong Youping, and Zhu Miaofen

Hangzhou Dianzi University, Hangzhou, China, 310018
chenguojin@163.com

Abstract. This paper studies the modeling method based on the unified constraint systems for complex products' designing and analyzing. The method describes the complex product's multi-domain optimization in a unified constraint model. The unified constraint expression for the product's multi-domain simulation and optimization model can implemented by the mapping mechanism transferring the physical models into the mathematical models. Aiming at designing and analyzing for the complex mechanical products of the multi-field mixture and the sub-hierarchy, the paper studies the product's associated constraints of the different areas' hierarchical relationship, the constitutive constraints expressing the product areas' constitutive relations, the body constraints describing the relationship between the association and the constitution, and the discrete constraints representing the discrete events to extract the commonality in different areas for these four constraints on the basis of the geometric constraints' representation and the physical systems' modeling. In connection with the commonality in the constraints, the products' model formulation in the various fields is unified in the constraint level based on equations using the various structural elements' constraints based on a unified expression of mathematical equations.

Keywords: constraint, multi-domain, modeling and simulating, homotopy iteration method, Modelica.

1 Introduction

Modern complex physical systems widespread in the aerospace, electrical power, automobile, shipping, engineering machinery and other industries, such as aerospace vehicles, robots, cars, loaders, etc. They are becoming increasingly sophisticated and are usually the synthesis composed of mechanical, electrical, hydraulic, control, and magnetic subsystems in different areas [1-3]. Multi-field coupling and continuous-discrete mixture of complex products are today's notable features, and multi-disciplinary cross is a development trend of the complex product's designing and analyzing. In the 1980s, the American Sobieski [1] first proposed the idea for optimal design of complex systems in many fields (Multidisciplinary Design Optimization, referred to as MDO). It is the method that the complex systems and subsystems are designed through the full exploration and utilization of the interacting coordination mechanism in the engineering systems. Its basic idea is that in solving a complex

K. Li et al. (Eds.): LSMS/ICSEE 2010, Part II, CCIS 98, pp. 177–183, 2010.

system, the system is decomposed into several subsystems belonging to different disciplines, making full use of the various disciplines of knowledge, modeling theory and analysis tools, applying the effective designing optimization strategy, taking full account of the interaction between the various disciplines, through the realization of concurrent designing to shorten the design cycle and to get the whole system's optimal solution, so as to develop the more competitive product. However, the traditional single-field optimization designing theory and methods are difficult to resolve the optimization problems of the complex multi-domain physical systems. Thus, the collaborative optimization theory and methods for the multi-domain physical systems have been gradually turned into the most popular model-based means for the optimization design of the complex systems [4].

2 Constraint Relationship for Complex Multi-domain Product's Model

A unified multi-domain simulation and optimization design is that under the goal of a given system the designing variables, constraint conditions and solving goals of the various areas are expressed by the unified statement equations to obtain the optimal solution of the complex systems [5-6], as shown in Figure 1. For the multi-domain physical systems, the model units of the machinery, electronic, hydraulic, control systems are respectively established, and for the common model units in these areas the reusable units' libraries are set up. Users can build the digital prototypes and the unified multi-domain constrained simulating and optimizing models through the visual interface operation and the model units. And the multi-domain constraint relationship models of complex products are based on the unified models. The product's multi-domain constraint models of simulating and optimizing are uniformly expressed by the mapping implementation mechanism transferring the physical model into the mathematical model. The parameter sensitivity of the unified constraint models is analyzed to reduce the size of constraints. By the principle that the overall comprehensive satisfaction is as the greatest as possible, the multi-field constraints in the component level are coordinated to achieve the compatibility and singularity analysis for the product's constraints. And the large-scale multi-domain simulation and optimization models after processing are solved using the homotopy iteration method. Finally adopting the unified multi-domain modeling language based on Modelica, the solving system of simulation and optimization is expressed by the unified multi-domain constraints.

According to the classification method of the product model, the product constraints are also divided into the product-level constraints, the part-level constraints and the component-level constraints. The sets and relevancies at the same level constitute the constraint networks of the level, such as the product-level constraint networks, the part-level constraint networks and the component-level constraint networks. The sets and relevancies of constraint networks in all levels constitute the overall product models of constraint relationship, as shown in Figure 2. The constraint relations have the associated constraints, the constitutive constraints and the body constraints. The associated constraints express the link between the product structural

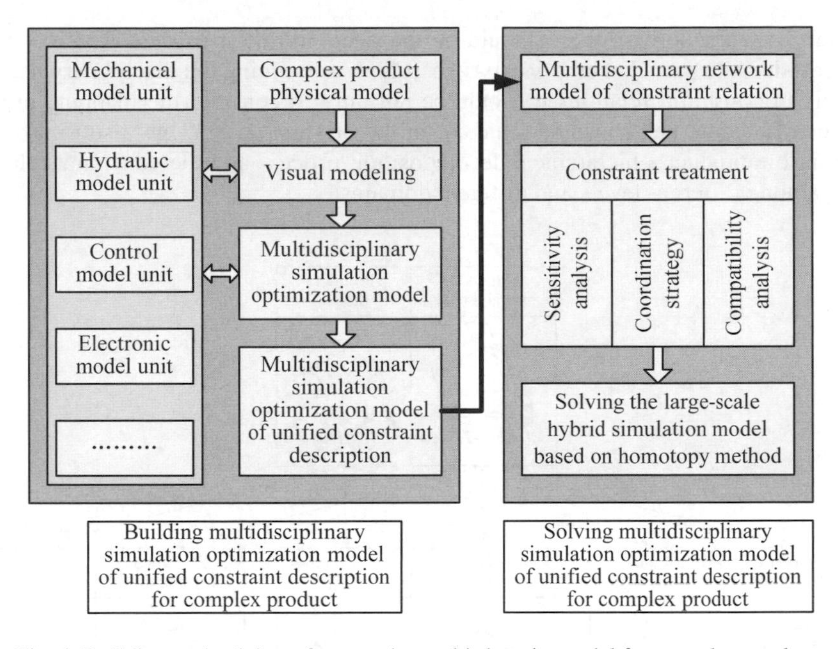

Fig. 1. Building and solving of constraint multi-domain model for complex products

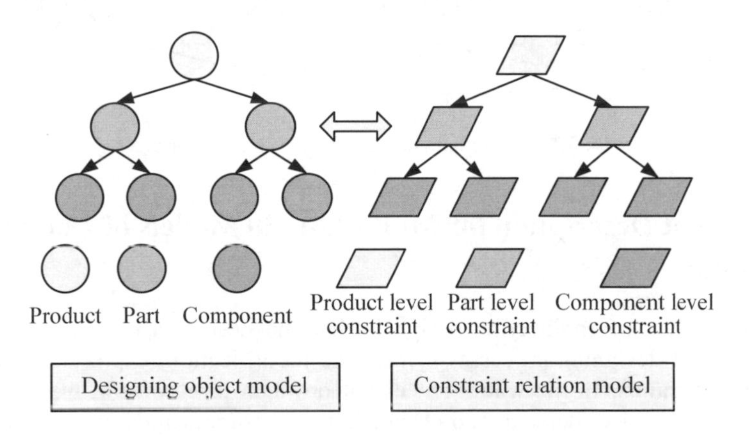

Fig. 2. Overall product models of constraint relationship

elements within a particular area or among different areas in the complex products. The constitutive constraints represent the related physical rules in the product elements' fields. The body constraints establish the relationship between the associated constraints and the constitutive constraints.

The same product is only concerned about different aspects. So there must be the interconnected relationships. The constraint relation networks in all subject areas constitute the multi-domain constraint relation networks for the products, as shown in Figure 3. In each subject area, the product's design undergoes the two-way iterative mapping processes in the functional domain, the behavior domain and the structural domain. In each domain there should be the decomposition process from coarse to fine (products to parts to components). Therefore, in building the disciplinary network models of constraint relations it should be not only to consider the mapping process between function, behavior and structure in the product design, but also to take into account the product's hierarchical decomposition process and the constraint relationships among different levels and different domains.

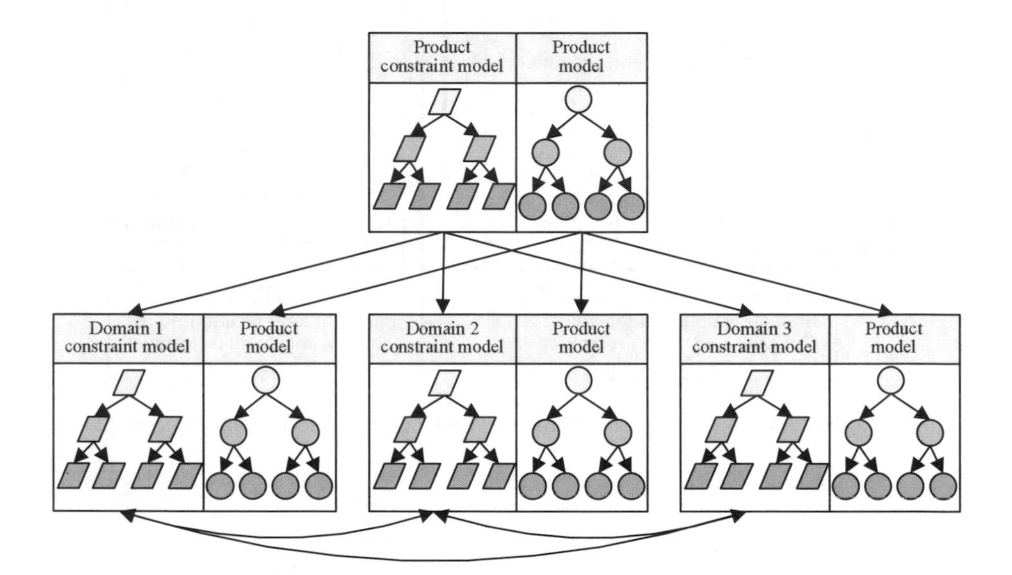

Fig. 3. Multi-domain constraint relation networks

3 Constraint Description for Multi-domain Models of Complex Products

An important idea of engineering design is that the design process is facing constraints. A lot of designing processes involve the identification, expression and satisfaction of constraints. In the whole process of product development, the constraints are continuously added, deleted, and changed. The constraint must be accurately expressed by the systematic and consistent manner. The constraints can be inspected and manipulated in different ways by engineers or computer personnel. Firstly, the constraints can be tested according to the decision-making accuracy, adequacy and integrity. Secondly, the constraint information can help designers to understand the complexity of designing problems. Thirdly, the constraints can be automatically changed. For the product's life-cycle simulation and optimization models based on Modelica, the constraints in the product's forming process should be rationally expressed. In order to building accurately the constraint model, two types of constraints need to be considered, that is, a single constraint and a set of constraints.

3.1 Expression of a Single Constraint

A single constraint has different expression modes, such as domain constraints, equations, inequalities and rules. The domain constraints express the scope of design parameters, while the equations, inequalities, and the rules describe the relationship between the design parameters.

(1) There are two domains, discrete and limited or continuous and unlimited.
(2) Equality constraints and inequality constraints.
(3) Rule-constraints, including the conditions and conclusions.

3.2 Expression for a Group of Constraints

For a group of constraints, there are three different types of modelling methods, as shown in Figure 4.

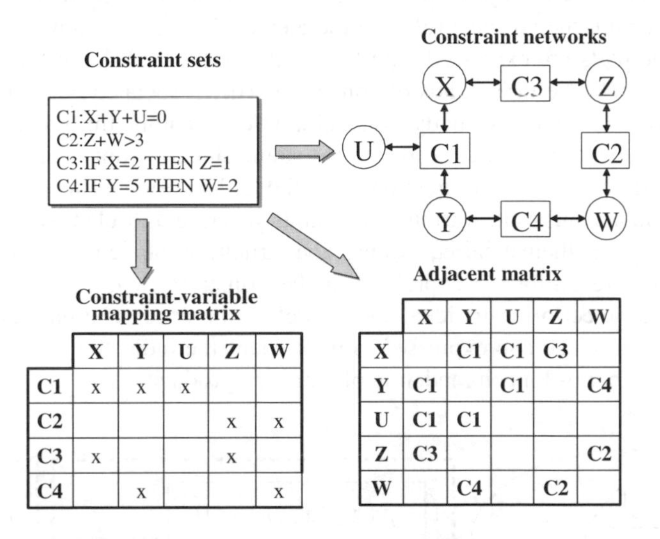

	X	Y	U	Z	W
C1	x	x	x		
C2				x	x
C3	x			x	
C4		x			x

	X	Y	U	Z	W
X		C1	C1	C3	
Y	C1		C1		C4
U	C1	C1			
Z	C3				C2
W		C4		C2	

Fig. 4. Constraint description

(1) Constraint networks.
(2) Constraint-variable mapping matrix.
(3) Adjacent matrix.

The constraint networks are a collection of constraints. Through the shared variables the constraints are interrelated. The constraint networks are very useful to detecting the constraints' transmission and collision. When some values are assigned to a variable set, these values pass through the networks. The values of the input variables in the equality constraints can derive the values of the output variables. The inequality constraints can be used to detect the constraints' collision in the constraints' transmission. The objective of the constraint-variable mapping matrix is to study the relationship between constraints and variables. In the matrix, the rows represent the constraints and the columns represent the variables. Similar to the designing matrix, each element in the matrix is two values, empty or "￥". The element A_{ij} in the "￥" symbol expresses that the variable *j* appears in the constraint *i*. In order to decompose the constraints, the poly family algorithm can be used to decompose a matrix into the independent sub-matrices interrelated.

4 Unified Expression for the Product's Simulation and Optimization Model of Multi-domain Constraints

The product's design and performance analysis must comply with the mechanism of constraint satisfaction. According to the principle, the three elements for expressing the elements and their relationships should be refined by profoundly grasping the structural elements' essential characteristics in different areas and their relationships with the outside world, beyond the physical boundaries of the areas. The three elements are the port, the physical rule and the element's body, as shown in Figure 5. In the figure the ports are expressed with the small circles, and the elements are linked through the ports. The three kinds of constraints (the associated constraints, the constitutive constraints and the body constraints) representing the complex product's design and analysis are reintegrated into the hierarchical common description by using the description based on equations and the object-oriented technology. For the constraint commonality, the various constraints of structural elements are uniformly expressed using mathematical equations. The structural elements' models in various product fields are uniformly expressed in the constraint level based on equations. Adopting the object-oriented idea, the model's tolerance relations and associated relations are organized and expressed to implement the indicative physical modeling and the unified multi-domain modeling of complex products.

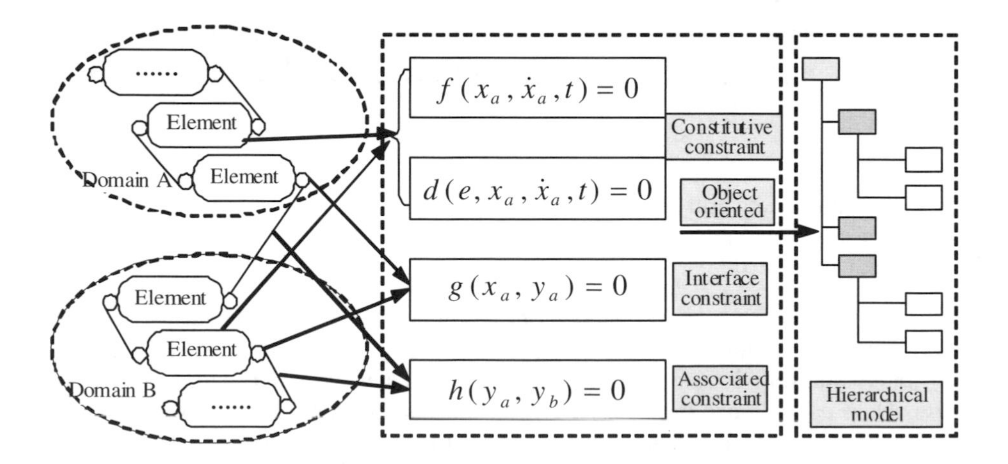

Fig. 5. Unified constraint expression for multi-domain simulation model

5 Conclusion

The same product is only concerned about its different aspects, so there must be the interconnected relationships. The associations among the constraint networks in all subject areas form the constraint relationships for the multi-domain products. In each subject area, the product design undergoes the two-way iterative mapping processes among the functional, behavior and structural domain. It is a key how to map the constraints among the function, behavior and structure in the product's designing process into a unified multi-domain simulation model. Aiming at the designing and analyzing for the complex mechanical products of the mixed multi-field and sub-hierarchy structures, the paper analyzes the associated constraints expressing the product's hierarchies among different areas, the constitutive constraints characterizing the physical constitution relations of the product areas, the body constraints describing the relations between the associated constraints and the constitutive constraints, and the discrete constraints representing the discrete events on the basis of the representation of the geometric constraints and the physical system's modeling. The unified modeling method for the multi-domain physical systems based on the level of the equation's constraints is proposed to lay the foundation for the collaborative optimization of the multi-domain physical systems.

Acknowledgements. This work is supported by National Natural Science Foundation of China (No.60873106 and No.60903087).

References

1. Sobieskij, S.: On the Sensitivity of Complex, Internally Coupled System. AIAA Journal (28), 153–160 (1990)
2. Zeyong, Y., et al.: Study on Multidisciplinary Design Optimization of Aero-engine. Engineering Science (9), 1–10 (2007)
3. Bao-gui, W., Hong-zhong, H., Ye, T.: Functional Digital Prototype-based MDO & Simulation. Journal of System Simulation (19), 861–864 (2007)
4. Elmqvist, H., Mattsson, S.E., Otter, M.: Object-oriented and Hybrid Modeling in Modelica. The Journal of European systemautomatics (35), 1–10 (2001)
5. Yizhong, W., Min, L., Liping, C.: Development of Hybrid Modeling Platform for Multi-domain Physical System. Journal of Computer-Aided Design & Computer Graphics (18), 120–126 (2006)
6. Jian-Wan, D., Li-Ping, C., Fan-Li, Z., Hua, H.: Consistency Analysis of Complex Declarative Simulation Models. Journal of Software (16), 1868–1875 (2005)

An Automatic Collision Avoidance Strategy for Unmanned Surface Vehicles

Wasif Naeem and George W. Irwin

Intelligent Systems and Control
School of Electronics, Electrical Engineering and Computer Science,
Queen's University Belfast, Belfast BT9 5AH, UK
{w.naeem,g.irwin}@qub.ac.uk

Abstract. Unmanned marine vehicles are useful tools for various hydrographical tasks especially when operating for extended periods and in hazardous environments. The autonomy of these vehicles depends on the design of robust navigation, guidance and control systems. This paper concerns the preliminary design of an automatic guidance system for unmanned surface vehicles based on standardised rules defined by the International Maritime Organisation. A guidance system determines "reasonable" and safe actions in order to complete a task at hand. Thus, autonomous guidance can be regarded as the mechanism that brings self-reliance to the whole system. The strategy here is based on way-point guidance by line-of-sight coupled with a manual biasing scheme. Simulation results demonstrate the functioning of the proposed approach for multiple stationary as well as dynamic obstacles.

Keywords: Unmanned surface vehicles, obstacle detection and avoidance, navigation, guidance and control.

1 Introduction

Automatic unmanned systems are an integral part of everyday life, normally employed to perform repetitive chores quickly and efficiently which are too tedious for humans. Most are designed to operate in structured environments where the surroundings do not vary considerably. Developing a fully autonomous system which can work in any unstructured or unpredictable environment is a challenging task that requires robust guidance and control strategies. For unmanned mobile systems operating in fast changing surroundings, such as an automobile, aircraft or a humanoid robot, the automatic guidance system or path planner plays a cental role in bringing autonomy to the whole system. The guidance or mission commands are normally sent through a wired or wireless channel by a remote operator whose responsibility is to constantly oversee the system and act accordingly. The guidance system is thus essential to determine "reasonable" and safe actions that are required to accomplish a mission.

In what follows, a guidance system, or more precisely, an obstacle avoidance system (path planner) is developed for an unmanned (maritime) surface vehicle

K. Li et al. (Eds.): LSMS/ICSEE 2010, Part II, CCIS 98, pp. 184–191, 2010.

or USV. USVs are useful tools in tasks such as oceanography, weapons delivery, environmental monitoring, surveying and mapping. There are several worldwide USV programs both in the defence and civil sectors such as the *Delfim* USV for mapping applications [1] and *Protector* USV [2] for maritime assets protection. Most of these programmes rely on remote operator guidance for sending mission commands and to constantly overlook the vehicle's status either by observation or via a wireless video link [3]. This adds to the operating cost of each mission and is not practical for extended periods. In order to fully benefit from this technology, a reliable obstacle detection and avoidance system is thus mandatory, a fact confirmed by leading researchers and industrialists in the field [3,4]. It is also important that the USV behaves in a manner that is discernable by other ships in the vicinity. This attribute would aid in integrating the USVs with the ambient marine traffic. The coastguard regulations on prevention of collision at sea (COLREGs), defined by the International Maritime Organisation (IMO) [5], can usefully be integrated for this purpose.

The proposed approach employs a simple waypoint by line-of-sight (LOS) guidance strategy coupled with a manual biasing scheme. This is tested in simulations on the USV dynamic model described in Section 3. To this end, several mission waypoints are selected between the USV launching position and the destination. The vehicle is normally guided to stay on the direct LOS route when no obstacles are found. Assuming that a vision-based detection system is present onboard, a bias is added to the current reference heading angle should an object is found posing a threat. This manual bias deviates the course of the USV and thus the obstacle is evaded. Subsequently, the craft is again commanded to follow the direct LOS angle between its current position and the next waypoint. It is demonstrated that the addition of the bias angle generates evasive manoeuvres that satisfy the IMO requirements. Simulation results are presented showing USV trajectories between all the waypoints for the case of multiple stationary as well as a single dynamic obstacle. Section 2 explains the motivation of this research including a brief description of COLREGs. The *Springer* USV dynamic model employed here is briefly described in Section 3. The problem formation is then outlined in Section 4. Section 5 presents simulation results whilst concluding remarks follow in Section 6.

2 Motivation and Background

Recent statistics have shown that 60% of casualties at sea are caused by collisions [6]. It has also been found that human error is a major contributing factor to those incidents. Furthermore, it is reported [7] that 56% of collisions at sea include violation of COLREGs. The infamous *Titanic* tragedy was in fact as a result of the unwillingness of the crew to change the speed of the vessel [8] as required by the rules of obstacle encounter at that time. Although these studies are compiled for manned ships, unmanned vessels without any form of onboard intelligence could even be more vulnerable. A review of related research has revealed that very few USVs are equipped with an onboard detection and avoidance system. In addition, only a handful of research programmes have considered developing COLREGs-based avoidance systems. Examples include those at MIT [9] at MIT using behaviour-based control and the work at the Space and Naval Warfare Systems Center in San Diego [10] employing a voting technique. Another collision avoidance method using fuzzy logic with reference to COLREGs was devised for the vessel traffic service (VTS) [11], but no experimental results were reported. Finally, a simulation study of COLREGs-based automatic collision avoidance for manned vessels at the Universities of Glasgow and Strathclyde [12] employed artificial potential field and speed vector for trajectory planning and collision avoidance.

In the absence of obstacles, the waypoint guidance scheme generally works very well. However, in practice, the real-world is full of unpredictable situations, so it is not possible to leave the unmanned vessel unattended during a mission. This paper introduces a simple, yet effective, technique for obstacle avoidance based on IMO regulations. The IMO rules or COLREGs suggest particular manoeuvres in various obstacle encounter settings. For instance, in the head-on collision scenario presented (Rule 14), both the vessels involved must turn towards their respective starboard sides. Also, Rule 15 defined the crossing situation, which is akin to the right of way rule in the automobile driving regulations.

The difficulty with the COLREGs is that they were written for humans and thus are subjective in nature. For instance, Rule 8(b) states that any change in the vehicle's course should be discernable by the ambient traffic and must not include a series of small changes. For a human captain, say a 28^0 starboard-side manoeuvre is no different than a 30^0 turn as long as it is avoiding the collision. This is clearly not optimal in any sense. Hence an automatic path planning system may also be useful for (the captain of) a manned vessel.

3 *Springer* USV

The *Springer* USV is a catamaran-shaped research vessel which was primarily designed to carry out pollutant tracking and environmental and hydrographic surveys in rivers, reservoirs, inland waterways and coastal waters. It is a low cost vehicle which is also intended to be used as a test bed for researchers involved in environmental data gathering, designing alternative energy sources, sensor and instrumentation technology and control systems engineering.

Each hull of *Springer* is divided into two watertight compartments containing some of the onboard sensors and electronics including battery packs. Pelicases are placed within the bay areas between the two cross beams, as depicted in Figure 1. These house the computers and the remaining onboard electronics and control circuitry. A GPS receiver and wireless router were also installed on the mount shown in Figure 1. The onboard computers are all linked through an *ad-hoc* wireless network providing an external intervention capability in the case of erratic behaviour or simply for monitoring purpose. For the interested reader, the detailed hardware development of the *Springer* USV is described in [13].

Fig. 1. *Springer* USV during trials at Roadford Reservoir, Devon

Springer's steering mechanism is based on differential thrust and the dynamic equations can be manipulated to generate the following single-input single-output state space model:

$$\mathbf{x}(k+1) = \begin{bmatrix} 1.002 & 0 \\ 0 & 0.9945 \end{bmatrix} \mathbf{x}(k) + \begin{bmatrix} 6.354 \times 10^{-6} \\ -4.699 \times 10^{-6} \end{bmatrix} u(k) \qquad (1)$$

$$y(k) = \begin{bmatrix} 34.13 & 15.11 \end{bmatrix} \mathbf{x}(k) \qquad (2)$$

where $u = n_d$ is the differential thrust input (excitation signal) in rpm given by Equation 3 in terms of the individual thruster velocities, n_1 and n_2. The controlled variable, $y = \psi$, is the output heading angle of the USV in radians.

$$n_d = \frac{n_1 - n_2}{2} \qquad (3)$$

It is obvious that when $n_d = 0$ i.e. $n_1 = n_2$, the vessel traverses in a straight line in the absence of external disturbances. The above dynamic model was obtained by applying system identification techniques to the input-output data acquired through trials carried out at a fixed speed.

4 Problem Formulation

In order for the automatic collision avoidance to work, a reliable detection system is mandatory. The detection system is responsible for keeping track of any changes in the vicinity of the vessel and reporting to the avoidance (guidance) module. For many reported applications, it is normally assumed that the location of the obstacles is known in advance. A map of the environment is also available which defines location of all the fixed infrastructure present around a sea port or a harbour. It is assumed that a camera and LIDAR (light detection and ranging) system is present onboard which can reliably detect any obstacles and provide their distance from the USV. The vision processing software detects the vertices of an object whose distance from the ship is accurately calculated using a LIDAR. Although this method of detection has some obvious disadvantages, it is only being used in this paper to demonstrate the viability of the proposed approach.

It is common to employ a virtual safety zone around the obstacles as well as the ship being controlled which must not be breached at any time unless necessary. The size of these zones depend on the dynamics (minimum turning radius and speed etc.) of the vessel. Here, a circular safety zone called the circle-of-rejection (COR) is assumed around each obstacle. A circle-of-acceptance (COA) is also assumed around each waypoint which flags the arrival of the ship at each one and the mission planner then selects the next waypoint. The COA is normally taken to be twice the length of the vessel being commanded.

The methodology adopted here has two distinct planning stages. Firstly, the vehicle must never enters the safety zone around the obstacle. Secondly, in order comply with COLREGs, the vehicle must pass by from the starboard side of the obstacle. This is true for both stationary and mobile obstacles. As explained earlier, the proposed approach employs a simple waypoint guidance by LOS coupled together with a manual biasing scheme. This strategy changes the current heading angle of the vehicle towards the starboard side when the distance of the ship to the obstacle is less than or equal to, the radius of COR thus avoiding the obstacle in accordance with coastguard regulations.

5 Simulation Results

Simulations have been carried out both for static and dynamic obstacles. As stated previously, the detection system determines the vertices of the obstacles and a COR is defined around each of them. Based on the waypoint guidance strategy, the vessel follows the LOS angle between its current position and the next waypoint. If there is a breach of the COR, the added bias in the current heading angle alters the course of the vessel towards the starboard side in order to avoid the collision as well as complying with the COLREGs. When the obstacle is fully avoided, the vehicle heads back to the current LOS to the next waypoint. The parameters, COR, COA and bias angle are chosen as $50m$, $10m$ and 75^0 respectively. In addition, a simple PID controller integrated with the path planner maintains the desired heading. The PID autopilot ensures that the vehicle stays on course as required by the guidance system.

The USV was assumed to have been launched at $(0, 0)$ where it eventually docked after completing the mission. There were seven waypoints and four obstacles (three rectangle and one triangle-shaped) in a field of 700 by 700 metres. The co-ordinates of the waypoints were chosen randomly taking care that no waypoint be located within any of the obstacle's boundary. The obstacle avoidance simulation is presented in Figure 2(a) which shows the USV's trajectory through all the waypoints.

From the plot, the effect of vessel dynamics is evident. Several evasive actions have evidently been generated by the path planner. From waypoint 1 to 2, the craft had to navigate away from the obstacle twice before arriving at the waypoint. There was a starboard side turn on the way to waypoint 3, whereas the trajectory from waypoint 5 to 6 consisted of several avoidance manoeuvres including a very sharp starboard turn. The path taken by the USV from waypoint 6 to 7 also contained COLREGs-compliant manoeuvres to avoid running into the obstacle. Note that waypoints 2, 3 and 6 are very close to the boundary of the COR and hence a breach was unavoidable. The vehicle finally docked at the launching point.

Next, a single mobile obstacle, comprising a ship initially considered to be moving in the South-Westerly direction at a fixed speed of $1m/s$ was examined. A COR of radius $50m$ was assumed around the obstacle. The USV launch coordinates were $(100, 100)$ which was also the final docking location. It is clear that the initial USV orientation would have been towards the North-East direction and therefore on a direct collision course with the oncoming ship. In order to create an interesting scenario, the direction of the oncoming ship was altered towards the North-East after the USV evaded its first encounter. This provided a practical situation or could also be regarded as two dynamic obstacles encountered during a mission. The complete USV route depicted in Figure 2(b) shows two evasive actions from waypoint 1 to 2 and from waypoint 4 to 5. In both cases, the USV passed on from the right-hand side of the moving ship and avoided the collision. The remaining trajectory consisted of approximately straight line or LOS paths. The relative speed limitation of the USV with the obstacle is a potential problem with this simulation as it may require a large COR so that appropriate action can be taken well in advance.

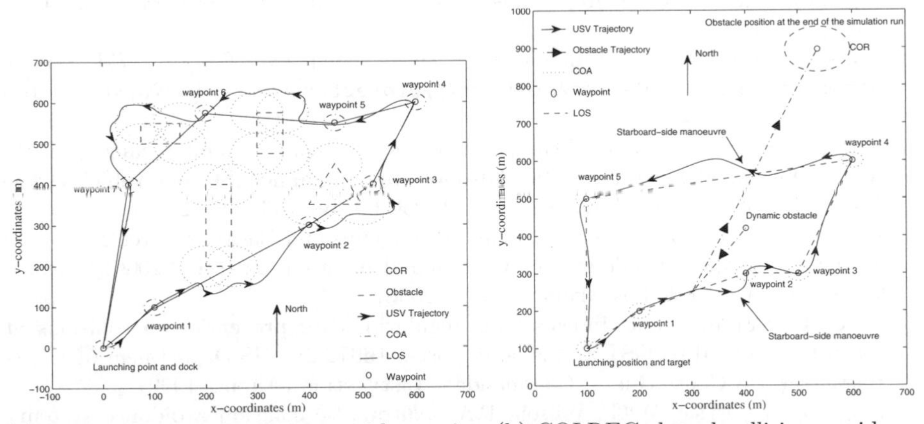

(a) COLREGs-based collision avoidance simulation for multiple static obstacles

(b) COLREGs-based collision avoidance trajectory for a dynamic obstacle

Fig. 2. COLREGs-based simulation analysis of collision avoidance in the presence of static and dynamic obstacles

6 Concluding Remarks

Preliminary simulation results have been presented on the development of an obstacle avoidance strategy for USVs. A simple manual biasing scheme was implemented together with the waypoint by LOS guidance technique. The highlight here is the integration of standardised IMO regulations or COLREGS in the path planning. The dynamics of an actual USV were also incorporated, providing realistic trajectories which are closely followed by the autopilot. In the proposed strategy, the USV must enter within the COR before the heading bias is introduced which diverts its heading towards the starboard side. It should be noted that manned vessels could also benefit from autonomous path planning, thus helping to eliminate the subjective nature of human decision making and safeguarding the onboard personnel. In the future, other motion planning strategies will be investigated for COLREGs-compliance using evolutionary algorithms such as genetic algorithms and particle swarm optimisation. Furthermore, automation of additional IMO rules will be carried out and the performance analysed in the presence of sea disturbances.

Acknowledgements

The authors would like to acknowledge the MIDAS Research Group at the University of Plymouth, UK for allowing the use of *Springer* model for this study.

References

1. Pascoal, A., et al.: Robotic ocean vehicles for marine science applications: the European ASIMOV project. In: Oceans 2000 MTS/IEEE Conference and Exhibition, vol. 1, pp. 409–415 (2000)
2. Rafael. Protector - Unmanned Surface Vehicle. World Wide Web, `http://defense-update.com/products/p/protector.htm` (accessed April 12, 2010)
3. Corfield, S.J., Young, J.M.: Unmanned Surface Vehicles - Game Changing Technology for Naval Operations. In: Advances in Unmanned Marine Vehicles, The Institution of Electrical Engineers, ch. 12, vol. 69, pp. 311–326 (2006)
4. Caccia, M.: Autonomous Surface Craft: Prototypes and Basic Research Issues. In: 14th Mediterranean Conference on Control and Automation (MED 2006), Ancona, Italy, pp. 1–6. IEEE, Los Alamitos (June 2006)
5. U. C. G. Commandant. International regulations for prevention of collisions at sea, 1972 (72 COLREGS). Technical Report M16672.2D, US Department of Transportation, US Coast Guard, Commandant Instruction (August 1999)
6. Jingsong, Z., Price, W.G., Wilson, P.A.: Automatic collision avoidance systems: Towards 21st century. Department of Ship Science 1(1), 1–5 (2008)
7. Statheros, T., Howells, G., McDonald-Maier, K.: Autonomous ship collision avoidance navigation concepts, technologies and techniques. Journal of Navigation 61, 129–142 (2008)
8. Lord, W.: A night to remember. Holt, Rinehart & Winston, New York (1955)

9. Benjamin, M.R., Curcio, J.A., Newman, P.M.: Navigation of unmanned marine vehicles in accordance with the rules of the road. In: IEEE International Conference on Robotics and Automation (ICRA), Orlando, FL, pp. 3581–3587. IEEE, Los Alamitos (May 2006)
10. Larson, J., Bruch, M., Halterman, R., Rogers, J., Webster, R.: Advances in autonomous obstacle avoidance for unmanned surface vehicles. In: AUVSI Unmanned Systems North America, Washington DC (August 2007)
11. Kao, S.L., Lee, K.T., Chang, K.Y., Ko, M.D.: A fuzzy logic method for collision avoidance in vessel traffic service. Journal of Navigation 60, 17–31 (2007)
12. Xue, Y., Lee, B.S., Han, D.: Automatic collision avoidance of ships. IMechE Proceedings Part M: Journal of Engineering for the Maritime Environment 223(1), 33–46 (2009)
13. Naeem, W., Xu, T., Sutton, R., Tiano, A.: The design of a navigation guidance and control syatem for an unmanned surface vehicle for environmental monitoring. IMeChE Transactions Part M, Journal of Engineerig for the Maritime Environment (Special Issue on Marine Systems) 222(2), 67–79 (2008)

Research on Fire-Engine Pressure Balance Control System Based Upon Neural Networks

Xiao-guang Xu and Hong-da Shen

Department of Electrical Engineering, Anhui Polytechnic University, Wuhu 241000 China
xuxg@163.com

Abstract. The pressure produced by the water coming out of the fire-engine pump outlet is controlled by the rotate speed of the fire pump. However, this RS is controlled through fire-engine accelerator voltage which is controlled by the ECU. In order to control and keep the fire-engine pressure balanced, it is necessary to take pressure, rotate speed and current rate as input parameters and control voltage as output parameter through BP neural network control system. Related researches indicate that BP neural network is appropriate for building the system whose target is to keep the pressure balanced. And, some modification can be done to the standard BP neural network algorithm. These modified BP neural network algorithms are BP neural network with momentum factors and self-adapting learning speed which can improve the response speed and performance of this control system dramatically.

Keywords: pressure balance neural networks momentum factors.

1 Introduction

As you know, Fire-Engine plays a vital role in fire industry. With the highly development of the fire-Engine technology, the design level and the manufacturing technology of fire Engine in our country has approached the international cutting-edge level. Simultaneously, there are a number of gaps and insufficiencies between the level of our country and the international cutting-edge level. At the present time, many staple water tankers usually discharge water through the water hose and fire water monitor. Therefore, firemen can turn on and off the outlet of the water hose or fire water monitor according to changing fire behavior. Under the circumstance that the state of switch changed suddenly, the pressure of the outlet will increase significantly and generate big recoil due to lack of timely adjustment to power of the fire pump. Such recoil will make great wallop to firemen and cause some safety misadventure. Fire-Engine Pressure Balance Control System will reduce the damage brought by fire pump pressure jump effectively under the condition that the working pressure of the fire pump is allowed.

Fire-Engine Pressure Balance Control System is built based on the pressure of the fire-engine water outlet, current rate of the outlet, fire pump rotate speed and so forth. Through adopting neural network control algorithm and building the pressure balance control system, it is possible to keep the pressure of the fire-engine water outlet balanced at a given range. In this system, pressure, current speed, rotate speed and voltage control signal are taken as input parameters and output parameters separately. Pressure control signal controls the accelerator pedal through automobile device electronic

K. Li et al. (Eds.): LSMS/ICSEE 2010, Part II, CCIS 98, pp. 192–200, 2010.

control unit (ECU) to control rotate speed of the generator of the fire-Engine further. Fire pump gains corresponding kinetic energy from the automotive engine through power takeoff to alter the rotate speed of the fire pump and reduce the pressure of the fire pump according to the size of outlet. From this, it can be seen that pressure of the fire pump depends on the rotate speed and the current rate of the outlet of the fire pump so that these three parameters are relevant. However, this system mainly controlled object is the fire-engine accelerator which is controlled through ECU. Therefore, Fire-Engine Pressure Balance Control System does control the accelerator of the fire-engine through the ECU.

It seems difficult to build a definite function relation between the accelerator control voltage in this system and air throttle which is controlled through ECU (a valve which is used to control the size of the accelerator), so the system can not use the control method which is similar to the method used to control the fire pump pressure.

Currently, the problem on the fire pump pressure device is often solved by putting the hand control gas pedal into use. When it comes to increase or decrease the fire pump pressure, it seems relatively simple to apply manually operation to change the size of the accelerator and convenient to operate live. But it should be paid attention that such operation often cause big safety problems because different automobiles from different factories may have various ECU characteristics and their manually operations have difference as well. When firemen are operating the control button of the accelerator, the pressure of the fire pump may become or big or small. Much more severely, pressure of pedal may increase intensely in short time and make big recoil to the firemen. For example, when working aloft, such big recoil may do great harm to firemen even have trips or some other dangers when they are operating the accelerator.

However, adopting neural networks algorithm will solve the problems mentioned above. Through building a neural network model and selecting appropriate algorithm will build relationship between the pressure, rotate speed and current rate of the fire pump and accelerator control voltage, it becomes easy to reduce the pressure of fire pump outlet effectively.

Currently, the widely-used neural network model is BP neural network algorithm. However, BP neural network algorithm has some deficiencies. With the development of research on neural network, many updated algorithms come into being including BP neural network with momentum factors or self-adaptive learning rate and LM algorithm which improve the performance of BP neural network in different fields.

This paper applies the standard BP neural network algorithm and updated BP neural network to building fire-Engine pressure balance system model. This system model is three-input-single-output system which has an uncertain relationship. From this system, it is obvious to see the feasibility and validity when applying the BP neural network to this system. At the same time, it will be seen that different BP neural network algorithms and different performance parameters may bring relatively great improvement to the standard BP neural network.

2 Fire-Engine Pressure Balance Control System Model

Fire-Engine Pressure Balance Control System controls the size of the fire-engine accelerator on the basis of the outlet pressure of the fire pump, current rate, rotate speed signal of the fire pump and output accelerator voltage signal which will be given to fire-engine ECU to realize certain control function in order to keep the pressure of the fire pump balanced. This system consists of sensor unit, transmitting unit and controller unit. Sensor unit involves pressure sensors, current rate sensors, rotate speed sensors which are responsible to collect detailed data which are important to the whole system; the main task of transmitting unit is to transform the data collected by sensor unit to standard electric signal and transmit the signal to controller unit; then the controller unit does convert the signal transmitted to controller unit by transmitting unit to digital signal, carry out operations in the internal part of the controller and output accelerator control signal which will be transmitted to fire-engine ECU to be used for accelerator. With the changing size of accelerator, the rotate speed of the engine will change. Next, fire pump will change its own rotate speed through fire-engine power takeoff, thus change the pressure of the fire pump. The structural drawing of this system model is shown in the Fig 1.

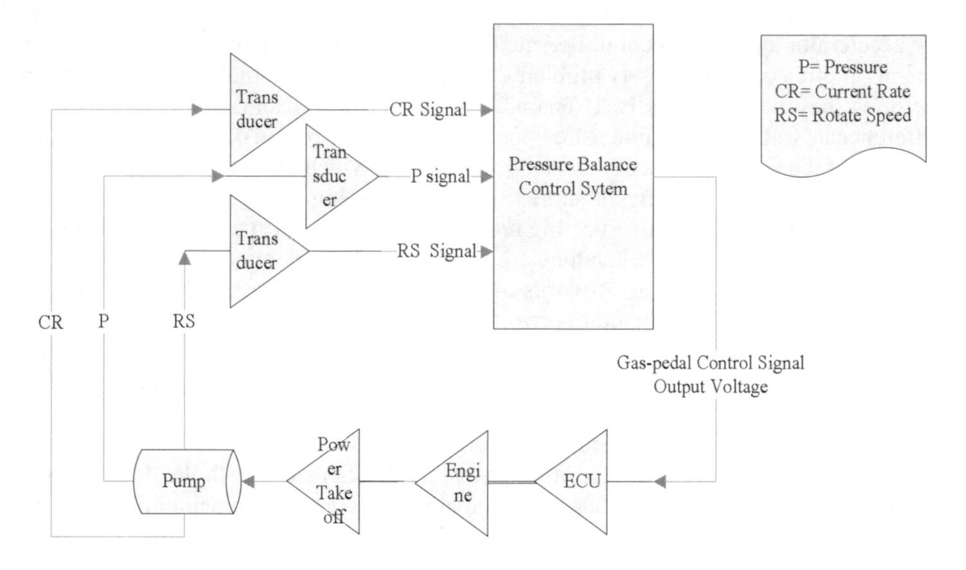

Fig. 1. Fire Pressure Balance Control System Model

In this system, the core units are the pressure balance controller which receives pressure, current rate and rotate speed signal and outputs accelerator voltage signal. The main target of this system is to keep the pressure balanced or maintain the pressure within the range that fire-engine users set before. When those external conditions changed, for instance, the on and off of the fire-engine outlet, it is required that the pressure variation should be maintained within the permissible range.

The controller mentioned above is chosen to be built by neural network to realize the pressure balancing control because the output accelerator voltage signal in response of pressure, current rate and rotate speed of the inputs can not be described in mathematical equations.

3 Pressure Balance Controller Design Based on BP Neural Network

3.1 BP Neural Network Construction

BP neural network is a kind of multilayer feedforward network algorithm based on Error Back Propagation (BP neural network algorithm). The transfer function of neurons in BP neural network is often "`Sigmoid`" function(S-function) which can realize any nonlinear mapping between input and output. This kind of network has simple structure and fast learning speed. As long as the appropriate weight value and threshold value can be assured, good performance of nonlinear system will be achieved easily which makes it widely used in the function approximation, pattern recognition, system identification and control field. However, it is difficult to fix the specific network structure and parameters. And slow learning speed, premature convergence and sensitivity of selected parameter may become big problems. For that reason, researchers do much modification to the standard BP neural network algorithm, for example, BP neural network algorithm with momentum factors and self-adaptive learning speed which adopting heuristic method [3].

Neural network topological structure influences directly the generalization ability of the network. Once the specific network structure and the number of variables have been fixed, the learning ability of the network and search space will be limited. If the reinforce of network classification capacity of complex problems becomes necessary, adopting the multilayer feedforward network seems a good method. That is to say, adding hidden layer between the input layer and output layer can improve the quality of the network effectively. Such multilayer network involves input layer, hidden layer and output layer. As it is imagined, it becomes simple to approach a continuous function if the transfer function of the hidden layer is continuous.

The Pressure Balance Control System Based On BP Neural Network Algorithm has a three-layer network structure in which the hidden layer contains 19 neurons. To fix the number of ordinary hidden layer neuron takes a collection of experiments. This system talked in this paper has a hidden layer which contains 19 neurons. At the same time, both the output and convergence curve are relatively perfect. According to the practical system demand of output, this neural network set one output which is in response to accelerator control voltage value. BP neural network model is shown in the Fig 2.

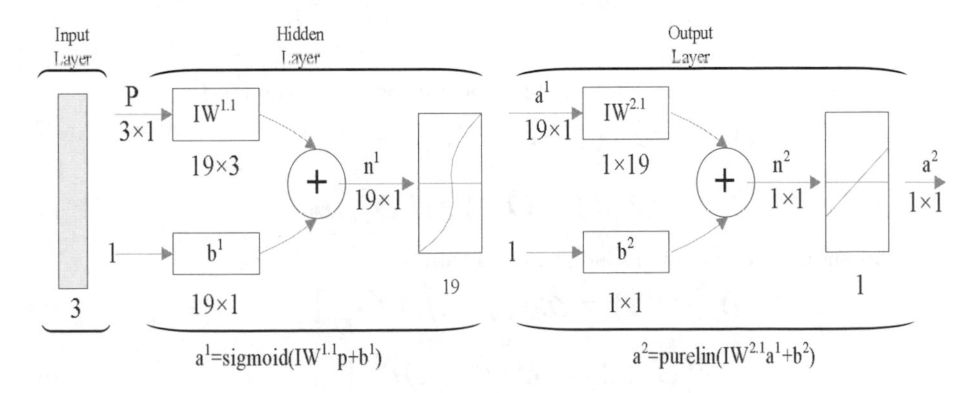

Fig. 2. BP neural network with a single hidden layer

In the Fig 2, the transfer function of hidden layer is sigmoid function. P in the Fig2 represents the input vector which includes the practical variables of this system; pressure, current rate and rotate speed of the fire pump. Therefore, the input vector shapes like [P1, P2, P3]. Three inputs and 19 neurons add up to 57 hidden layer weight values. These weight values and 19 threshold values are parameters which are expected to train. a^1 in Fig2 is a 19×1 vector of hidden layer output. W^1 is hidden layer weight value and b^1 is hidden layer threshold value. W^2 is output layer weight value and b^2 is output layer threshold value. W^2, b^2 are 1×19 and 1×1 vectors separately, a^2 is output layer actual output. in other words, the final output of the network.

3.2 BP Neural Network Algorithm

With adopting supervised learning and least mean-square error rule to train the weight values and threshold values, 20 groups of training samples will be achieved through detect sensors and related experiments which is used to meet the requirement on the precise output and convergence curve. The whole sample group stands for pressure, rotate speed and current rate of the fire pump separately. Simultaneously, target output is made up of the corresponding system output voltage. Gas pedal control voltage is taken as network output and also system feedback signal in order to maintain the pressure balance. BP neural network can be realized based on the following steps [5-7].

1) Setting initial values of all weight values and threshold values $\omega_j(0)$, $\theta_j(0)$. All these values can be limited between 0 and 1 randomly and taken as initial weight values and initial threshold values.

2) Input learning sample

Training samples including three input variables: pressure, RS and CR are collected in the related experiments. The more the samples are, the more accurate the output curve will be. Among those data collected in the experiments, they are divided into two groups: one is training sample, another one is considered as testing group.

3) Calculating the actual output of the network and actual state of hidden units:

$$O_{pj} = f_j(\Sigma\, \omega_i O_i - \theta_j) \tag{1}$$

In this formula: transfer function f is sigmoid function, that is

$$f(x) = 1/[1 + \exp(-x)] \tag{2}$$

4) Calculating the training error

Input layer and hidden layer errors function are formula 3-1and 3-3

$$\delta_{pj} = O_{pj}(1 - O_{pj})(t_{pj} - O_{pj}) \tag{3}$$

$$\delta_{pj} = O_{pj}(1 - O_{pj}) \sum k \delta_{pk} \omega_k \tag{4}$$

5) Amending weight values and threshold values

$$\delta_i(t+1) = \omega_i(t) + \eta \delta_j O_{pj}] \tag{5}$$

$$\theta_i(t+1) = \theta_i(t) + \eta \delta_j] \tag{6}$$

In this formula: η represents learning rate.

6) After p changes from 1 to P, it is easy to judge that whether the actual precision can satisfy the requirement of E..

$$E = \sum E_p \tag{7}$$

$$E_p = \sum (t_{pj} - O_{pj})^2 / 2 \tag{8}$$

Here, E<ε,and ε represents the actual precision. If the demand can be satisfied, the training process can be stopped, otherwise, jump to Step 3.

7) Stop training and end.

3.3 BP Neural Network Updated Algorithm

Standard BP neural network updated algorithm aims at the weight values and threshold values of the network added the momentum factors. In fact, this algorithm contributes to generate new weight values and threshold values. Standard BP neural network algorithm ignores the original system gradient direction when calculating the weight values and threshold values in the direction of negative gradient sometime. This deficiency leads the network system to converge at a slow speed and unexpected oscillation during the training process. Yet, BP neural network algorithm with momentum factors influence momentum factor of this time by momentum factor generated during the previous learning process.

That is:

$$\delta_i(t+1) = \omega_i(t) + \eta\delta_j O_{pj} + \alpha[\omega_i(t) - \omega_i(t-1)] \tag{9}$$

$$\theta_i(t+1) = \theta_i(t) + \eta\delta_j + \alpha[\theta_i(t) - \theta_i(t-1)] \tag{10}$$

Where α represents momentum factors which are often positive.

From formula 3-9、3-10, it is obvious to see momentum factors and error momentum factors are opposite in sign, which serves to decrease the momentum factors of this time when the previous adjustment is overshooting. Adversely, if the previous adjustment turns undershoot, momentum factors and error momentum factors have the same sign, which take effect of accelerating adjustment. Facts indicate that BP neural network algorithm with momentum factors can obtain faster convergence speed and less opportunity premature convergence. That is to say, more stable network and faster convergence speed will be approached if the BP neural network with momentum factors is adopted. Moreover, another advantage of the BP neural network with momentum factors lies on accelerating convergence speed effect which is brought by momentum factors when the gradient stays unchanging. In other words, BP neural network algorithm (MOBP) can improve the quality and performance of the network and makes the network more effective and efficient.

4 Simulation and Results

In order to test the real effect of the pressure balance control system based on BP neural network, this paper chooses 20 groups of sample data to train the standard and updated BP neural network. Other 5groups will be used to test the network. Each group consists of 3 variables. According to the requirement that every variable should be limited in certain range, pressure varies from 0.4MP to 2.0MP and output voltage varies from 0.2V to 4.8V. RS and CR values are collected by the sensors installed in the fire-Engine which is used for experiment. Specific sample data is shown in the following Table 1.

Table 1. Sample Data

Pressure (MPa)	RS（Kr/min）	CR（T/h）	Output Voltage (V)
0.40	2.010	5.03	0.25
0.54	2.544	4.71	0.47
0.63	3.060	4.86	0.72
0.72	3.523	4.91	0.93
0.83	4.249	5.12	1.20
0.95	4.530	4.77	1.41
1.00	5.019	5.02	1.67
1.05	5.300	5.05	1.92
1.15	5.504	4.79	2.16
1.21	6.099	5.04	2.43
1.29	6.371	4.94	2.64
1.37	6.500	4.74	2.88
1.57	7.253	4.62	2.13
1.64	7.395	4.51	3.36
1.68	7.528	4.48	3.60
1.70	7.789	4.58	3.85
1.73	7.959	4.60	4.02
1.89	7.996	4.23	4.39
1.96	7.989	4.08	4.58
2.00	8.000	4.00	4.80

In the MATLAB, calling the neural network toolbox to build a bi-layer feedforward neural network, training max step length epoch=8000, error performance target value goal=0.01, learning rate lr=0.01, the weight values and threshold values of hidden layer are set between 0 and 1 randomly adopting tansig function as hidden layer transfer function, purelin function as output layer transfer function.

Through simulation experiment, it is possible to obtain convergence curve and error curve of standard BP neural network and updated BP neural network, as they can be seen in Figure 3 and Figure 4. Figure 3 describes the convergence curve or error curve of standard BP neural network algorithm. The training function of this algorithm is traingd(). From the curve shown in the Figure 3, it is obvious that BP neural network works well and the step length taken to reach the specific target error value for error curve is 6022 which turns out, therefore, that BP neural network algorithm can build appropriate relation between pressure, RS, CR and gas pedal control voltage to construct the pressure balance control system.

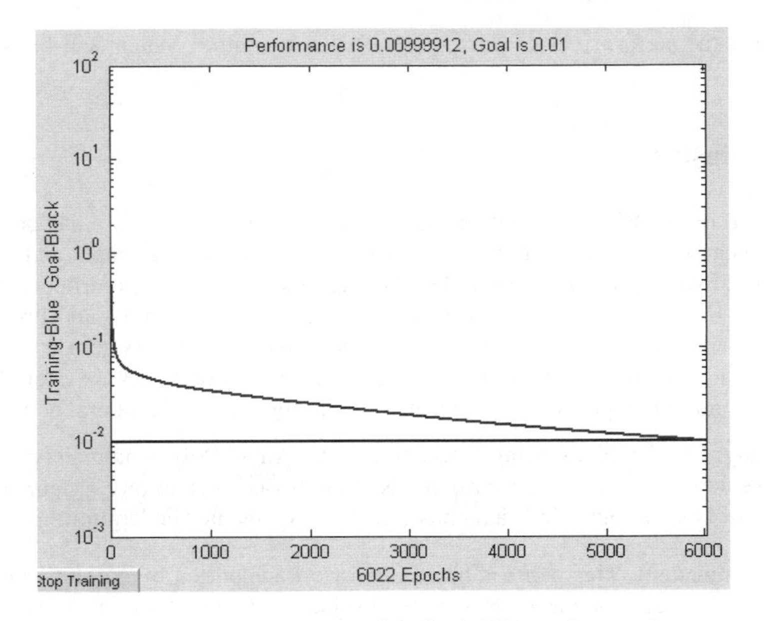

Fig. 3. convergence curve of the basic BP neural network

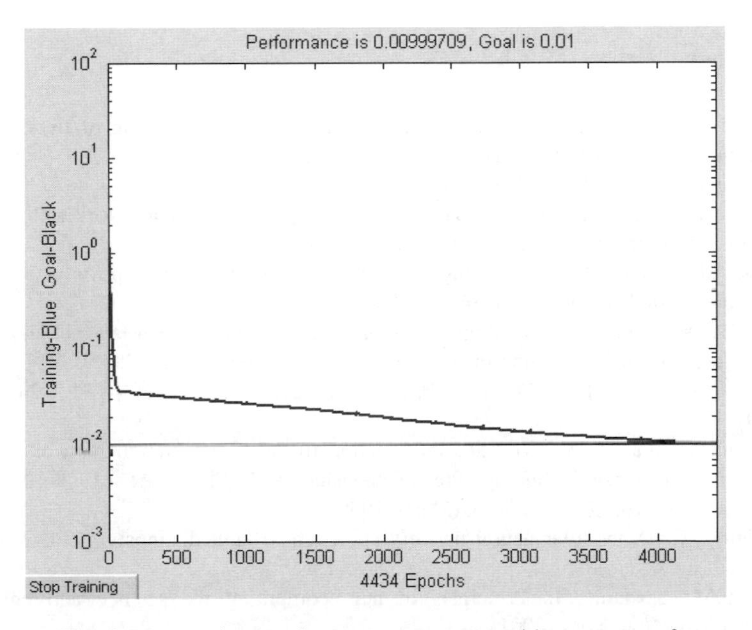

Fig. 4. BP neural network convergence curve with momentum factor

Simulation of system model adopting the updated BP neural network with momentum factors and self-adaptive learning rate yields convergence curve and error curve in the Figure 4. It can be seen that this kind of network has faster convergence speed because the actual error takes 4434 steps to reach the target error. Due to significant decrease in cycle step, the performance of this pressure balance control system will be improved dramatically.

5 Conclusion

This paper applies BP neural network algorithm to design the pressure balance control system because it is difficult to build the definite relation between input and output of the system by using the PID controller. Through the simulation experiment, it is obvious that BP neural network can realize pressure balance control and updated BP neural network algorithms can improve the performance of this system dramatically. These updated algorithms not only accelerate the convergence of the system but also solve local minimum problem which is usual in application of BP neural network.

Biography: Prof.Xu Xiao-guang is teaching at the Anhui Polytechnic University. His main research field is multi-sensor data fusion, network of field bus system and computer control technology. He is a member of China Computer Federation.

Acknowledgement. This paper is based upon work supported by a grant from Anhui Province College Science Research Projects. Grant No.KJ2009B048 and No.KJ2008B098.

References

1. Feng, H., Lin, D.S.: China's situation and the development direction of fire-engine. Fire Technique and Products Information (11), 69–71 (2003) 何锋, 董松林. 我国消防车的现状及发展方向. 消防技术与产品信息 (11) 69-71 (2003)
2. Ho, K.L., Hsu, Y. Y., Yang, C. C.: ST LF using a multilayer neural net work with an adaptive learning algorithm. IEEE Trans. on PS 7(1), 141–149 (1992)
3. Parlos, A.G.: An accelerated learning algorithm for multi player perceptron networks. IEEE Trans. on Neural Networks 5(3), 86–88 (1994)
4. Hong, L., Qiu-fang, T., Hui, L.: Application of BP a lgor ithm in the ba lance of underactuated manipulator. Journal of Beijing Institute of Machinery 24(3), 17–21 (2009) 厉虹,田秋芳,李慧. BP算法在欠驱动机械臂平衡控制中的应用. 北京信息科技大学学报. 24(3), 17–21 (2009)
5. Shouren, H., Shaobo, Y., Kui, D.: Introduction to Aritificial Neural Networks. National University of Defense Technology Press, Changsha (1996) 胡守仁,余少波, 戴葵. 神经网络导论. 长沙: 国防科技大学出版社 (March 1999)
6. Rodriguez, C.: A modular neural network approach to fault diagnosis. IEEE Trans. on N ns 7(2), 326–340 (1996)
7. Ham, F.M., Kostanic, I.: Principles of neurocomputing for science and engineering. McGraw-Hill, New York (2001)

The Summary of Reconstruction Method for Energy Conservation and Emission Reduction of Furnace

Xiaoxiao Wang[1] and Xin Sun[2]

[1] School of Mechatronics Engineering and Automation,
Shanghai University, yanchang road. 149,
200072 Shanghai, P.R. China
[2] Shanghai University, yanchang road. 149, 200072 Shanghai, P.R. China

Abstract. With the raising of the third session of the CPPCC national committee first proposal" promote the development of china's law-carbon economy on the pro- postal", the R&D and result. This article studies deeply and reconstructs cosmically the structure of the furnace, aiming at the energy saving method of the furnace's energy saving application, in order to achieve the saving of fuel as could be under the temperature of material outputting and ensuring the safety. What has been discussed above still need more proofs during the practice, achieving the best result through various kinds of technic improvement and co-operation.

Keywords: low-carbon economy, furnace, energy saving.

1 Introduction

Nowadays, the chemical industry in our country for most part is still of high energy consumption and of high pollution, causing huge sources exhausting and strongly pollution. According to the forecast from the State Council Development Research Center: China will consume coal resource 2.8 billion tons and crude oil 0.6 billion tons in 2020, which will rise more than 2.5 times than that in 2000. It is estimated that China petrol import in 2020 will be up to above 250 million tons, and import de- pendency rate achieve 70%. At the same time, the carbon emission will reach 1.94 billion tons. Faced the great challenge of energy supply and living environment now, the chemical engineering industry should start with improving their produce methods to survive, gradually jumping the track, which take the" high increase, high consumption, high pollution and expanding without detail consideration" as the main industry mode.

Energy saving is the core content of China Energy Strategy. To solve the energy and resource problems, the basic method is to rely on science and technology improvement, by adding the S&T content. Lowing the rely on the energy and lowing the environment pollution at the same time. This article lists the recycle of the waste heat of furnace, the technical-reconstruction of the body of furnace, the furnace inside bilge clean and series of technology for reconstruction. Hope that one or more of the tech- nologies can apply on the furnace at the same time and achieve the best effect of energy saving.

K. Li et al. (Eds.): LSMS/ICSEE 2010, Part II, CCIS 98, pp. 201–205, 2010.

2 Optimizing the Control System

The present control system of furnace transfer the traditional simulate instrument to field control system(FCS). Using the abundant hardware and software functions of computer can accomplish the adjustment, direction, enactment, record, alarm, display of dynamic chart for the process variable and so on.[1] It has new tactics for major parameter technical control , switch aptitude control, interlock advanced, etc empolder application aspects, which benefits the safely of equipment, stability, op- timization running on the obvious energy saving effect.

Effective automatic control tactic of furnace system can reduce the depression of the efficiency of furnace which caused by man-made. Install the essential monitoring instrument for pressure, temperature flow on the furnace main body and the distribution system, and combined with the automatic control system to accomplish that according to the need of the user achieve that the furnace system stay at the best running estate. Of course, excellent control arithmetic advantage now appear: not only to get the well control effort ensuring for the manufacture, but also to protect the equipment better.

3 Abatement of Electric Energy Consumption

The reactor of the chemical industry and its system run with a lot of electricity. So saving the electric energy is considerable important, too. The frequency change and speed adjusting is now recognized as one of the energy saving technique of motor which have series of advantages including high-precision, wide speed range, high operation efficiency, high power factor, easy operation and so on. It becomes one of the main methods for change-tradition industry, variable flow improvement, auto- mastic level during produce process improvement, product quality improvement, technology improvement launching.

In the past, furnace induced draught fan adapted butterfly valve control, heating up the temperature being decided by the fuel flow gushed into the furnace, fuel quantum being inferred by he fuel flow, putting into computer. The computer adjust the opening of the butterfly valve by how much the fuel quantum, providing wind reasonable. When the proportion between wind and fuel is most appropriate, the effect is the best. For the butterfly valve is equipped in the wind channel, even opening fully, there still be some throttle. When it runs in winter with high load, some furnace, always feels that there is a lack of wind. While it runs in summer with low, controlling the wind by butterfly valve, the need of the wind just need a very little butterfly valve opening. But the power of the fan still being the same ,butterfly valve throttle then cause a huge loss of electricity. That is the wasting of resources.

Transducer changes the rotating of the fan by adjusting the output frequency, in order to achieve governing of the goal for AC machines (Eg. As Fig1), reconstructing the transducer technology of the blower of the furnace system. Transducer accomplishes" provide what is needed" by reducing the output power. The square law characteristic uses for the fan and pumps load can reach 50% of fractional energy saving. The load uses for other technics also can get 10%--40% energy saving. After rotating of the equipment has cut down[2], it can reduce the abrasion to extend its using time, and then gain a considerable and indirect economic benefit.

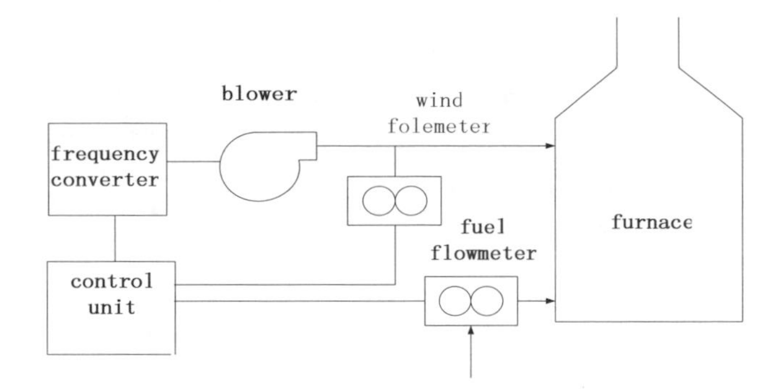

Fig. 1. Fan Equipment with transducer

4 Effective Recycle of the Remaining Heat

During the running, the heat efficiency of furnace is often 8%--10% lower than the planned efficiency. The huge amount of heat eliminated from chimney of the furnace, not only decrease the heat efficiency, but also cause the heavy pollution to the environment. To solve this problem, heat pipe air preheater is installed to recycle the remaining heat, which can decrease the smoke-eliminated temperature, improve the burning efficiency, reduce the carbon of remaining content, promote the efficiency of furnace, decrease the CO2 emission, achieving the great effect of energy saving.

The structure principle of heat pipe air preheater completely differs from pipe air preheater. It is the assembly of the heat pipe components, taking advantage of its characters. Transfer the inner gas flow of the PAP to hea-exchange outside, gas outside exchange heat up and down respectively indirectly through a wall. Its gas downside of the heat releasing and upside of the heat absorbing are both achieved by changing (Boiling, condensation)in its tube media(Eg. as Fig2). Its lowest metal wall temperature could designed approaching the temperature of hot side fluid, and the both sides have great isothermal performance. The testing material shows that the gap between the walls if heat pipe the 2(cold and hot)sides is no more than 5℃ during its running. The higher wall temperature of the heat pipe is higher than the gas's water dew point even the acid dew point, so that the heating surface avoids or reduces it opportunity to be corroded by the sulfur content of flue gas. When the dust fells on the wall with high temperature it won't become hard and keep in loose state. It can more into the dust collector following the gas, and it can also be cleaned with other easy ways.[3]

For that the strengthen of burning stability radiation and heat transfer could reduce the loss of the chemical incomplete combustion; On the other hand, air preheater use the remaining heat of the gas further decrease the loss of gas-eliminated, so that the heat efficiency of the furnace can be increased. According to the experience, when the air temperature rises 1.5% in the preheater, the gas-eliminated temperature reduces 1℃ .After the air preheater installed in the furnace's channel of gas , if the air temperature rises 150℃—160℃ then the gas-eliminated temperature could reduce 110℃—120℃. The heat efficiency of furnace is promoted 7%--7.5%, and the fuel could be saved 11%--12%.

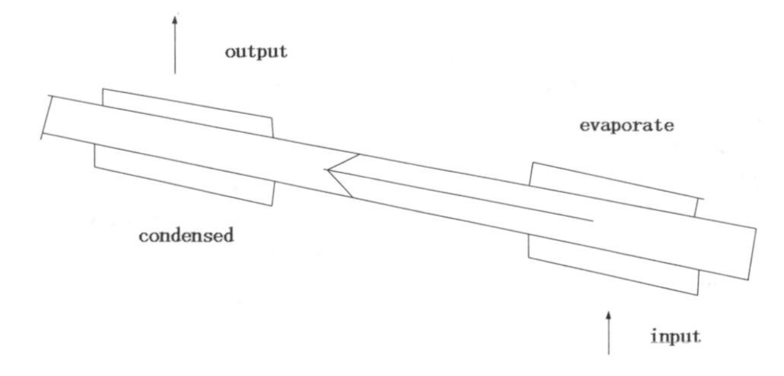

Fig. 2. Brief picture of heat pipe transfer principle

5 Reduce the Heat Loss in the Furnace

Blackbody Technology is a blackbody principle according to infrared physics, studied the "blackbody origin" concept to invent an industry standard called blackbody component. The abundant blackbody on the wall like the second transmitter of volleyball, absorbing the duffusion-like heat ray in the furnace and controlling the relaunching process. Let the heat ray ready as fast as possible, both speeding up the heat transfer and reducing the facial cooling. The insulation resistance produced by the blackbody component can play the role of calming the temperature and pressure fluctuation caused by inverting. After the enhanced treatment of the wall, its air tightness enhanced, which can both ease the oxidation rate inner and prevent the wall leakage effectively.[4] It can obviously increase the scouring force of the airflow by furnace's wall on the fuel furnace. Obviously, the furnace has environmental function, which reduce the CO2 emission and the heat pollution to the atmosphere. A comparison test before and after the reconstruction of blackbody technology in some factory shows that its 40 tons gas eliminating temperate has decreased 70℃.

The problem the blackbody technology solved is to change the position of heat rays in the furnace and to promote the irradiance, no relating to which heat produce the rays. So it can be use in electric, gas, oil, coal whatever heat furnace. All kinds of furnaces which need heating in the hearth (box-style, desktop cars, pit type resistance furnace, vacuum furnace, controlled atmosphere furnace and its production lives)[5]. Sted rolling heating furnace, forging furnace, ceramic kiln furnace and glass tank furnace can adopt the blackbody technology to attain the pleased effects of long-term quality, high efficiency, low exhausting, and environmental protection. The range of it is very wide. Blackbody enhances radiation heat transfer energy-saving technology practical produce shows that the energy of furnace and heat–dealer can decrease 20%--30% on the existing basis.

6 Dirt Block Prevention in the Furnace

6.1 Application of Organic Fluorine and Resin

The dirt prevention principle of organic fluorine resin layer way take the fluorine layer as the major material adding other metals, non-metals oxide and ceramic ponder.

In preventing dirts and block dirts aspects use the inertia of the fluorine layer, the scale out performance of metal, non-metal oxide, and crude medium scale material hard surface layer adhesion in transit. It starts from the face to prevent and block the dirt.

6.2 Electrostatic Water Treatment Technology

Because the position of hydrogen atom and oxygen atom in the water is asymmetric ,there is polarity. Under the action of the electrostatic field, water molecules will be directed to the negative by positive neatly arranged to form chains. When water containing dissolved, salt ions, these ions will be surrounded by the water dipoles, and make it not move freely, inhibiting the precipitation of the forming dirt of calcium and magnesium ions, and then achieve the function of preventing the dirt. After the electrostatic handling , there will be little oxygen released in the water. Set up a micro-oxidation on the paradise wall system preventing its corrosion.[6]

For the scale of electronic force between molecule destroyed by the oxygen released by the system, changing its crystal structure, making the hard dirt loose. The distance between the water molecule turns big under the electrostatic field, enhancing the capacity of salt ions with water, so that it helps to increase the dissolution rate of the water dirt, making it general erosion, and then felling down as fragments. That causes the dirt cleaning.

7 Conclusions

Through the discussion of the methods for energy saving technology of furnace, A legible thoughts for energy saving of furnace has provided. In order to achieve the goal of energy saving, we still need further study and develop more advanced and abundant new technology, using them in chemical produce industry of furnace system, and finding different reconstruction plan for different furnace system.

References

1. Li, M.Z.: The discuss and research of control strategy for furnace. Techniques of automation & applications 27(11), 83–85 (2008)
2. Chang, G.Y., Wang, S.H., Zhang, D.F.: The application of Variable-frequency regulating speed technology in the furnace fan energy-saving (7), 61–62 (2006)
3. Zhuang, J., Zhang, H.: Heat pipe technology and engineering applications. Chemical industry press (2000)
4. Lan, J.: A Portable Black Body Furnace W -th Fixed Middle Temperature Points. Auto-mated instrumentation 22(8), 23–24 (2001)
5. Barton, W.A., Miller, S.A., Veal, C.J.: Drying Technology 17(3), 497–522 (1999)
6. Zhu, X.X., WaNg, H.: Static water treatment system. Techniques of automation & applications (5), 38–39 (2006)

Osmotic Energy Conversion Techniques from Seawater

Yihuai Hu[1] and Juan Ji[2]

[1] Shanghai Maritime University, 1550 Pudong Avenue, Pudong New Area, Shanghai, P.R. China
[2] Qingdao Harbor Vocational and Technical College, 65 West Chongmingdao RD, Qingdao, P.R. China
yhhu@shmtu.edu.cn

Abstract. This paper firstly introduces the principles of Pressure Retarded Osmosis (PRO) and Reverse Electro Dialysis (RED). It is concluded that the RED method is more suitable than the PRO method for power generation from seawater. Theoretical analysis of RED stack is made for the design of a test RED compartment. Tests were carried out to study the relationship between produced energy power density, test compartment width and ion exchange membrane area. Experimental conclusions are drawn out at the end of the paper.

Keywords: Osmotic Energy, Reverse electro dialysis, Power generation.

1 Introduction

Traditional fossil fuels have a number of drawbacks, such as emissions of greenhouse gases, depletion of finite sources and dependence on fuel oil-exporting regions in the world. With global warning and oil price rising, more and more interests are put on renewable energies including solar energy, wind energy, biomass energy, hydro energy and ocean energy. Where a river flows into a sea, a lot of energy could be released due to the different salt concentration between seawater and fresh water. This is called osmotic energy existing in sea mouth of big rivers. The global energy output from the estuaries is estimated at 2.6 TW [1], which represents approximately 20% of the present worldwide energy demand. Two membrane-based energy conversion techniques are developed from the osmotic energy: PRO and RED.

2 Principles of Osmotic Energy Conversion

2.1 Principle of PRO Technique

In this osmotic process two solutions with different salt-concentrations are involved (often freshwater and seawater). A semi-permeable membrane is put between these two solutions to separate the freshwater and salt-water. This kind of membrane only lets small molecules like water molecules pass. Then water aspires to decrease the salt concentration on the side of the membrane that contains seawater, while the water streams through the membrane and amount of water on the salty side will increase creating an osmotic head pressure. This pressure could be utilized for power generation by

K. Li et al. (Eds.): LSMS/ICSEE 2010, Part II, CCIS 98, pp. 206–212, 2010.

using a turbine and a generation as shown in Fig.1. The amount of fresh water that passes through the membrane depends on the salt concentration in the salt water before the osmotic process begins.

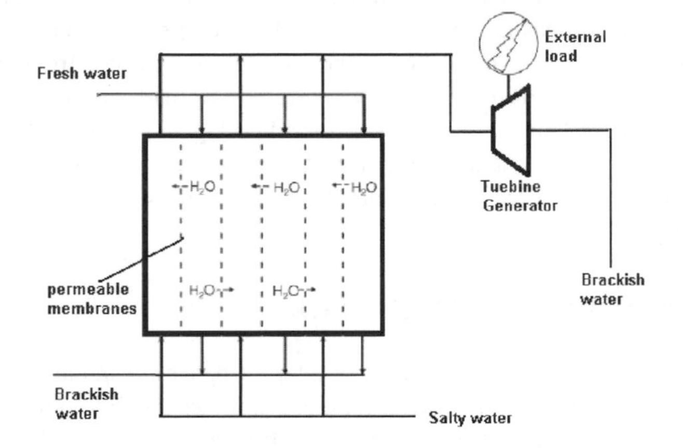

Fig. 1. Scheme of PRO technique

2.2 Principle of RED Technique

In this osmotic process two kinds of ion exchange membrane are involved and stacked in an alternating pattern between a cathode and an anode. Fresh water and salty water is filled alternating between the membranes. The anion exchange membranes only let negative ions (such as chloride ion) pass, while the Cation exchange membranes only let positive ions (such as sodium ion) pass.

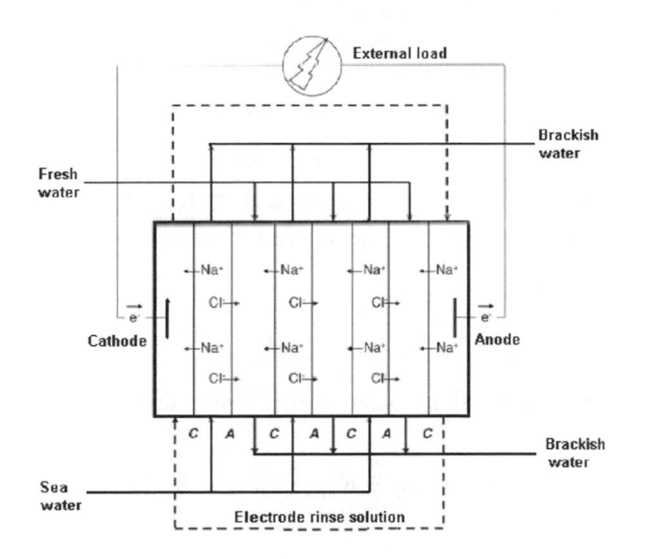

Fig. 2. Scheme of RED technique

In this way chemical potential difference will be established between the outer compartments of the membrane stack on the Cathode and Anode. As a result, an electron could be transferred from the anode to the cathode via an external electric circuit as shown in Fig.2. The voltage obtained between the Cathode and Anode depends on the member of membranes in the stack, the absolute temperature, the ratio of concentrations of the solutions, the internal resistance of cells and the electrode properties.

Comparison between PRO and RED techniques was made in literature [4]. It was concluded that the RED technique has several advantages over the PRO technique in the respect of membrane maintenance, membrane leakage, membrane strength and system complexity. RED technique seems more suitable for power generation from seawater than PRO technique.

3 Theoretical Analysis of RED Stack

A typical RED stack consists of a number of alternating cation and anion ion exchange membranes. The compartments between these membranes are fed with fresh water and salty water. The generated electromotive force of this RED cell could be given by:

$$E = (\alpha_{CEM} + \alpha_{AEM})\frac{RT}{zF}\ln(\frac{\alpha_c}{\alpha_d}) \tag{1}$$

Where:

E——Electromotive force generated by the diffusion of Na+ and Cl- ions;

α_{CEM} ——Perm-selectivity of the cation exchange membrane;

α_{AEM} ——Perm-selectivity of the anion exchange membrane;

R——Gas constant (R= 8.31 J/(mol·k));

T——Temperature of the solution (K);

z——Valency (where z=1 for Na+ and Cl-);

F——Faraday constant (F=96485.3 C/mol);

α_c ——Activity of ion Na+ and Cl- in salty water

α_d ——Activity of ion Na+ and Cl- in fresh water.

In an ideal RED stacks there is no ionic shortcut current. The internal resistance of RED stack R_i depends on each cell resistance r, number of cells N and resistance of electrode R_{cl}. That is:

$$R_i = N \cdot r + R_{cl} \tag{2}$$

Where the cell resistance r is the sum of membrane resistances and compartment resistances.

$$r = R_{CEM} + R_{AEM} + R_f + R_s \tag{3}$$

Where:

R_{CEM} ——Resistance of the cation exchange membrane;

R_{AEM} ——Resistance of the anion exchange membrane;

R_f ——Resistance of fresh water compartment;

R_s ——Resistance of salty water compartment.

$$R_{CEM} = \rho_{CEM} \cdot A \tag{4}$$

$$R_{AEM} = \rho_{AEM} \cdot A \tag{5}$$

$$R_f = \frac{l_f}{f_v \cdot k_f \cdot A} \tag{6}$$

$$R_s = \frac{l_s}{f_v \cdot k_s \cdot A} \tag{7}$$

Where:

ρ_{CEM} ——Area resistance of the cation exchange membrane (Ω/m^2);

ρ_{AEM} ——Area resistance of the anion exchange membrane (Ω/m^2);

A ——Area of the cell (m^2);

l_f, l_s ——Thickness of fresh water and salty water compartment respectively (m);

k_f, k_s ——Specific conductivity of fresh water and salty water respectively (s/m);

f_v ——Void factor of the volume occupied by the solutions.

To minimize the internal resistance of the stack R_i, the thickness and area of the compartment should satisfy the following equation as:

$$\frac{l_c}{l_d} = \frac{k_c}{k_d} \tag{8}$$

$$A = \sqrt{\frac{1}{\rho_{CEM} + \rho_{AEM}} \cdot \left(\frac{l_c}{k_c} + \frac{l_d}{k_d}\right)} \tag{9}$$

4 Experimental Study

As the salinity of sea water in the East Sea is between 30g/L and 35g/L, while the salinity of fresh water in the Yangtze River is between 2.5g/L and 12g/L, the salinity of the tested salty water and diluted water were set as 30g/L and 3g/L respectively. The temperature of solution was 298.15K. Two kinds of ion exchange membranes were used. One is a heterogeneous ion exchange membrane and the other is a homogeneous ion exchange membrane of type DF120. The area resistance of DF120 membrane is between 5.9 and 6.5 Ω/cm2 .The perm-selectivity of it is between 96.2 and 96.5%. Two titanium-coating plates were used as electrodes with good electric catalytic activity, catalytic selectivity and high current efficiency. A sensitive lvium potentiostat of type J0409 was used with measuring range of –300 ~ 0 ~ +300 μA, 2.5 grade precision and internal resistance of 108 Ω.

The tests were done with four single-cell homogeneous ion exchange membrane stacks and one multi-cell heterogeneous ion exchange membrane stack. Structural dimensions of these stacks are shown in Table 1.

Table 1. Structural dimensions of the tested RED stacks

Test stacks	Thickness of fresh water compartment (cm)	Thickness of salty water compartment (cm)	Area of membranes (cm×cm)
1# single-cell stack	4 cm	22 cm	15 cm × 20 cm
2# single-cell stack	4 cm	12 cm	15 cm × 20 cm
3# single-cell stack	4 cm	22 cm	10 cm × 15 cm
4# single-cell stack	4 cm	12 cm	10 cm × 15 cm
Multi-cell stack	8 cm	8 cm	80 cm × 40 cm

Measures were carried out with the Ivium potentiostat. From the measured I (U) and calculated internal resistances, the maximum electric power were got as shown in Fig.3, Fig.4 and Fig.5.

From the experiment results, we could see that.

1. Output current of the multi-cell stack is only 50μA, obviously smaller than that of single-cell stacks as shown in Fig.3. This is because the usage of the heterogeneous ion exchange membrane (IEM) in the single-cell stacks. Area resistance and total area

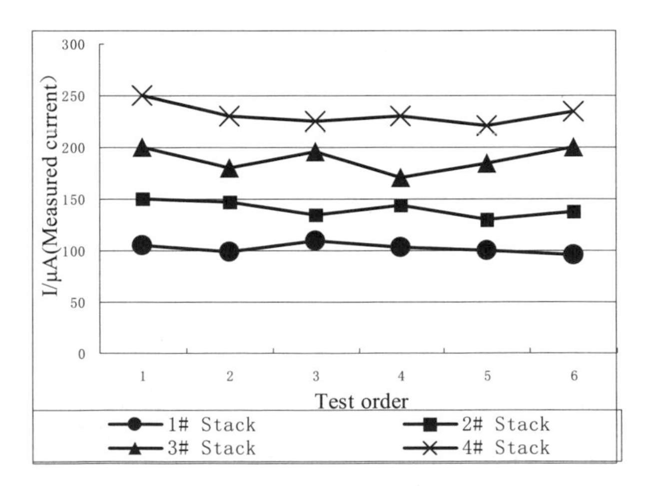

Fig. 3. Measured currents of the tested single-cell stacks

of the heterogeneous IEM is larger than that of the homogeneous IEM, which makes smaller output current in the RED stack.

2. Comparing Fig.3 Fig.4 and Fig.5, it could be found that smaller thickness of compartment produces higher current and bigger power with the same kind of membrane. This may be due to the bigger perm-selectivity of the solutions under narrower compartment. It is also found that smaller membrane area makes higher current and bigger power. This is due to smaller membrane resistance with smaller area.

3. These tests verify the feasibility of electrical power generation with RED stacks. To produce bigger electrical power, the RED stacks should be made as compact as possible.

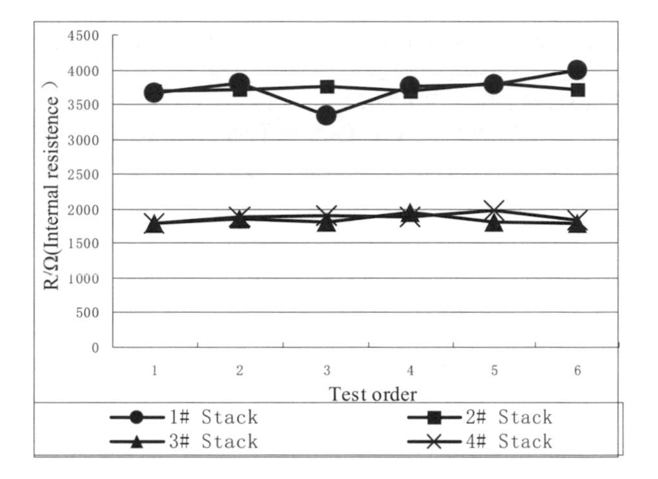

Fig. 4. Calculated resistances of the tested single-cell stacks

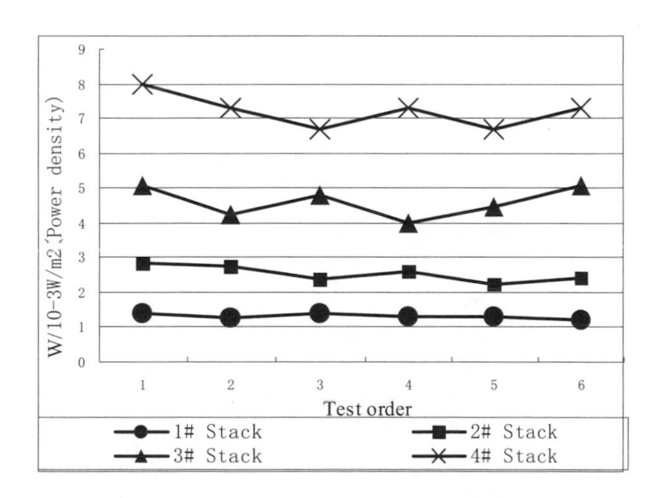

Fig. 5. Measured power densities of the tested single-cell stacks

References

1. Zhu, T.: Development and application of ocean energy, 1st edn. Chemical Industry Press (2004)
2. Ji, J.: Experimental study on electrical power generation from seawater osmotic energy. Dissertation for M. S., Shanghai maritime university (2008)
3. Yan, X.: Osmotic power generation. Solar energy, No. 1, pp. 6–7 (1999)
4. van den Ende, K., Groeman, F.: Blue Energy. KEMA consulting (October 2007), http://www.Leonardo-energy.org
5. Monney, N.: Ocean Energy Resources. Press of the American society of mechanical engineers (1977)
6. Wick, G.L., Schmitt, W.R.: Harvesting Ocean energy. The Unesco Press (1981)
7. Post, J.W.: Evaluation of pressure-retarded osmosis and reverse electro-dialysis. Journal of Membrane Science (288), 218–230 (2007)
8. Mingxiang, L.: Chemical power, 1st edn. Press of Tianjin University (1992)

Vibration Monitoring of Auxiliaries in Power Plants Based on AR (P) Model Using Wireless Sensor Networks

Tongying Li and Minrui Fei*

School of Mechatronical Engineering & Automation, Shanghai University, Shanghai 200072, China
litongying@shu.edu.cn, mrfei888@x263.net

Abstract. There are many auxiliaries with high rotating speed in a power plant, such as pumps, fans, motors and so on.To warrant their safe and reliable operation, their state of vibration has to be monitored. But because of their scattered location, the traditional way of online monitoring with shielded cable connections is costly and work expensive and the precision, reliability and safety of itinerant measurements are unable to meet the requirements of customers. A novel method of vibration monitoring for auxiliaries in power plant based on wireless sensor networks (WSN) has therefore been proposed to realize vibration data acquisition, on-line-detection and data analyzing in this paper, which meets the requirements of auxiliaries with less expenditure and warrants safe operation in the long run. The multi-sink topological structure of WSN can improve the transmission efficiency of multi-hop network to meet the vibration test requirements of low-latency, high frequency sampling and high data throughput. The sensor node can schedule its sampling and communication time to minimize sampling frequency and communication traffic, reduce the energy consumption and prolong the lifetime of WSN according to the prediction value of the sample data probability mode.

Keywords: Wireless Sensor Networks, Vibration, multi-sink topological structure, sample data probability mode, dynamic scheduling, energy management.

1 Introduction[2]

In the large thermal power plant there are a large number of high-speed rotating machines. In addition to the host generator, there are a lot of auxiliaries, such as turbine pump, electric pump, fans, motors and so on. The vibration monitoring of these auxiliaries is an important means to warrant safe and reliable operation of these auxiliaries. At present, in the general power plant the online turbine supervisory instrumentation(TSI) is installed in the main equipment and the large auxiliaries, such as turbine pump and so on. Sensors are installed on the facilities and the field vibration signal is sent to the monitoring meter in the central control room by laying of shielded cable. Real-time monitoring is realized by monitoring the vibration of the unit. If the vibration signal of the measured points is too large, the contact signal will be emitted to forcibly shut down the equipment to warrant the security of the unit. Behind TSI some plants also install

* Corresponding author.

K. Li et al. (Eds.): LSMS/ICSEE 2010, Part II, CCIS 98, pp. 213–222, 2010.

vibration analysis and fault diagnosis system, such as FFT analysis, orbital analysis, waterfall chart analysis and so on. The unit's vibration causes are long-term real-time analyzed under different conditions, to identify the reasons of the unit's vibration, to predict the unit's lifespan and to determine it's maintenance time, that is lifespan estimation, which is widely used and is riper in power plants. However, small-scale rotating machines such as fans and motors are widely scattered in the plant in great number and are far from the central control room. If we adopt the above method, it means great amount of cable laying work and the cost is quite high. Therefore, generally the online long-term monitoring protective equipment is not installed in the field and the point inspection personnel in regular patrol record the data using the portable vibration meter. With the growing capacity of power plant unit and increasing automation, the requirement of the plant running safety performance is higher and higher. Ground damage and equipment obsolescence in many power plants caused by excessive vibration of fan, pump, etc lead to the serious consequence of the generator's shutdown. Now more and more customers are aware of the seriousness of the problem. Therefore the online test equipment is installed in many new power plants and some stronger companies. However, people generally believe that the cost is too high. Some other companies increase the number of inspection in order to reduce the risk of equipment failure by the point inspection personnel in regular patrol. It reduces the cost, but there are its own shortcomings as follows:

(1)Relatively portable vibration meter is less precise. It can only measure the amplitude, but it can not measure the waveform and analyse the spectrum to identify the problem.

(2)Auxiliaries are scattered in every corner of the power plant. Some are located at the point which personnel should not reach and some are located in the hot or dangerous area, where inspection personnel should not have the point inspection for their security when the auxiliaries are running. Therefore, the point inspection is of great difficulty and labor intensity.

(3)Because vibration is a vector, vibration values measured in different directions are very different. Power plant standard of vibration measurement is required in 45° horizontal or vertical direction. However, in practice it is very difficult for the inspection personnel to control the view direction and the data results measured by the different inspection personnel are different.

For the above-mentioned reasons, there is an urgent need for the vibration monitoring system of low cost, long-term monitoring, high precision and high reliability to monitor the vibration of the power plant's auxiliaries. In order to meet the needs of the customers, this paper proposes a new vibration monitoring and diagnosing system which is very suitable for monitoring and diagnosing the power plant auxiliary vibration based on WSN.

2 The Scheme of WSN for Vibration Monitoring [2-4]

Vibration is an important characteristic parameter of the running auxiliaries in power plant and the vibration monitoring is an important means of the auxiliary state monitoring

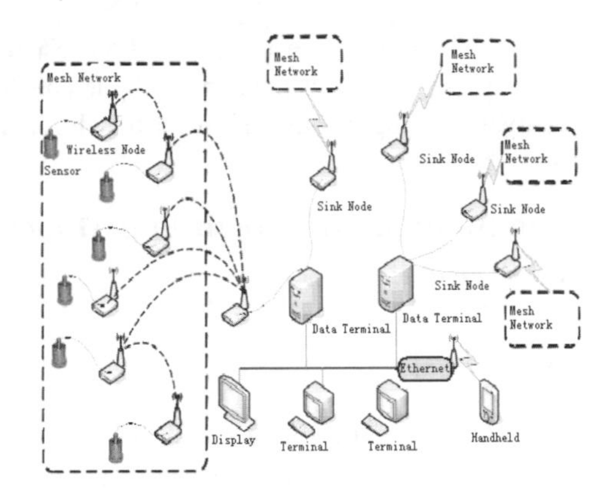

Fig. 1. The scheme of WSN for vibration monitoring

and fault diagnosis. People can obtain large amounts of detailed and reliable information at any time, place and any environmental conditions from WSN. Development of WSN technology provides a new way to monitor the auxiliaries' vibration in the power plant. To build a wireless, distributed vibration monitoring system using WSN monitoring mode can solve the conflict of the data acquisition range, precision and monitoring costs and achieve the free expansion and large-scale monitoring of the system at lower cost. At the same time, the wireless transmission of WSN can improve data collection methods of the monitoring process. Using the processing capacity of the sensor nodes, front-end data processing and data fusion synchronize the data collection and analysis to form sensor network mode of state maintenance which has preliminary self-analysis and self-diagnostic capabilities. This can not only improves the efficiency of data acquisition and processing but also effectively reduces the communication flow between the collection points and the system of back-end analysis and diagnosis. The scheme of WSN for vibration monitoring is shown in fig.1.

Mesh network is a new type of WSN which only allows a node to communicate with its adjacent nodes of short distance, in which all nodes are equal. The vibration monitoring process of auxiliaries in power plant requires higher transmission stability and higher data throughput of the network. At the same time the number of sensor nodes is small, network mobility is not strong after deployment and nodes need simple and reliable self-organization algorithm. In view of this, WSN uses multicast clustering network mode of upper multiple base station nodes(sink nodes) combined with bottom mesh network. Sensor nodes collect the vibration signals of auxiliaries. Under normal conditions the sensor nodes are usually in sleep or park. According to actual needs, the sink node will wake up the sensor nodes to query vibration conditions on it's own initiative. Then the sensor node will complete data collection in collaboration and send back the data to the sink node. If a vibration characteristic parameter exceeds the threshold, the sink node will require the sensor nodes to uninterruptedly send vibration data of the auxiliary region. On the top of the monitoring network the multiple base station nodes' processing capability is strong without regard to power supply problems because of cable connection of their ends. This realizes topology control of multiple data collection points and increases data transfer rate to meet the vibration test requirements of low-latency, high frequency sampling and high data throughput.The following is specific data collection process analysis and algorithm design for the above monitoring network.

3 Sampling and Communication Scheduling Modeling Based on Simple Probability Model

3.1 Modeling of Sample Data[9]

In the above-mentioned WSN of vibration monitoring, each bottom mesh network is corresponding to a sink node. Each sensor node in mesh network completes sampling and modeling collaborative tasks under the sink node scheduling.

As the resources of sensor nodes that can be used are limited, the modeling algorithm running on the sensor nodes should be as simple as possible and the memory size to be used also should be small. Therefore, a relatively simple AR(P) model is used for modeling of sample data. The following is the method of sample data modeling.

As the sampling data of the sensor node is a time series, we can use the probability model to denote the time series. For the limited resources of sensor nodes, we use the easily implemented AR(P) model[9] to denote the sampling data of the sensor nodes. The form of AR(P) model is given as follows:

$$X_t = \phi_1 X_{t-1} + \phi_2 X_{t-2} + \cdots + \phi_p X_{t-p} + \alpha_t (t \geq p), \alpha_t \sim NID(0, \delta_t^2) \tag{1}$$

Where, X_t is the sampling data at moment t, φ the regression coefficient, α_t a white noise sequence, which obeys the standard normal distribution. P-order model, denoted by AR(P), uses a linear combination of the front P sampling data and a random white noise to predict data at the moment t.

AR(P) model can be used for prediction. For example, for AR(1) model, it can be for k-step forward prediction with 95% confidence level at moment t given as follows:

$$X_t(k) \pm 1.96 \delta_t^2 (1 + \varphi_1^2 + \varphi_1^4 + \cdots + \varphi_1^{2(k-1)})^{1/2}$$

Where, $X_t(k)$ can be recursively calculated by formula (1), to make P=1.

Using the P-order AR(P) model to model, N sampling values $X_1, X_2, \ldots, X_N$ are to be collected by sensor nodes of the mesh network and then through least squares calculate the coefficients $\varphi_1, \varphi_2, \ldots, \varphi_P$ order:

$$Y = (X_{P+1}, X_{P+2}, \cdots, X_N)^r$$

$$\theta = (\varphi_1, \varphi_2, \cdots, \varphi_P)^r$$

$$A = \begin{bmatrix} X_P & X_{P+1} & \cdots & X_{N-1} \\ \vdots & \vdots & \vdots & \vdots \\ X_1 & X_2 & \cdots & X_{N-P} \end{bmatrix}^r$$

$$\varepsilon = (\alpha_{P+1}, \alpha_{P+2}, \cdots, \alpha_N)^r$$

Then the formula (1) can be expressed as Y=Aθ+ε and the coefficient matrix θ can be calculated through least squares method:

$$\theta = (A^r A)^{-1} A^r Y \tag{2}$$

The task of formula (2) is divided into multiple sub-tasks and each node implements one of them. Order $A_t=(X_t,X_{t-1},\ldots,X_{t-p+1})^r$, which denotes the sample sequence of length p at cut-off time t. Pay attention to $A_{t+1}=Push(Drop(A_t),X_{t+1})$. That is first to remove the sample X_{t-p+1}, second to add the new sample data X_{t+1} and then to get the transpose matrix A^r of matrix A, which is given as follows: $A^r=(A_p,A_{p+1},\ldots,A_{N-1})$.

In order to calculate the formula (2), we conduct respectively the following calculation:

$$A^r A = \sum_{k=p}^{N-1} A_k A_k^r \tag{3}$$

$$A^r Y = \sum_{k=p}^{N-1} X_{k+1} A_k \tag{4}$$

The formula (3) and (4) decompose the large matrix operation of the formula (2) into the vector operations of the length P without taking up much storage space. In practice, the values of P is usually not very large, for example, P=5, many actual sequences can be fitted[9].

In order to maintain energy balance of sensor nodes in WSN, the above calculation task is divided into several sub-tasks in accordance with the amount of residual energy of each node in the mesh network. The sub-tasks for each node to be completed is sampling within a specified time limited by task, calculating the partial sum of formula (3) and (4) with the sampling data and returning the above two partial sums to the sink node. Finally the sink node completes the formula (2) calculation. Task partitioning and scheduling is also completed by the sink node.

3.2 Sample Data Modeling Algorithm[5]-[8]

As the AR(P) model suppose the time series with mean 0, the modeling algorithms of sample data is first to estimate the mean of the time series, then is the sample value minus the estimated value of the mean at each sensor node, that is 0 average processing. The algorithm includes two parts, which is executed in the sink node and in each sensor node of mesh network. The used symbols are given as follows:

M: the number of participating computing nodes within a mesh network.

$Power_i$: Residual energy of node i.

N: The total number of samples.

N_i: The number of samples assigned to node i.

L_i: The starting sampling time of the designated node i.

μ_t: The calculated sample mean at moment t.

T_m: Sampling time that the sink node needs to calculate the initial average values.

X_j: The sampling value at the moment j.

P: the orders of AR(P) model.

ARA_i: The calculated partial value of A^rA at the node i in formula 3.

ARY_i: The calculated partial value of A^rY at the node i in formula 4.

A_i: the vector A_k in formula (3) and (4) at node i.

S_i: Sum of each node's sampling value that is used to calculate the average value.

The modeling algorithm of sample data is designed as follows:

Algorithm 1. Modeling algorithm within WSN.

Input: N samples data

Output: The matrix θ of AR(P) model and the varianceδ^2 of white noise sequence.

3.2.1 Algorithm Executed in the Sink Node

(1) $\mu_t = (\sum_{t-T_m \le j \le t} X_j) / |\{X_j | t - T_m < j < t\}|$;

//Caculate the average values the sampling data within T_m period and initialize $\dot{\mu_t}$.

(2) $N_i = Max(N * Power_i / \sum_{j=1}^{m} Power_j, P+1)$;

//Assign the number of sampling data according to the node residual energy and the numbe of sampling data is at least P+1.

(3) $L_i = (\sum_{j=1}^{i-1} N_j) - P + 1, where, 1 < i < m, L_1 = 1$;

// Calculate the starting sampling time of each sensor node.

(4) The sink node informs the results of these steps to the sensor nodes involved in the computation within the mesh network and then enter the sleep status.

(5)In accordance with the scheduling program, the sink node will return into the work state and receive the results when a sensor node within mesh network returns the results of subtasks. When all subtasks are finished, the following calculation is performed:

(6) $A^r A = \sum_i^m ARA_i$ //Implement the formula (3)

(7) $A^r Y = \sum_i^m ARY_i$ //Implement the formula (4)

(8) $\theta = (A^r A)^{-1} A^r Y$

(9) $\mu = (\sum_{i=1}^{m} S_i) / N$

//According to the return the partial sums of the sensor nodes calculate mean.

(10)The sink node informs each sensor node the parameter θ within the mesh network and receives their return residual values ε^2 and $Power_i$;

(11) $\sigma_\alpha^2 = (\sum_{i=1}^{m} \varepsilon_i^2)/m$

//Calculate variance of the white noise sequence in the AR(P) model.

3.2.2 Algorithm Executed in the Sensor Nodes within the Mesh Network

According to scheduling scheme of the sink node, in the time interval t_i=[L_i,L_i+N_i], the sensor node i perform the following steps:

Step 1. Calculate the modeling sub-tasks assigned to the sensor node i:

a)ARA_i=ARY_i=0;

b)While(t_i≤L_i+p-1)

c)Begin

d) $A_i[t_i] = X_{t_i} - \mu$; // the first P sampling data is processed with zero mean

e) t_i= t_i+1;

f)End

g)S_i=0;

h)While(t_i≤L_i+N_i)

i)Begin

j) ARA_i=ARA_i+$A_{ti-1}A^r_{ti-1}$; //Implement Formula (3)

k) ARY_i=ARY_i+(X_{ti}-μ)*A_{ti-1}; //Implement Formula (4)

l) At_i=Push(Drop(A_{ti-1}),X_{ti});

m) t_i= t_{i+1};

n) S_i=S_i+X_{ti};

o)End

Step 2. The sensor node i returns ARA_i, ARY_i and S_i to the sink node and then turns into sleep status;

Step 3. According to the scheduling program, when the sink node sends the parameter matrix the sensor node i will be waked up;

Step 4. The sensor node i calculate and return ε_i and $Power_i$;

$$\varepsilon_i = X_{L_i+N_i} - \sum_{k=L_i+N_i-1}^{L_i+N_i-p} \varphi_k X_k$$

4 Realization of Sampling and Communication Scheduling

Through algorithm 1, each sink node establishs the AR(P) model on the sampling data of the monitoring target.With the AR(P) model and the user-specified precision the sensor nodes are dynamically woken up to sample and transmit the data so as to minimize the energy consumption. The basic idea of vibration sampling and time scheduling algorithm is given as follows: With the AR(P) model prediction capability, we can predict the future data and dynamically adjust sampling time interval according to historical data and the user's precision requirement. If the predicted value can meet the precision requirement, we need not start the actual sampling and data transmission of the sensor nodes and can properly extend the sensor nodes's sleep time. If the forecast error is over the precision range, on the one hand we need the real sampling and data transmission, on the other hand, we must properly adjust the AR(P) model.

For AR(P) model, the l-step forward prediction value at the moment t can be calculated by the following formula:

$$X_t(l) = \sum_{k=1}^{P} \varphi_k X_t(l-k)$$

The corresponding 1-α confidence interval is given as follows:

$$X_t(l) \pm z_{\alpha/2}\sigma_\alpha (1 + G_1^2 + \cdots + G_{l-1}^2)^{1/2} \tag{5}$$

Where, $\varphi(z_{\alpha/2} = 1\text{-}\alpha/2), \sum_{i=1}^{j} \varphi_i G_{j-i}, G_0 = 1$, φ is standard normal distribution function, G_j is called Green Function[9].

According to the above sampling and scheduling model the sampling and scheduling algorithm is designed as follows. In the algorithm, when a sensor node within mesh network is awakened, it first samples, checks the error between the predicted value calculated by the formula (5) and the current sampling value, and then determine the next sampling time, whether to transmit data and adjust the model, etc according to the above error. The involved symbols of the sampling and time scheduling algorithm are given as follows:

T: The initial sampling period

t_0: The start time of the sensor node's sampling

len: Specify the number of a sensor node's samples.

t: The current time

T_1: The sampling period after adjustment.

X_t: The predicted value at the moment t.

x_t: The sampling value at the moment t.

error_bound: The threshold allowed by the error.

n_error: The number of errors which exceed the threshold within the sampling period developed by the system.

Algorithm 2 The vibration sampling and communication scheduling algorithm of WSN

1)T_1=T; // the initial sampling period of the sensor node

2)After each time interval T1, the sensor node will be waked up. Suppose the current time is t, the following steps are excuted by the sensor node :

3)If ($t<t_0$+len*T) //The current time is still within the specified sampling time

4)Begin

5) If $|X_t-x_t|$>error_bound

6) Begin

7) n_error++;

8) if n_error>threshold of the system

9) T_1=max(T,T_1-T/2);

//Shorten the sampling period and obtain more accurate sampling data.

10) return x_t; // Return the real sample data.

11) End

12) Else

13) T1=T1+T/2; //Increase the sampling period.

14)End

15)Else //The specified sampling task has been completed.

16) Go to the sleep mode

5 Conclusion

The above method of vibration monitoring based on WSN is a novel method of monitoring. Its main feature is wireless and online. It can collect waveforms to analyse with high precision and reliability. In this paper, the vibration measurement is limited to auxiliaries of rotating machinery in the power plant and it can be gradually applied to the host vibration monitoring after we accumulate some experience. It thus has wide prospects of application.

Acknowledgements

Supported by National High Technology Research and Development Plan of P.R.China under Grant 2007AA04Z174 and International Cooperation Project of Shanghai Municipality under Grant 08160705900.

References

1. Zhu, X.L.: Research on vibration monitoring in power plant. Southeast University Journal 23(8), 131–136 (1992)
2. Rauch, D.: Towards the Sensor Web: A Framework for Acquistion, Storage and Visualization of Wireless Sensor Networks Data. Master's thesis, Distributed Systems Group, Department of Computer Science, ETH Zurich, Zurich, Switzerland (February 2008)
3. Kung, H.Y., Hua, J.S., Chen, C.T.: Drought forecast model and framework using wireless sensor networks. Journal of Information Science and Engineering 22(4), 751–769 (2006)
4. Pierce, F.J., Elliott, T.V.: Regional and on-farm wireless sensor networks for agricultural systems in Eastern Washington. Computers and Electronics in Agriculture 61(1), 32–43 (2008)
5. Li, T.Y., Fei, M.R.: Information fusion in wireless sensor network based on rough set. In: Proceedings of 2009 IEEE International Conference on Network Infrastructure and Digital Content (IEEE IC-NIDC 2009), pp. 129–134 (2009) (EI: 20101012756412)
6. Li, T.Y.: Fuzzy Query of Data Warehouse. Jiangnan University (2002)
7. Li, T.Y.: Query for Distributed Database based on Rough Sets. Applications of the Computer Systems (1), 26–29 (2005)
8. Li, T.Y.: Fuzzy Optimization Query for Web-based Database. Computer Engineering 28(5), 97–99 (2002)
9. Sudhakar, M.P.: Time Series and System Analysis with applications. John Wiley and Sons, New York (1983)

Performance Prediction for a Centrifugal Pump with Splitter Blades Based on BP Artificial Neural Network

Jinfeng Zhang, Shouqi Yuan, Yanning Shen, and Weijie Zhang

Research Center of Fluid Machinery Engineering and Technology Jiangsu University, Zhenjiang 212013, China
zhangjinfeng@ujs.edu.cn

Abstract. Based on MATLAB, a BP artificial neural network (BPANN) model for predicting the efficiency and head of centrifugal pumps with splitters were built. 85 groups of test results were used to train and test the network, where the Levenberg—Marquardt algorithm was adopted to train the neural network model. Five parameters Q, Z, β_2, D_i, b_2 were chosen in the input layer, η and H were the output factors. Through the analysis of prediction results, the conclusion was got that, the accuracy of the BP ANN is good enough for performance prediction. And the BP ANN can be used for assisting design of centrifugal pumps with splitters. Meanwhile, the method of CFD flow field simulation was also used to predict the head and power for a centrifugal pump with splitters, and compared with that from the BPANN model. The comparison of prediction results and experimental value demonstrated that the prediction values acquired through numerical simulation and BPANN were uniform with the test data. Both methods could be used to predict the performance of low specific speed centrifugal pump with splitters.

Keywords: BP artificial neural network, centrifugal pumps with splitters, Performance prediction.

1 Introduction

Yuan [1-3] revealed that the splitter blade is one of the techniques to solve most hydraulic problems of low specific speed centrifugal pumps. Miyamoto et al.[4] examined the influence of splitter blades by measuring the velocity and pressure in un-shrouded and shrouded impeller passages. Asuaje et al.[5] studied the splitter blades effect on the performance of a centrifugal pump through both numerical simulation and experiments. Experimental studies were made to investigate the effects of splitter blade length on deep well pump performance for different blade numbers [6, 7]. And the authors used an ANN to predict the performance of deep well pumps with and without splitter blades [8, 9]. We also conducted an optimization and PIV test to study the influence of splitter blades on the flow of centrifugal pump impeller [10].

In this study, a BPANN has been used for predict the performance of centrifugal pumps with splitters. 85 groups of test results were used to train and test the network, where the L-M algorithm was adopted to train the neural network.

K. Li et al. (Eds.): LSMS/ICSEE 2010, Part II, CCIS 98, pp. 223–229, 2010.

2 Test and Impellers

As shown in Fig.1, most of the tests were carried on the open test rig, with accuracy of II class. 85 groups test results were used to train and test the ANN model and predict performance[10-12].

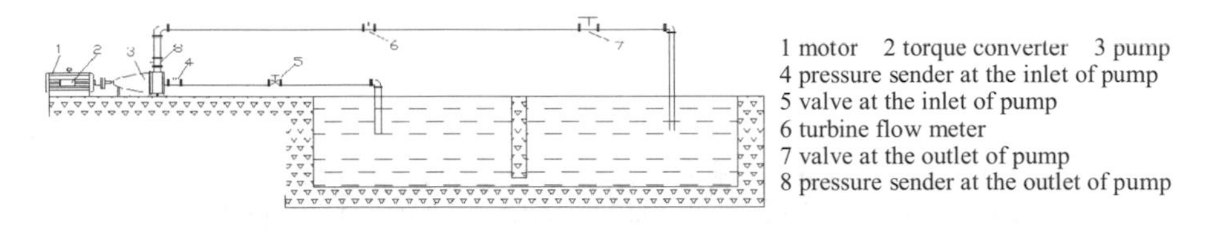

Fig. 1. Sketch of open pump performance test rig

Parts of impellers were shown in Fig.2. The model pump No. is IS50-32-160, with designed flow rate: Q= 25m^3/h, head: H=32m, rotation speed: n=2900r/min, specific speed: n_s=47, rated efficiency: η= 56%, horse power: P=3 kW.

Fig. 2. Some impellers machined by the technology of Fused Deposition Modeling(FDM)

3 Performance Predictions by BP Artificial Neural Network

3.1 BP Algorithm and ANN Based Model

The BPANN in this study use the hyperbolic tangent sigmoid activation function in the input and hidden layer and the linear (*purelin*) activation function in the output layer. According to the former research on the centrifugal pumps with splitters [10], five parameters are chosen as the input parameters: flow rate Q, blade number Z, blade outlet angle β_2, splitter inlet diameter D_{in}, blade outlet width b_2. And the performance parameters head H and efficiency η are chosen the output parameters. And the number of the neutrons was selected by many times tries, by comparison, 20 neutrons was determined. So the selected BP ANN model of a three-layer with 5 inputs, 20 hidden neurons and 2 outputs has been shown in Fig. 3.

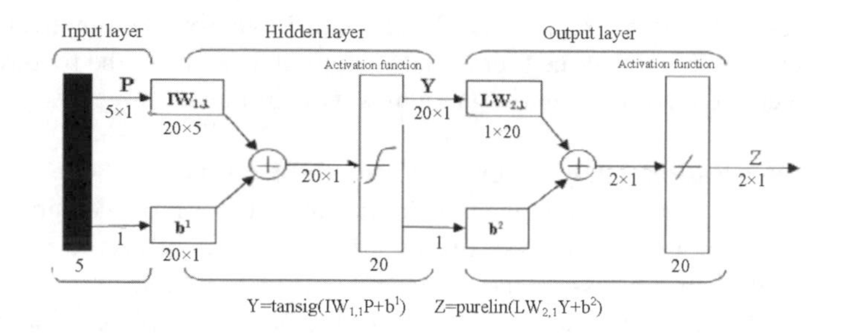

Fig. 3. a BP networks model with 5 inputs, 20 hidden neurons and 2 outputs

3.2 Performance Prediction by BP ANN

3.2.1 Model Training

The function *premnmx* was used for normalizing the inputting and outputting data, the function *tramnmx* was used to pre-treat the data, and the function *postmnmx* was used to realize the normalization data restoration. The network function *newff* was established to build the BP neural network, and the function *trainlm* was chosen as a training function, and the learning algorithm is L-Malgorithm.

The train function was applied to train the established network, and the training times was set up as 400, learning efficiency was set as 0.05, the network's target error is 10^{-3}, after 38 times training, the training achieve the required accuracy.

3.2.2 Network Simulation Testing and Analysis

The function *sim* was used to simulation tests, and 12 test data were randomly selected as the test samples. The results of simulation show that the maximum relative deviation of the predicted efficiency is 10.10%, the average relative deviation is 5.97%, and the minimum relative deviation is 1.19%, and those corresponding deviations of the predicted head are 8.71%, 4.78% and 0.44% respectively.

Fig.4 shows the predicted efficiency and head by the BP ANN model, and compared with the original experimental data, respectively. The data predicted by BP ANN show

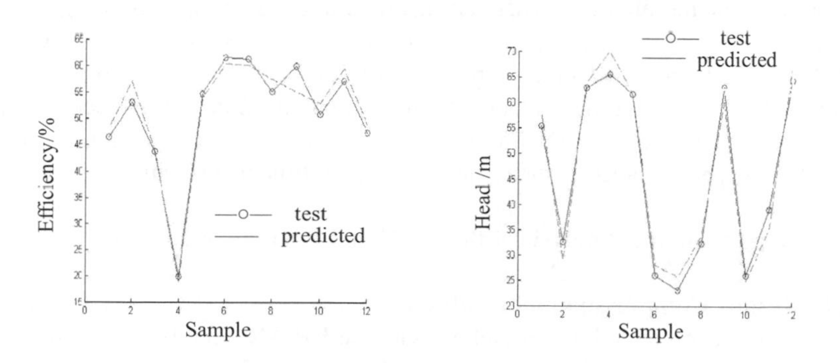

Fig. 4. Test and ANN predicted results for efficiency and head

the similar trend with test, except several data with large error. From error analysis, conclusions can be drawn that the trained network model can meet the requirement for the performance prediction of centrifugal pumps with splitters.

3.2.3 Linear Regression Analyses for the Predicted Results

The function *postreg* was used to analyze the linear regression of the network output results. Fig.5 is the regression analysis results, whose vertical axis is the change rate of the predicted value; the abscissa is the change rate of the test value.

The value of Pearson Correlation Score(R) for the predicted efficiency and head are 0.978 and 0.989 respectively, which prove that the correlation between the predicted and test value and the generalization capability are very good.

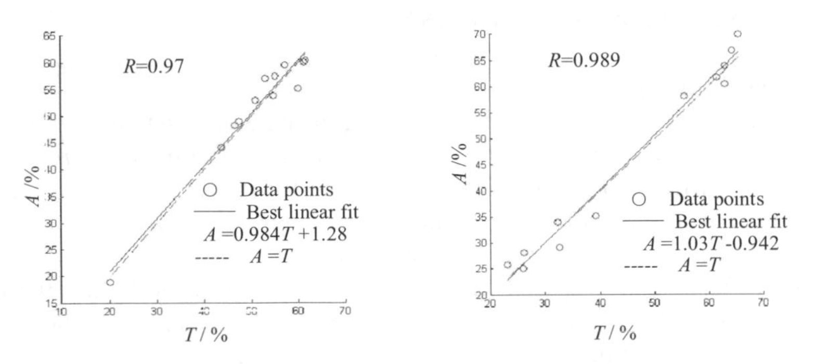

Fig. 5. Linear regression analysis for ANN predicted efficiency and head

4 Performance Predictions by CFD and ANN under Different Operation Conditions

4.1 Performance Predicted by CFD and BP ANN

During CFD performance prediction, The steady turbulent numerical simulations were finished in Fluent6.3, and the performance of power and head were predicted. The 3D models for calculation were generated in Pro/E and the computational grids for different schemes were formed in Gambit, which is the pre-processor of Fluent. And for turbulence modeling, a RNG k-ε turbulence closure in the RANS approach had been adopted. The details on the fundamental mathematical models had been given in the reference [10]. And the BP ANN model introduced in the former section was used also to predict the performance of the same pumps as in Fluent simulation.

4.2 Comparison and Analysis of the Predicted Performances

4.2.1 Results Comparison and Analysis

Fig.6 shows the predicted power and head by the BP ANN model, and compared with the original experimental data, respectively. The data predicted by Fluent and BP ANN

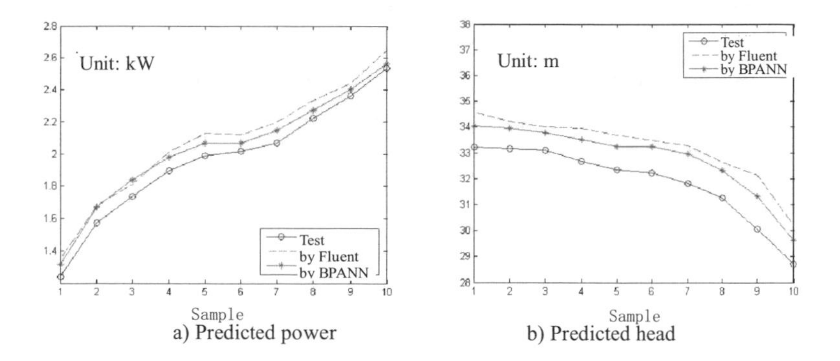

a) Predicted power
b) Predicted head

Fig. 6. Power and Head comparison between predicted value and test data

model show the similar trend with the test data, and the performance predicted by BP ANN are more closer to the test data.

4.2.2 Relative Deviation of the Predicted Value

The average relative deviation for Fluent and ANN predicted power are 5.69% and 3.78% respectively ; and the value for predicted head are 3.87% and 2.97% respectively. And the maximum and minimum relative deviations for them are all listed in Table1. The two prediction methods have the similar trend, and the BP ANN method has relatively smaller deviation.

Table 1. Relative deviation of the predicted value

Relative Deviation	Predicted Head		Predicted Power	
	Fluent	BP ANN	Fluent	BP ANN
the maximum relative deviation	6.96%	4.16%	8.70%	6.11%
the average relative deviation	3.87%	2.97%	5.69%	3.78%
the minimum relative deviation	2.75%	2.05%	3.26%	1.14%

4.2.3 Linear Regression Analyses for the Predicted Results

The linear regression of the predicted results by Fluent and BP ANN were analyzed by using the function *postreg* in MATLAB, as show in Fig.7 and Fig.8.

The value of R for Fluent predicted and test power is 0.998, and the R for ANN predicted and test power is 0.999. Meanwhile, the value of R for Fluent predicted and test head is 0.984, and the R for ANN predicted and test power is 0.995. Which prove that the correlation between the predicted and test value and the generalization capability are very good, which can provide reference value for the performance prediction of centrifugal pumps with splitters.

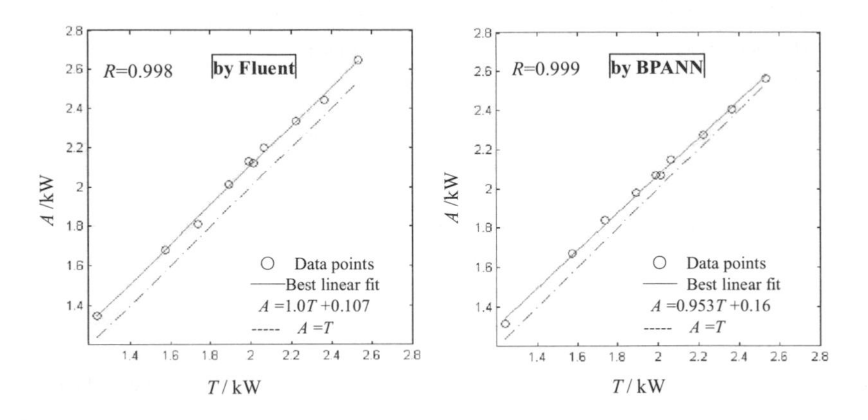

Fig. 7. Linear regression analysis of predicted power results by Fluent and ANN

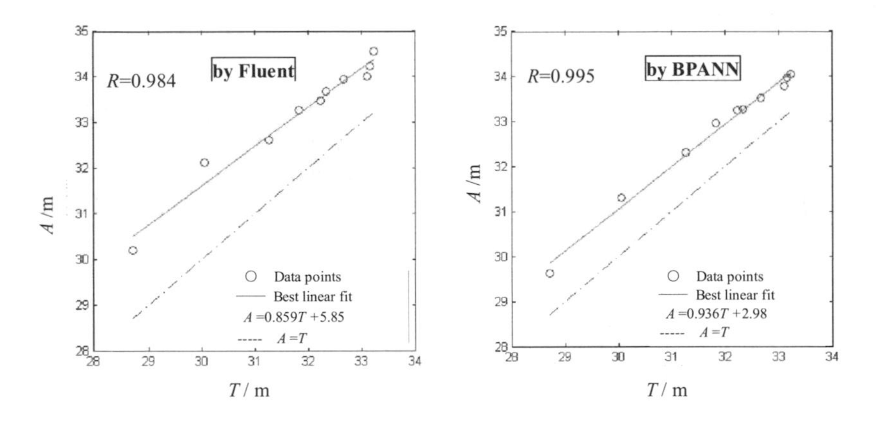

Fig. 8. Linear regression analysis of predicted head results by Fluent and ANN

5 Conclusions

(1) A BPANN model with 5 inputs, 20 hidden neurons and 2 outputs for predicting performance of centrifugal pumps with splitters were built. The average relative deviation for the efficiency and head are 5.97% and 4.78% respectively, and the predicted data has the similar trend with test data, and the value of R for the predicted efficiency and head are 0.978 and 0.989 respectively, which prove that the correlation between the predicted and test value and the generalization capability are very good, and the trained network model can meet the requirement for the performance prediction of centrifugal pumps with splitters.

(2) Meanwhile, the method of CFD was also used to predict the head and power for a centrifugal pump with splitters, and compared with that from the BPANN model. The prediction values acquired through CFD and BPANN were uniform with the test value.

Both the two methods could be used for the performance prediction of low specific speed centrifugal pump with splitters.

Acknowledgments

This study is supported by National Outstanding Young Scientists Foundation of China (No.50825902), National Natural Science Foundation of China (No.50979034), and Natural science Foundation of Jiangsu Province (No.BK2009218) and the Foundation for senior person with ability in Jiangsu University of China (No.08JDG040). The authors would like to thank the others of our research team for their working.

References

1. Yuan, S.Q.: Advances in Hydraulic Design of Centrifugal Pumps. In: ASME, Fluid Engineering Division Summer Meeting, Vancouver, BC, Canada, pp. 1–15 (1997)
2. Yuan, S.Q.: The Theory and Design of Low Specific-Speed Centrifugal Pumps. Mechanical Industry Press, Beijing (1997) (in Chinese)
3. Yuan, S.Q., Chen, C.Q.: CAO Wuling: Design Method of Obtaining Stable Head-Flow Curves of Centrifugal Pumps. In: ASME. Pumping Machinery Meeting, FED, vol. 154, pp. 171–175 (1993)
4. Miyamoto, H., Nakashima, Y., Ohba, H.: Effects of Splitter Blades on the Flows and Characteristics in Centrifugal Impellers. JSME Int. J. Ser. 2: Fluid Eng. Heat Transfer Power Combustion Thermo Physical Properties 35, 238–246 (1992)
5. Asuaje, M., Bakir, F., Noguera, R., et al.: 3-D Quasi-Unsteady Flow Simulation in a Centrifugal Pump. Influence of Splitter Blades in Velocity and Pressure Field. ASME Fluids Engineering Forum - FEDSM 2004 - HT-FED2004-56600 (2004)
6. Gölcü, M., Pancar, Y., Sekmen, Y.: Energy Saving in a Deep Well Pump with Splitter Blade. Energy Conversion and Management 47, 638–651 (2006)
7. Gölcü, M., Pancar, Y.: Investigation of Performance Characteristics in a Pump Impeller with Low Blade Discharge Angle. World Pumps 9, 32–40 (2005)
8. Gölcü, M.: Neural Network Analysis of Head-Flow Curves in Deep Well Pumps. Energy Conversion and Management 47, 992–1003 (2006)
9. Gölcü, M.: Artificial Neural Network Based Modeling of Performance Characteristics of Deep Well Pumps with Splitter Blade. Energy Conversion and Management 47, 3333–3343 (2006)
10. Yuan, S.Q., Zhang, J.F., Yuan, J.P., et al.: Research on the Design Method of the Centrifugal Pump with Splitter Blades. In: Proceedings of the ASME, Fluids Engineering Division Summer Meeting, Vail, Colorado USA (August 2009)
11. Chen, S.S., Zhou, Z.F., Ge, Q., et al.: Orthogonal experimental study on centrifugal pump with deviated splitter vanes. Journal of Yangzhou University Natural Science Edition 8, 45–48 (2005) (in Chinese)
12. Luo, C.L., Yuan, Y.C.: The Problem of the Efficiency and Unsteady Performance Curve for Low-Specific Speed Centrifugal Pump. Pump Technology 3, 20–27 (1983) (in Chinese)

Short-Term Traffic Flow Prediction Based on Interval Type-2 Fuzzy Neural Networks

Liang Zhao

College of Electrical Engineering, Henan University of Technology,
Zhengzhou 450007, China
zhaoliang270@gmail.com

Abstract. In this paper, a TSK interval type-2 fuzzy neural network is proposed for predicting the short-term traffic flow. The proposed fuzzy neural network is adaptively organized from the collected short-term traffic flow data. The whole process includes structure identification and parameter learning. In structure identification, the hierarchical fuzzy clustering algorithm performs the training traffic flow data set in order to generate the network structure. After the structure identification is finished, the BP algorithm is adopted to perform the parameter learning. Then the trained fuzzy neural network is employed the collected short-term traffic flow test set and the prediction result verifies that the TSK interval type-2 fuzzy neural network has high prediction accuracy.

Keywords: traffic flow prediction, TSK interval type-2 fuzzy neural network, structure identification, parameter learning.

1 Introduction

In intelligent transportation systems, traffic flow prediction is one of most important elements of traffic monitoring and control. In many important traffic management aspects, such as traffic flow dynamic induction, congestion management, emergency warning and management, traffic flow prediction is indispensable. In terms of the current technological means, it is impossible that the future traffic flow states are completely accurate predicted. The cause of traffic flow is very complicated, which is the interaction result of a large number of uncertain factors. In this paper, interval type-2 TSK fuzzy neural network (IT2FNN) is employed to predict the short-term traffic flow.

So far, there are many methods used to conduct short-term traffic flow forecasting, such as dynamic traffic distribution method [1-2], time series analysis method [3], historical data average method [4-5], etc. In the past several years, fuzzy neural networks have been applied to the traffic flow prediction [6]. In [7], fuzzy time series is employed to predict short-term traffic flow. In [8-9], the neural network prediction method was systematically expounded. Recent studies have shown that the predictive ability of artificial neural network is better than those of Kalman filter model and ARIMA model in information processing. Although the fuzzy neural network has high predictive ability, it is not suitable for handling traffic flow data, which contain a

K. Li et al. (Eds.): LSMS/ICSEE 2010, Part II, CCIS 98, pp. 230–237, 2010.

large number of uncertain information. Therefore, IT2FNN is employed to build the short-term traffic flow model.

In this paper, we propose IT2FNN short-term traffic flow prediction method. IT2FNN is more complicate FNN which has a more powerful predictive ability. The whole modeling process can be divided into two stages: structure identification and parameter learning. In the structure learning phase, the subtractive clustering determines size of fuzzy rule base at first, and FCM algorithm determines the initial parameters of the precondition and consequence of the IT2FNN. In order to obtain higher prediction accuracy, we use BP algorithm to adjust the network parameters of the precondition and consequence. To confirm its prediction capability, we apply the actual traffic flow data collected to verify the established prediction model.

The whole paper is organized as follows. Section 2 describes the IT2FNN prediction model. Section 3 introduces how to build short-term traffic flow prediction model. Section 4 validates the effectiveness of our proposing model. Section 5 concludes remarks and points out the future research direction.

2 TSK IT2FNNs

In this section, we introduce the IT2FNN structure shown in figure 1. IT2FNN is a new fuzzy neural network system whose rule base represents expert knowledge or experience. In figure 1, the input vector is x and in the membership function layer, an IT2 Gauss membership function is employed, which centers m are certain and variances σ are uncertain. The detailed introduction about the IT2FNN structure and reasoning process can be illustrated in [10]-[11].

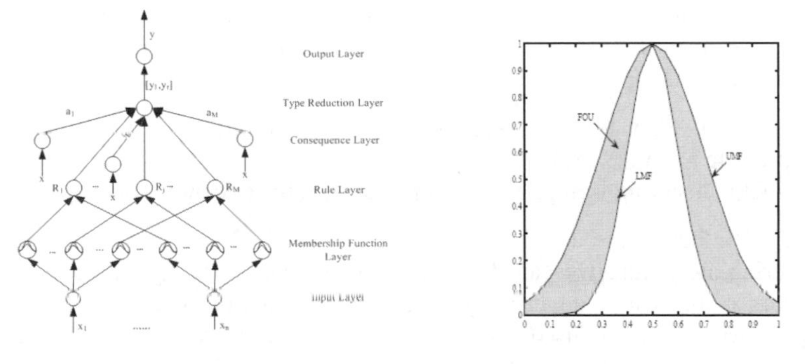

Fig. 1. IT2FNN

Fig. 2. IT2 Gauss membership function

The IT2FNN can realize the following IT2 fuzzy inference process.

Rule j: If x_1 is $\tilde{A}_{1j}$ and x_2 is $\tilde{A}_{2j}$ and … and x_n is $\tilde{A}_{nj}$, then $y = C_{0j} + C_{1j}x_1 + \cdots + C_{nj}x_n$.
Where $C_{ij} = \left[c_{ij} - s_{ij}, c_{ij} + s_{ij}\right]$.

In following section, we will discuss the learning algorithm of the IT2FNN.

3 The Learning Algorithm of IT2FNNs

Similar to its type-1 counterpart, the learning process of interval type-2 fuzzy neural network can be divided into two steps: (1) structure learning; (2) parameter learning. Structure identification includes the precondition and consequence identification of interval type-2 fuzzy neural network. Its precondition structure identification corresponds to the input space division, which determines the number of the fuzzy rules and linguistic values of each input variable. Meanwhile, the consequence structure identification corresponds to determine the parameters of the consequence linear function. Parameter learning includes the accurate adjustment of free parameters of precondition and linear function parameters of consequence. The following section describes the learning algorithm of IT2FNN.

3.1 Structure Learning

The structure learning is equivalent to the division to the input/output sample space. Usually the lattice method is applied to divide the input/output space. However, it will cause the "curse of dimensionality". In order to avoid the "curse of dimensionality" problem, clustering methods have been developed to deal with input/output space division and generate the appropriate number of fuzzy rules. Currently, there are many kinds of clustering algorithms, such as classification tree, nearest neighbor clustering, subtractive clustering, k-means clustering, fuzzy c-means clustering and self-organization competition law. This paper adopts the hierarchical fuzzy clustering algorithm to determine the IT2FNN structure.

3.1.1 Hierarchical Fuzzy Clustering Algorithm

Before making calculations, the number of sample data n, the neighborhood radius r_a and r_b and the error allowed value ε should be determined in advance according to the specific questions.

(1) The subtractive clustering is used to determine the number of clusters of the sample data $\{x_1, x_2, \ldots, x_n\}$.
(2) The FCM clustering is applied to determine the cluster center $\{m_1, m_2, \ldots, m_M\}$.

For the IT2FNN, the sample data is $\{(x_1, y_1), \ldots, (x_n, y_n)\}$. Based on this hierarchical fuzzy clustering algorithm, we can determine the number of fuzzy rules, the initial center vectors of the Gaussian primary membership function and the initial linear function parameters of the consequence c_{0j} , $j = 1, 2, \ldots, M$. We will determine the uncertain variances $\overline{\sigma}_{ij}$ and $\underline{\sigma}_{ij}$ and the parameters of consequence line function c_{ij} and s_{ij} , $i = 1, 2, \ldots, n$, $j = 1, 2, \ldots, M$.

3.1.2 Initial Parameters Determination

After the hierarchical fuzzy clustering, we will determine the remaining initial parameters.

(1) The center matrix M is re-sequenced from top to bottom starting from the first row. The subtracting results of the adjacent elements of the matrix M are compared and the greater numbers shall be the initial variance. Finally the initial variance matrix $\sum$ can be obtained by the inverse sorting.

(2) The each element of the initial variance matrix $\sum$ will be introduced by the uncertain parameters λ to get initial uncertainty variance matrix.

$$\overline{\Sigma} = \frac{\Sigma}{\lambda} \qquad \underline{\Sigma} = \lambda\Sigma \tag{1}$$

(3) The line function parameters C will be introduced by the uncertain parameters γ. Therefore, the consequence variances are determined according to the consequence centers.

$$s = \gamma c \tag{2}$$

After structure identification, we determine the number of fuzzy rules in fuzzy rule base and the initial parameters of precondition and consequence. Certainly they are very rough. In order to further improve the modeling accuracy, it is necessary to further fine-tune the initial parameters.

3.2 Parameters Learning

The initial parameters determined in structure learning are relatively rough. In order to improve the modeling accuracy, the BP algorithm is adopted to further fine-tune the parameters of the precondition and consequence. The aim that the BP algorithm adjusts the parameters reduces the errors between the network output and actual output to minimize. That is

$$E = \frac{1}{2}\left[y_d(k) - y(k)\right]^2 \tag{3}$$

Here the Gauss functions which have certain centers and uncertain variances are determined as the primary membership functions. The ideal output sample data are expressed as $y_d(k)$ and $y(k)$ is said actual output data. The upgrade rules of the adjustable parameters of the IT2FNN are derived as follows.

$$\begin{aligned} \overline{\sigma}_{ij}(k+1) &= \overline{\sigma}_{ij}(k) - \eta^{\overline{\sigma}} \frac{\partial E}{\partial \overline{\sigma}_{ij}}(k) + \beta^{\overline{\sigma}} \Delta \overline{\sigma}_{ij}(k) \\ \Delta \overline{\sigma}_{ij}(k) &= \overline{\sigma}_{ij}(k) - \overline{\sigma}_{ij}(k-1) \end{aligned} \tag{4}$$

$$\begin{aligned} \underline{\sigma}_{ij}(k+1) &= \underline{\sigma}_{ij}(k) - \eta^{\underline{\sigma}} \frac{\partial E}{\partial \underline{\sigma}_{ij}}(k) + \beta^{\underline{\sigma}} \Delta \underline{\sigma}_{ij}(k) \\ \Delta \underline{\sigma}_{ij}(k) &= \underline{\sigma}_{ij}(k) - \underline{\sigma}_{ij}(k-1) \end{aligned} \tag{5}$$

$$\begin{aligned} m_{ij}(k+1) &= m_{ij}(k) - \eta^{m} \frac{\partial E}{\partial m_{ij}}(k) + \beta^{m} \Delta m_{ij}(k) \\ \Delta m_{ij}(k) &= m_{ij}(k) - m_{ij}(k-1) \end{aligned} \tag{6}$$

$$c_{ij}(k+1) = c_{ij}(k) - \eta^c \frac{\partial E}{\partial c_{ij}}(k) + \beta^c \Delta c_{ij}(k)$$
$$\Delta c_{ij}(k) = c_{ij}(k) - c_{ij}(k-1) \tag{7}$$

$$s_{ij}(k+1) = s_{ij}(k) - \eta^s \frac{\partial E}{\partial s_{ij}}(k) + \beta^s \Delta s_{ij}(k)$$
$$\Delta s_{ij}(k) = s_{ij}(k) - s_{ij}(k-1) \tag{8}$$

Here the gradient descent method is adopted to adjust the free parameters of precondition and consequence. Though it may fall into a local minimum solution, the other global optimal algorithms, such as genetic algorithm, particle swarm optimization and ant colony algorithm, which can theoretically converge to the global optimal solution, are parallel search and their efficiency makes them unable to meet our requirements. The intelligent optimal algorithms can not are used as the parameters learning algorithms.

4 Experiment Result and Analysis

In order to validate the effectiveness of the proposed model, raw traffic flow sampled data which are downloaded at the Traffic Research Center's of Berkeley University website (http://www.ce.berkeley.edu/daganzo/index.html) is employed in this experiment. With collecting traffic flow data, the sample period is 2 min from 14:00 to 20:00. Selecting average data of five lanes at station 3 as the experiment data, we obtain 195 traffic flow data. The input sample data are $(O(k), O(k-1), O(k-2))$ and ideal output sample data are $O(k+1)$. The performance evaluation function is the root mean square error (*RMSE*):

$$RMSE = \sqrt{\frac{1}{N}\sum_{i=1}^{N}\left[O(k+1) - O_d(k+1)\right]^2} \tag{9}$$

Here $O(k+1)$ is deal output of TSK IT2FNN. We select 100 sample data as training samples, remaining data will be used to testing data. The whole training process includes structure identification and parameter learning.

For the IT2FNN, the structure learning parameters are set as $\zeta = 0.7$, $r_a = 0.6$, $r_r = 0.5$, $r = 0.5$, $m = 2.0$, $Max = 100$, $MinT = 10^{-5}$. Using the selected 100 training data with input vector $[O_p(k), O_p(k\text{-}1), O_p(\text{k-}2)]^T$ and ideal data $O_p(k+1)$. After the structure identification, we can get 3 fuzzy rules. In the parameters learning, the learning rates are shown as follows. The learning rate $\eta^{\bar{\sigma}} = \eta^{\underline{\sigma}} = \eta^m = \eta^c = \eta^s = 0.01$. The momentum terms are $\beta^{\bar{\sigma}} = \beta^{\underline{\sigma}} = \beta^m = \beta^c = \beta^s = 0.01$. The Maximum iteration number is $MaxI = 500$. Table 1 shows short-term traffic flow prediction performance.

Table 1. The performance comparison of the IT2-1 FNN, IT2-2 FNN, IT2-3 FNN and T1 FNN

	IT2-1 FNN	*IT2-2 FNN*	*IT2-3 FNN*	*T1 FNN*
Training	3.2302	3.2336	3.3056	3.3056
Testing	2.3385	2.2956	2.3118	2.3118

After learning, the actual output $O(k)$ and ideal output $O_d(k)$ of the IT2FNN are shown in figure 4. The evolution curve of BP algorithm is shown in figure 5.

In the learning process, the upper and lower bounds of variances $\overline{\sigma}_{ij}$ and $\underline{\sigma}_{ij}$ are decided by λ_{ij}. Similarly, the radius of consequence interval linear functions are determined by γ_{ij}. The parameters λ_{ij} and γ_{ij} are random numbers, which have great effect on the learning results. When we randomly select 5 parameters λ_{ij} and γ_{ij} , the different learning results are shown in Table 2.

Table 2. Prediction performance comparison of the IT2FNN

	Group 1	*Group 2*	*Group 3*	*Group 4*	*Group 5*
Training	3.3047	3.3041	3.3056	3.2302	3.3027
Testing	2.3016	2.3080	2.3366	2.3385	2.3005

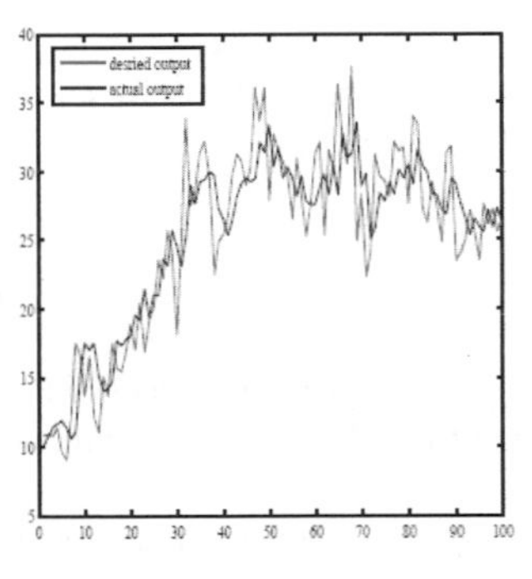

Fig. 3. Prediction result of the IT2FNN

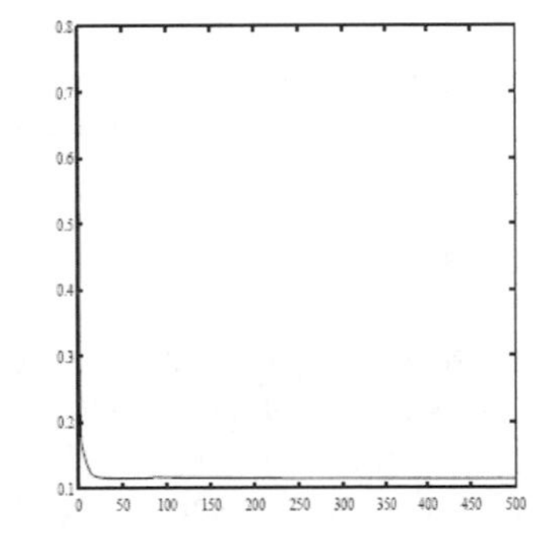

Fig. 4. Evolution curve of BP algorithm

The remaining 75 sample data are employed to evaluate the learning effectiveness. The testing result has already been shown in Table 1. The evolution curve of testing result illustrates in figure 6.

From the table 2, we draw a conclusion that the parameters λ_{ij} and γ_{ij} have great effect on the training results. Therefore, it is vital how to select parameters λ_{ij} and γ_{ij}.

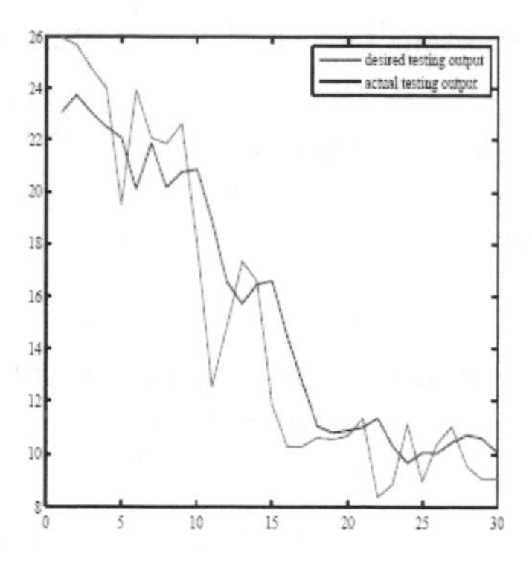

Fig. 5. Testing result of the IT2FNN

5 Conclusions

Based on IT2FNN and its learning algorithm, we propose a new method of short-term traffic flow prediction. The whole training process can be divided into two stages, i.e., structure identification and parameters learning. Firstly, we determine structure parameters of TSK IT2FNN by hierarchical clustering algorithm. Then BP algorithm is employed to adjust the precondition and consequence parameters. Finally, we verify that our proposed method is effective and valuable.

Acknowledgments. This work is supported by High-Level Talents Research Foundation of Henan University of Technology (No.2009BS060).

References

1. Sen, A., Sööt, S., Ligas, J., Tian, X.: Arterial Link Travel Time Estimation: Probes, Detectors and Assignment Type Models. In: Preprints of papers at the Transportation Research Board 76th Annual meeting, vol. 21 (1997)
2. Akiva, B., Cascetta, M., Gunn, E.H.: An Online Dynamic Traffic Prediction Model for an Inter-Urban Motorway Network. In: Gartner, N.H., Improta, G. (eds.) Urban Traffic Networks: Dynamic Flow Modeling and Control, pp. 83–122. Springer, New York (1995)
3. Chen, H.B., Muller, S.G.: Use of Sequential Learning for Short-term Traffic Flow Forecasting. Transportation Research, Part C 9, 319–336 (2001)
4. Smith, B.L., Demetsky, M.J.: Traffic Flow Forecasting: Comparison of Modeling Approaches. Journal of Transportation Engineering 123, 261–266 (1997)
5. Kaysi, I., Ben-Akiva, M., Koutsopoulos, H.: An Integrated Approach to Vehicle Routing and Congestion Prediction for Real-time Driver Guidance. In: Transportation Research Record 1408, TRB, Washington, D.C, pp. 66–74 (1993)
6. Zhao, L., Wang, F.Y.: Short-Term Fuzzy Traffic Flow Prediction Using Self-Organizing TSK-Type Fuzzy Neural Network. In: IEEE International Conference on Vehicular Electronics and Safety, pp. 294–299 (2007)

7. Zhao, L., Wang, F.Y.: Short-Term Traffic Flow Prediction Based on Ratio-Median Lengths of Intervals Two-Factors High-Order Fuzzy Time Series. In: IEEE International Conference on Vehicular Electronics and Safety, pp. 287–293 (2007)
8. Smith, B.L.: Forecasting Freeway Traffic Flow for Intelligent Transportation System Applications. Ph. D Dissertation, Department of Civil Engineering. University of Virginia, Charlottesville
9. Williams, B.M.: Modeling and Forecasting Vehicular Traffic Flow as a Seasonal Stochastic Time Series Process. Ph.D Dissertation. Department of Civil Engineering, University of Virginia, Charlottesville
10. Zhao, L.: Direct-Inverse Modeling Control Based on Interval Type-2 Fuzzy Neural Network. In: 29th Chinese Control Conference (accepted)
11. Zhao, L.: TSK Interval Type-2 Fuzzy Neural Networks for Chaotic Time Series Prediction. In: 8th World Congress on Intelligent Control and Automation (accepted)

Neural Network and Sliding Mode Control Combining Based Reconfigurable Control

Gongcai Xin, Zhengzai Qian, Weilun Chen, Kun Qian, and Li Li

Department of Aerial Instrument and Electric Engineering,
The First Aeronautical Institute of Air Force, Xin Yang, 464000, China
Xingc2752@163.com

Abstract. This paper introduces a neural network adaptive control and sliding mode control combining reconfigurable control for aircraft. The control law is based on the nonlinear dynamic inversion. Sliding model control and fuzzy neural network adaptive control are used to compensate twice the inversion error induced by aircraft actuator failures. Thereby the robustness of dynamic inversion control is greatly improved, and the modeling accuracy has been solved. The simulation shows that this method is feasible.

Keywords: neural network, dynamic inversion, sliding mode control, adaptive control.

1 Introduction

The purpose of flight control is to complete the tasks of a variety of flight modes, which is completed by controlling the aircraft attitude and trajectory, and aircraft safety should be guaranteed in such tasks firstly. In the course of flight, if aircraft has control surface failure (such as stuck, damaged and so on), it will produce lethal effects to flight safety. It is a research focus on the aviation field about how to use aerodynamic redundancy of more control surface shape of the aircraft, self-repair, enhance the failure adaptability. In this way, the failure aircraft also can maintain a certain handling qualities and fly safely. Multi-layer perceptron neural network was applied to the arm motion control of robot by Lewis [1] firstly, and he proposed a new weight adjustment algorithm. Research team led by Professor Calise [2], [3] introduced neural network to flight control and proposed neural network-based adaptive inversion tracking control structure. McFarland designed the adaptive controller of agile missile with RBF [4] structure and SHL [5] structure neural network respectively. In recent years, the United States have carried out RESTORE [6], [7] plans to study adaptive reconfigurable control scheme with neural networks for the new generation of fighter, and validated the effectiveness of this control scheme through the flight test of non-vertical tail aircraft.

In this paper, the dynamic inversion theory, neural networks, sliding mode control and adaptive control combined method are used to test tentatively about the nonlinear control of the aircraft. The results indicate that good learning of adaptive neural network, nonlinear approximation properties, and robustness and transient nature of the sliding mode control are used to make the error fast convergence based on the nonlinear

K. Li et al. (Eds.): LSMS/ICSEE 2010, Part II, CCIS 98, pp. 238–243, 2010.

dynamic inversion control. In this way it can solve the problem of model accuracy and improve tracking performance of flight control system.

2 Dynamic Inversion Implementing Feedback Linearization

Inverse system method is one feedback linearization [8] method, and its basic idea is: for a given system, first of all, the object model produces "an ath order integral inverse system " [9] of the original system which can be achieved by feedback method, and object compensation becomes a decouple standardized system (known as pseudo-linear system) with linear transitive relation; And then, use a variety of design theory of linear systems to complete the synthesis of pseudo-linear system. The essence is to offset the non-linearization of the controlled object with nonlinear inversion and nonlinear function and to get global linearization.

2.1 Control Method and Structure Design

Consider the following second-order nonlinear system:

$$\ddot{x} = f(x, \dot{x}, \delta) \tag{1}$$

$$y=h(x) \tag{2}$$

Where, x is state variable; δ is control variable; y is output; $x, \dot{x} \in R^n, \delta \in R^m$, $m \geq n, y \in R^p$. The control task is to get the control variable δ which satisfies the requirements of the output y. Now, gets derivative for equation (2)

$$y' = \begin{bmatrix} y_1^{(r_1)} \\ y_2^{(r_2)} \\ \vdots \\ y_p^{(r_p)} \end{bmatrix} = F(x, \dot{x}, \delta) = v \tag{3}$$

Then $$\delta = F^{-1}(x, \dot{x}, v) \cdot \tag{4}$$

For any output subsystem:

$$y_i^{(r_i)} = v_i \quad \text{(i=1, 2, …, p)} \tag{5}$$

Here, v_i is the pseudo-control variable.

By equation (5), we can see, non-linear features are turned into the linearization. However, it is the calculation under ideal circumstances. But inverse calculation errors and external disturbances still exist. Therefore, equation (3) can be turned into $\ddot{x} = f(x, \dot{x}, \delta) + \Delta(x, \dot{x}, \delta)$, $\Delta(x, \dot{x}, \delta)$ is the system random error. On the other hand, if the mathematical model is not accurately knowable, let the derivative of $F(x, \dot{x}, \delta)$ equal $F'(x, \dot{x}, \delta)$, so the equation (4) can be changed into $\delta' = F'^{-1}(x, \dot{x}, v)$, we can see, the nonlinearity can not be fully offset. And its control block diagram is shown in Figure 1. Where $F^{-1}(x, v)F(x, \hat{\delta}) = I$ ·

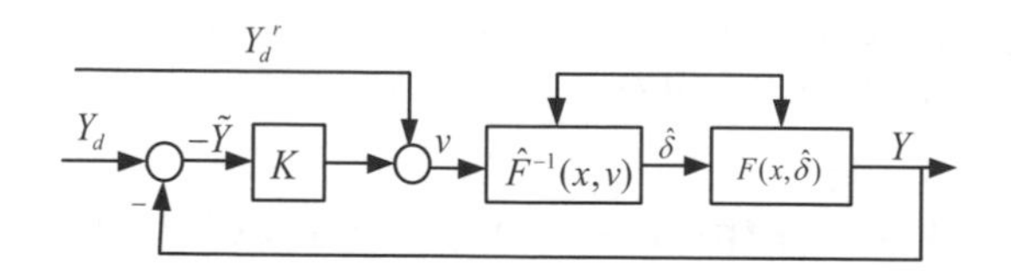

Fig. 1. Approximate dynamic inversion control

2.2 Shortage

The essence of the nonlinear dynamic inversion integration system is to counteract the nonlinearity of controlled objects by means of non-linear inversion and nonlinear function. When the object model is inaccurate, nonlinearity can not be counteracted reasonably, and system robustness cannot be guaranteed. Therefore, a suitable model error compensation is required to deal with model uncertainty. Adaptive control [2] is the primary means of dealing with model uncertainty, but it can only handle the uncertainty of object parameter.

To get a good flying qualities even if the aircraft is in an error state. We should eliminate its inversion error. So the error compensation for each channel should be carried out after inversion system method based systems are decoupled. On the one hand, it is necessary to compensate for external disturbances; On the other hand, it needs to deal with modeling errors.

3 Neural Network Adaptive Control

In this paper, Lewis proposed multi-layer perceptron neural network to reconfigure the inversion error, and neural network structure is shown in reference [1], its input-output mapping relationship can be written as

$$y_i = \sum_{j=1}^{N_2}[w_{ij} \cdot \sigma(\sum_{k=1}^{N_1} v_{jk}x_k + \theta_{vj}) + \theta_{wi}], \quad i = 1, 2, \cdots, N_3$$

where, y_i denotes the hidden layer activation function; v_{jk} denotes the connection weights between the input layer and hidden layer; w_{ij} denotes the connection weights between the hidden layer and output layer, θ_{vj}, θ_{wi} are neuron threshold; N_1, N_2, N_3 is the neuron number for the input layer, hidden layer, output layer respectively.

If the enough input information and the number of hidden layer neurons are given, neural network can approach continuous nonlinear functions with any accuracy. Therefore, for continuous uncertain nonlinear inversion error function $\Delta(x, \dot{x}, \delta)$ and the inversion system standard error $\varepsilon_N > 0$, there are N_2 and network weight matrices W^*, V^*, we have

$$\Delta(x, \dot{x}, \delta) = W^{*T}\sigma(V^{*T}\bar{x}) + \varepsilon, \quad \|\varepsilon\| \le \varepsilon_N$$

As long as the Adaptive items are taken as $v_{ad} = W^T\sigma(V^T\bar{x})$, we can get the approximate performance of nonlinear function by adjusting their weights, and minimize errors.

4 Neural Network Adaptive Control and Sliding Mode Control Integrated

From above, we can see, neural network adaptive control is a step-by-step learning process in the course of system running, this characterization determines that it can only deal with the steady or gradient uncertain model, and these uncertainties must also be expressed with a group of the unknown parameters, but it can not deal well with rapidly changing parameters and the dynamics systems which have not constructed model. Therefore, this paper introduces sliding mode control to improve the fuzzy neural network adaptive control.

The design ideas are: take the sliding surface of subsystem j: $s_j(\tilde{y}_j,t)=\tilde{y}=0$, let $\dot{s}_j=0$, we can get solution of equivalent pseudo-control $\hat{v}_j=\hat{\dot{y}}_j=0$. Define ρ_j^* as residual error, and take $\bar{v}_j=\rho_j\,\mathrm{sgn}(s_j)$ as sliding robust compensation term. Let $\tilde{\rho}_j=\rho_j-\rho_j^*$, under the pseudo-control condition,

$$\dot{s}_j=\Delta_j-v_{adj}-v_j-k_1\dot{s}_j=0$$

Define the Lyapunov function

$$V_j=\frac{1}{2}s_j^2+\frac{1}{2\lambda_j}\tilde{\vartheta}_j^T\tilde{\vartheta}_j+\frac{1}{2\eta_j}\tilde{\rho}_j^2$$

Then

$$\bar{V}_j=s_j[-\tilde{\vartheta}_j^T\rho(z)+\varepsilon_j-\rho_j\,\mathrm{sgn}(s_j)-k_1s_j]+\frac{1}{\lambda_j}\tilde{\vartheta}_j^T\tilde{\vartheta}_j+\frac{1}{\eta_j}\tilde{\rho}_j\dot{\tilde{\rho}}_j$$

The adjustment rules of $\tilde{\vartheta}_j,\rho_j$ are shown as follows:

$$\dot{\tilde{\vartheta}}_j=\dot{\vartheta}_j=\lambda_js_j\rho(z)\qquad \dot{\tilde{\rho}}_j=\dot{\rho}_j=\eta_j\,|\,s_j\,|$$

We can see, $\bar{V}_j\le 0$, so the system is wholly stable. Now, re-defined Lyapunov function $V_{j1}(t)=V_j(t)-\int_0^t(\dot{V}_j+k_1s_j^2)d\tau$, we can deduce the system error approach zero.

To eliminate the buffeting, compensation term of robust sliding mode as follows:

$$\bar{v}_j=\rho_j sat(\frac{s_j}{\Phi_j})$$

When the system begin to operate, the system trajectory leaves the boundary layer of the sliding surface, and adaptive neural network system based on sliding model compensation starts to work, then the system trajectory quickly back to the inner of boundary layer of the sliding surface, at this time, the adaptive control process is cut off. If the system trajectory leaves again, repeat this process until the sliding mode compensation term can fully compensate residual errors.

5 Control System Design

5.1 Structural Design

Control system block diagram is shown in Figure 2.

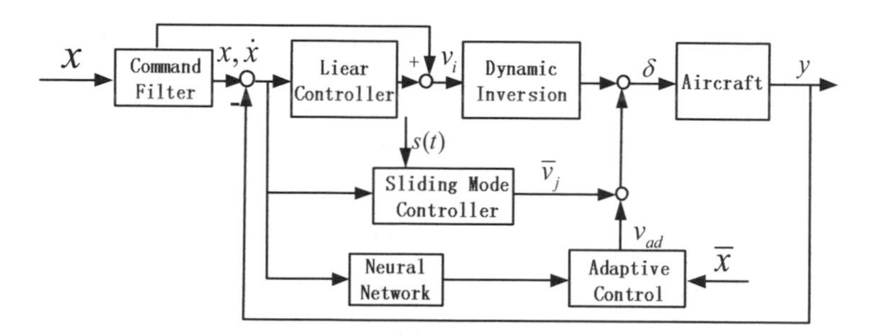

Fig. 2. Control system block diagram

5.2 Control Law Design

(1) For any output y_j (j = 1, 2, ..., p), derivative on δ until it explicitly appear in the expression;

(2) If $\sum_{j=1}^{p} r_j = n$, the system can be linearized; If $\sum_{j=1}^{p} r_j < n$, study the stability of the inside dynamics subsystems;

(3) To perform approximately linearization with dynamic inversion theory;

(4) Defines the sliding surface: $s(\tilde{Y},t) = \tilde{y}^{(r-1)} + k_r \tilde{y}^{(r-2)} + \cdots + k_2 \tilde{y} = 0$;

(5) Determine the adaptive control structure, find adaptive items by neural networks;

(6) Find sliding mode robust compensation item $\overline{v}_j$ by sliding mode control theory;

(7) Use Lyapunov function to educe adaptive adjustment rules, and eliminate the impact of parameter drift.

6 Simulation Results

To realize the stronger robustness and stability of the control strategy, we do simulation with simple model, the simulation graphs are shown in Figure 3, Figure 3(a) shows that the response curve of dynamic inversion control. Figure 3(b) shows that the response curve is greatly improved by adding neural network adaptive control. However, figure 3(c) shows that the transient state is improved obviously when adding sliding mode control. The solid line denotes the inverse error in figure 3(d), the dashed line denotes the total output of two compensation. We can see, through the compensation, the error has been fitted basically, and control accuracy is greatly improved. At last, it greatly improves tracking performance of flight control system. The control requirements of high-performance unmanned aerial vehicles can be satisfied basically.

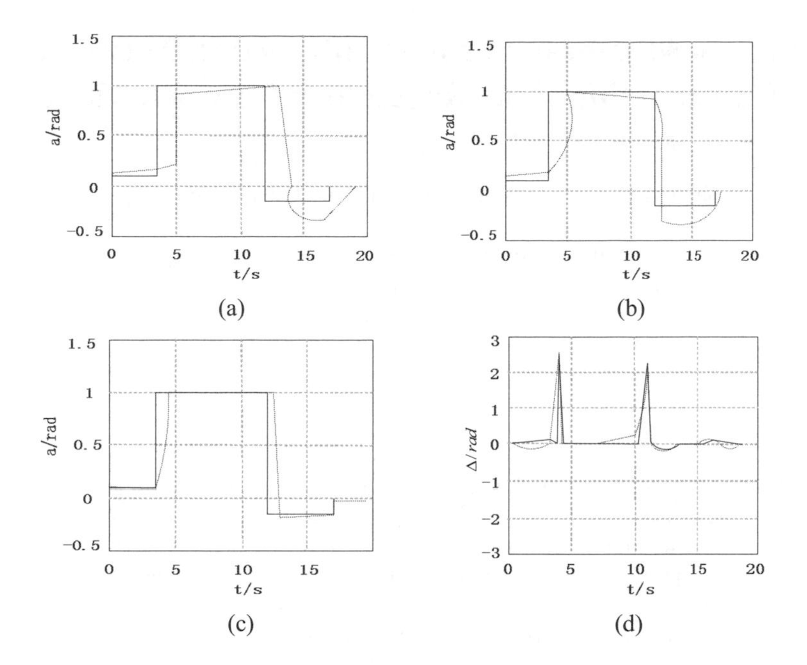

Fig. 3. Simulation diagram. (a) Simplified model of dynamic inversion control; (b) Neural network adaptive compensation; (c) Adaptive and sliding mode compensation; (d) Output and inverse error of two compensation items

References

1. Lewis, F.L., Yesildirek, A., Liu, K.: Multilayer neural-net robot controller with guaranteed tracking performance. IEEE Transactions on Neural Networks 7(2), 388–399 (1996)
2. Kim, B., Calise, A.J.: Nonlinear flight control using neural networks. Journal of Guidance, Control and Dynamics 20(1), 26–33 (1977)
3. Calise, A.J., Rysdyk, R.: Nonlinear adaptive flight control using neural networks. IEEE Control System Magazine 18(6), 14–25 (1998)
4. McFarland, M.B., Calise, A.J.: Adaptive nonlinear control of agile antiair missiles using neural networks. IEEE Transactions on Control System Technology 8(5), 749–756 (2000)
5. McFarland, M.B., Calise, A.J.: Neural networks and adaptive nonlinear control of agile antiair missiles. Journal of Guidance, Control, and Dynamics 23(3), 547–553 (2000)
6. Brinker, J.S., Wise, K.A.: Reconfigurable flight control for a tailless advanced fighter aircraft. AIAA-98-4107 (1998)
7. Calise, A.J., Lee, A., Sharma, M.: Direct adaptive reconfigurable control of a tailless fighter aircraft. AIAA-4108 (1998)
8. Isidoria: Nonlinear Control Systems. Springer, Berlin (1995)
9. Shu-xiang, L., Suo-feng, G.: Neural network based dynamic inversion control for a supermaneuverable aircraft. AcTa Aeronautica ET Astronautica Sinica 18(I), 26–30 (1997)

Study on Membrane Protein Interaction Networks by Constructing Gene Regulatory Network Model

Yong-Sheng Ding[1,2,*], Yi-Zhen Shen[1], Li-Jun Cheng[1], and Jing-Jing Xu[1]

[1] College of Information Sciences and Technology
[2] Engineering Research Center of Digitized Textile & Fashion Technology, Ministry of Education
Donghua University, Shanghai 201620, P.R. China
ysding@gordonlifescience.org,
ysding@dhu.edu.cn

Abstract. At present, about a quarter of all genes in most genomes contain transmembrane (TM) helices, and among the overall cellular interactome, helical membrane protein interaction is a major component. Interactions between membrane proteins play a significant role in a variety of cellular phenomena, including the transduction of signals across membranes, the transfer of membrane proteins between the plasma membrane and internal organelles, and the assembly of oligomeric protein structures. However, current experimental techniques for large-scale detection of protein-protein interactions are biased against membrane proteins. In this paper, we construct membrane protein interaction network based on gene regulatory network model. GRN model is proposed to understand the dynamic and collective control of developmental process and the characters of membrane protein interaction network, including small-world network, scale free distributing and robustness, and its significance for biology. The proposed method is proved to be effective for the study of membrane protein interaction network. The results show that the approach holds a high potential to become a useful tool in prediction of membrane protein interactions.

Keywords: membrane protein interaction network, gene regulatory network, small-world network, scale free distributing, robustness.

1 Introduction

The membrane protein, particularly the helical membrane protein interaction network, is an important part of the protein study. Genomic studies suggest that membrane proteins make up ~25% to ~33% of the predicted proteins in an organism. Though there exist some experimental techniques that detect interactions between individual transmembrane (TM) [1], mapping the membrane protein is difficult. It has so far been difficult to experimentally construct a genome-wide map of membrane protein interactions and TM interactions in yeast [1-3].

There is an increased interest in how lipids interact with membrane proteins and how these interactions lead to various cell membrane functions. Therefore, membrane

* Corresponding author.

K. Li et al. (Eds.): LSMS/ICSEE 2010, Part II, CCIS 98, pp. 244–252, 2010.

protein interaction models have been studied more and more deeply. ToF-SIMS is a unique technique to investigate these interactions by chemically identifying the location of each molecule [4, 5]. However, it is extremely difficult to characterize the native structure of cell membranes due to their innate complexity, i.e., the eukaryotic cell membranes consist of up to 500 different lipid species. Models of membrane systems, such as supported lipid bilayers and Langmuir-Blodgett (LB) monolayers, have been proven to be good mimics of cellular membranes [6, 7]. These simplified models can be used with different combinations of lipid and protein molecules to form a membrane, which represents a bottom-up approach to study individual interactions. Nevertheless, the studies above are just restricted in the membrane protein interaction.

Here we consider the gene regulatory network (GRN) model to study membrane protein interaction network [8-11]. In the GRN model, the timing of gene activation is critical to the execution of the regulatory program [12]. The topology of developmental GRNs specifies inputs into the regulatory system of each participating gene, and where this gene encodes a transcription factor, its outputs to target genes in the next tier of the hierarchical network. Thus any given domain of a GRN consists of prior or upstream, responding or downstream, and regulatory gene circuitry. In the operation of the GRN, time flows in the same direction as the causality determined in the GRN topology (except for feedback) [12]. Thus in terms of transcription dynamics, the measurable output of the GRN is a temporal sequence of cohorts of regulatory gene expressions. There is a one way logic relationship between overall GRN architecture and the temporal progression of transcription patterns: GRN topology predicts the kinetics of this progression, barring post-transcriptional modulations; however, it is almost impossible to infer network topology exclusively from dynamic expression data, except for linear cascades of such simplicity as are rarely seen in embryonic development.

2 Gene Regulatory Network Model for Membrane Protein Interaction Network

In the following, we construct membrane protein interaction networks based on the GRN model and study on its dynamic and collective characters.

The basic form of the GRN model formulation for membrane protein interaction network is as follows [8, 9]:

$$\frac{d[x]_{i,j}}{dt} = A - B \pm C \pm D \tag{1}$$

where A is composing, B is decay, C is transform, D is transfers, i is a membrane protein, and j is membrane protein surface (the interaction of membrane protein). Composing period represents the transfers of membrane proteins. For membrane proteins, there is only one composing period except multi-gene provides quite a few transcriptions. "Disintegration" class represents first order decay process certainly whether the given species disappears, even though it is very slow.

Using the model of differential equation, dynamics interaction is transferred to mathematics formula. These space state equations dominate the dynamics of membrane protein. Some periods like transfers flux are omitted for predigestion [8, 9].

The membrane protein dynamics formulation using hedgehog coding masterdom is as follows:

$$\frac{d[hh]_i}{dt} = T_{\max}\rho_{hh}\left(\frac{[EN]_i^{\nu_{ENhh}}}{K_{ENhh}^{\nu_{ENhh}} + [EN]_i^{\nu_{ENhh}}}\right) - \frac{[hh]_i}{H_{hh}} \tag{2}$$

$$\frac{d[HH]_{i,j}}{dt} = \frac{P_{\max}\sigma_{HH}[hh]_i}{6} - \frac{[HH]_{i,j}}{H_{HH}} - k_{PTCHH}[HH]_{i,j}[PTC]_{n,j+3} \tag{3}$$

$$\frac{d[PH]_{i,j}}{dt} = k_{PTCHH}[HH]_{n,j+3}[PTC]_{i,j} - \frac{[PH]_{i,j}}{H_{PH}} \tag{4}$$

Non-dimensionalization model and dimensionalization model are coherent because it is just a transformation. If the known scaling constants are used to be non-dimensionalization change in the model, the non-dimensionalization state will be translated into dimensionalization state easily. The non-dimensionalization model parameters are the results of the dimensionalization parameters, so the values of the known dimensionalization parameters can decide the values of the non-dimensionalization parameters uniquely.

The conversion formula of non-dimensionalization model [8, 9]:

$$t = T_0\tau,\ [x]_{i,j} = [x]_0 x_{i,j}(\tau) \tag{5}$$

In Eq. (5), τ is non-dimensionalization time, T_0 is eigentime constant about t and τ, $[x]$ is dimension muster degree, $x(\tau)$ is the substitution of non-dimensionalization, and $[x]_0$ is the characters muster degree of x. We choose $[x]_0$ as the most stable state muster degree. Videlicet, the most probably given dimension muster degree dominates the parameter x in the differential equation. In the most stable state, we only consider the main composition and the decay period. Otherwise, many equations can not be solved to be a simple and useful form. The derivative (composition, result splitting and so on) species characters muster degree is set to the main characters muster degree of membrane protein result.

To hh, the equation is as follows:

$$[hh]_0 = T_{\max}\rho_{hh}H_{hh} \tag{6}$$

To membrane protein, the equation is as follows:

$$[HH]_0 = P_{\max}\sigma_{HH}H_{HH}[hh]_0 = P_{\max}\sigma_{HH}T_{\max}\rho_{hh}H_{hh} \tag{7}$$

First, we propose PH and PTC as the coherent eigenconstant, so the relative species number can be evaluated easily. Through the substitution of the Eqs. (6) and (7), and then the renamed non-dimensionalization parameter group, Eqs. (2) ~ (6) are proposed as follows:

$$\frac{d\ hh_i}{d\tau} = \frac{T_0}{H_{hh}} \left(\frac{EN_i^{\nu_{ENhh}}}{\kappa_{ENhh}^{\nu_{ENhh}} + EN_i^{\nu_{ENhh}}} \right) - EN_i \tag{8}$$

$$\frac{d\ HH_{i,j}}{d\tau} = \frac{T_0}{H_{HH}} \left(\frac{hh_i}{6} - HH_{i,j} \right) - T_0 k_{PTCHH} HH_{i,j} \bullet [PTC]_0 PTC_{n,j+3} \tag{9}$$

$$\frac{d\ PH_{i,j}}{d\tau} = T_0 k_{PTCHH} [HH]_{n,j+3} PTC_{i,j} - \frac{T_0 PH_{i,j}}{H_{PH}} \tag{10}$$

This transformation is very convenient because each state parameter changes between 0 and 1. Therefore, the model is in the segment period of the most muster degree in each group, not expressed through absolute muster degree, however. In addition, in the group combining dimensionalization parameters and non-dimensionalization ones, there are one-third parameters cursorily estimated that they do not lose any facility [8, 9]. Therefore, non-dimensionalization parameters are usually more intuitionistic than dimensionalization parameters. For instance, the parameter κ_{ENhh} exists in Eq. (8), which decides how to expect EN activating hh transcription, as follows:

$$\kappa_{Xy} = \frac{K_{Xy}}{[X]_0} = \frac{K_{Xy}}{P_{\max} \sigma_X H_X T_{\max} \rho_\tau H_\tau} \tag{11}$$

In Eq. (11), Y is a catalyzing enzyme process, and transferring X is a conditioner. $T_{\max}$, $P_{\max}$, all σ and ρ are renamed as κ in non-dimensionalization equations. All κ parameters mean the same thing. The upper value (near 1.0) implies that the conditioner is weak, and the lower value (near 10^{-3}) implies a strong conditioner. The parameter ν itself is non-dimensionalization form, and can hold the line. Semi-decay can hold the dimension because minute is usually adopted to describe semi-decay. The best solution is setting it equal to the eigentime constant. Hence, we take T_0 equal to one minute.

Eqs. (8)~(10) denote isomerism dimerization, ligan binding, and other second order reactions. We propose a non-dimensionalization parameter as follows:

$$\chi_{PTCHH} = T_0 k_{PTCHH} [PTC]_0 \tag{12}$$

When the reaction is higher than second order, the parameters like χ_{PCTHH} are proposed meaningless in physics, but can be combined with other parameters. However, one unit can be easily converged to one or other k_{PTCHH} or PTC_0 for getting rid of dimension. They can still retain non-dimensionalization parameters after disappearing. When there is a scalar quantity value and its range consistent with the labeled dimensionalization copy, T_0 can combine with k_{PTCHH} and PTC_0 to make more convenient non-dimensionalization parameters: T_0, k_{PTCHH} and PTC_0 denote whether they reactive to emerge in a separate time unit.

3 Results and Discussion

3.1 Data Sets

In order to facilitate the study, the dataset constructed by [13] is used as the working dataset. They carried out a large-scale screen to identify interactions between yeast integral membrane proteins of Saccharomyces cerevisiae by using a modified split-ubiquitin technique. They identified 1,985 putative interactions including many interactions that had not been observed by either systematic or small-scale approaches, which involved 536 proteins among 705 proteins annotated as integral membrane proteins. Here, we select some special interactions of the 1,985 ones as our training and testing dataset. The yeast integral membrane proteins mainly include 3 classes and 21 characters.

3.2 Membrane Protein Interaction Network

In the simulation platform of membrane protein interaction network constructed by GRN model, the simulated network topology is set to 1,985 membrane protein interactions. The membrane protein interaction networks are generated by randomly reassigning node and edge labels for the networks of interest. After the membrane protein interaction network constructed by GRN model, we could calculate the relative coefficients using current equations [14, 15] to understand the dynamic and collective control of developmental process and get the membrane protein interaction network characters following the general protein interaction network we studied.

3.2.1 Small-World Network

Adjacency matrix A depicts the connecting relation between complex network, where the element A_{ij} depicts the connecting relation between the i node and the j node. The i node connects with the j node if $A_{ij}=1$. On the contrary, the i node does not connect with the j node if $A_{ij}=0$. The degree of the node i denotes the number of nodes connecting with the node i, which is defined as follows [14]:

$$K_i = \sum_{j=1}^{N} A_{ij} \tag{13}$$

where N is the sum of nodes in the network.

The membrane protein interaction network has the character of small world network. When the network has shorter average shortest path length and higher average ensemble coefficient, this network satisfies the small-world character.

The ensemble coefficient of node i is defined as follows [14, 15]:

$$C_i = \frac{2n_i}{k_i(k_i-1)} \tag{14}$$

where n_i denotes the number of borders between k_i neighbor nodes of node i.

The average ensemble coefficient of node i is defined as follows:

$$C = \frac{1}{N}\sum_{i=1}^{N} C_i \tag{15}$$

where N is the sum of nodes in the network.

The average shortest path length of the membrane protein interaction network is 1.6. As shown in Fig. 3, the average ensemble coefficient C_i decreases with the slope $\gamma = 0.013$ along with K's increasing, which indicates the membrane protein interaction network in the forecast results has small-world character.

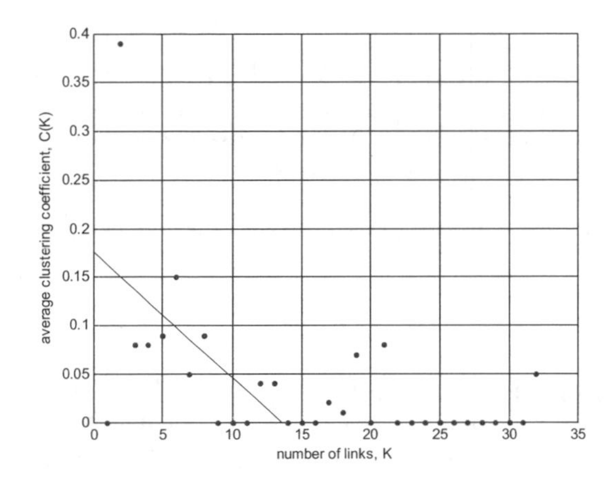

Fig. 1. The average ensemble coefficient of the membrane protein interaction network

3.2.2 Scale Free Distributing

The membrane protein interaction network also has scale free distributing character, which denotes the existing probability of the nodes with connecting degree k in the network satisfies exponential distributing, $P(k) \propto k^{-\gamma}$. To a biological network, we usually propose $2 < \gamma < 3$. From Fig. 4, we can get $\gamma = 2.75$, which denotes the membrane protein interaction network in the forecast results satisfies scale free distributing.

It was pointed out that the scale free network structure put out good robustness to the nodes taken out in the network. However, it can not resist the hub nodes taken out. The faster disturbance spreading speed and the shorter reaction time are relative with the small-world character [16].

3.2.3 Robustness

We propose that the average connecting coefficient calculating the membrane protein interaction network is to evaluate the network robustness, resisting disturbance ability namely. The calculation process is defined as follows:

$$H_i = \frac{\sum_{i=1}^{n}(n_i - 1)}{n} \tag{16}$$

By inference, we can get the average connecting coefficient defined as follows:

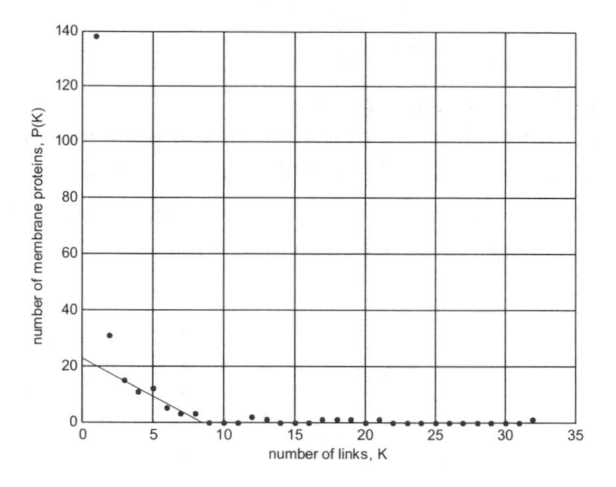

Fig. 2. The dimension distributing of the membrane protein interaction network

$$H = \frac{1}{N}\sum_{i=1}^{N} H_i \tag{17}$$

The results are shown in Fig. 5, where $H(k) \sim k^{-\gamma}$, $\gamma = 0.63$. The connecting coefficient presents the exponential distributing rule, which denotes that the membrane protein interaction network has better robustness and stronger resisting disturbance ability.

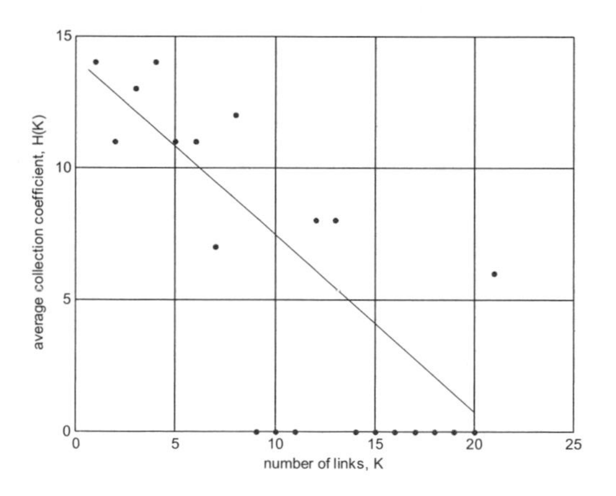

Fig. 3. The average correlative coefficient of the membrane protein interaction network

4 Conclusions

A thought provoking implication of the GRN model summarized to be used to construct the membrane protein interaction network is that the dynamics of gene regulatory circuits follows simply from the membrane protein network topology and the function of regulatory modules on their inputs. To understand the dynamic and collective control of developmental process and the characters of membrane protein interaction network itself, GRN model is proposed. The kinetic parameters are of course temperature dependant, but for a given system they are approximate constants which control the overall dynamics of regulatory life according to the GRN structure. The levels and identity of the transcription factors in a given membrane protein identify the regulatory state of the membrane protein at every point in development. The results show that the approach holds a high potential to become a useful tool in prediction of membrane protein interactions.

Acknowledgement

This work was supported in part by the National Nature Science Foundation of China (No. 60975059, 60775052), Specialized Research Fund for the Doctoral Program of Higher Education from Ministry of Education of China (No. 20090075110002), Project of the Shanghai Committee of Science and Technology (No. 09JC1400900, 08JC1400100), Shanghai Talent Developing Foundation (No. 001), Specialized Foundation for Excellent Talent from Shanghai, and the Open Fund from the Key Laboratory of MICCAI of Shanghai (06dz22103).

References

1. Chou, K.C., Shen, H.B.: MemType-2L: A Web server for predicting membrane proteins and their types by incorporating evolution information through Pse-PSSM. Biochemical and Biophysical Research Communications 360(2), 339–345 (2007)
2. Zhao, P.Y., Ding, Y.S., Chou, K.C.: Prediction of membrane protein interactions based on a fuzzy support vector machine classifier (to appear)
3. Ren, L.H., Ding, Y.S., Shen, Y.Z., Zhang, X.F.: Multi-agent-based bio-network for systems biology: Protein-protein interaction network as an example. Amino Acid 35(3), 565–572 (2008)
4. Zheng, L., Baker, M.J., Lockyer, N.P., Vickerman, J.C., Ewing, A.G., Winograd, N.: Investigating lipid-lipid and lipid-protein interactions in model membranes by ToF-SIMS. Applied Surface Science 255(4), 1190–1192 (2008)
5. Dario, L., Bernard, C., Stefanie, D.K.: Lipid membrane interactions of indacaterol and salmeterol: Do they influence their pharmacological properties? European Journal of Pharmaceutical Sciences 38(5), 533–547 (2009)
6. Toshinori, S., Haruyuki, I., Ena, O., Naoya, S., Hiroshi, U., Ryoichi, K.: Development of membrane chip system for study on membrane-protein interaction. Journal of Bioscience and Bioengineering 108(suppl. 1), S147–S164 (2009)
7. Peter, M.: Glycolipid transfer proteins and membrane interaction. Biochimica et Biophysica Acta 1788(1), 267–272 (2009)
8. Dsssow, G., Meir, E., Munro, E.M., Odell, G.M.: The segment polargity network is a robust development module. Nature 406(13), 188–192 (2000)
9. Dsssow, G., Meir, E., Munro, E.M., Odell, G.M.: Formulation of a model of the segment polarity network as a system of first-order ordinary differential equations using Ingeneue (2000), http://www.ingenue.org

10. Wang, Z.D., Lam, J., Wei, G.L., Fraser, K., Liu, X.H.: Filtering for nonlinear genetic regulatory networks with stochastic disturbances. IEEE Trans. Automatic Control 53(10), 2448–2457 (2008)
11. Wang, Z.D., Gao, H.J., Cao, J.D., Liu, X.H.: On delayed genetic regulatory networks with polytopic uncertainties: robust stability analysis. IEEE Trans. NanoBioscience 7(2), 154–163 (2008)
12. Ben-Tabou de-Leon, S., Davidson, E.H.: Modeling the dynamics of transcriptional gene regulatory networks for animal development. Developmental Biology 325(2), 317–328 (2009)
13. Miller, J.P., Lo, R.S., Ben-Hur, A., Desmarais, C., Stagljar, I., Noble, W.S., Fields, S.: Large-scale identification of yeast integral membrane protein interactions. Proc. Natl. Acad. Sci. 102(34), 12123–12128 (2005)
14. Watts, D.J.: Small Worlds. Princeton University Press, Princeton (1999)
15. Strogatz, S.H.: Exploring complex networks. Nature 410(6825), 268–276 (2001)
16. Pereira-Lea, J.B., Levy, E.D., Teichmann, S.A.: The origins and evolution of functional modules: Lessons from protein complexes. Philos. Trans. R Soc. Lond. B Biol. Sci. 361(1467), 507–517 (2006)

Author Index

Printing: Mercedes-Druck, Berlin
Binding: Stein + Lehmann, Berlin